Lecture Notes in Mathematics 1648

Springer
Berlin
Heidelberg
New York
Barcelona
Budapest
Hong Kong
London
Milan
Paris
Santa Clara
Singapore
Tokyo

R. Dobrushin P. Groeneboom M. Ledoux

Lectures on Probability Theory and Statistics

Ecole d'Eté de Probabilités
de Saint-Flour XXIV – 1994

Editor: P. Bernard

Springer

Authors

Roland Dobrushin †

Piet Groeneboom
Delft University of Technology
Department of Mathematics and
Computer Science
Mekelweg 4
NL-2628 CD Delft, The Netherlands

Michel Ledoux
Université Paul Sabatier
Département de Mathématiques
Laboratoire de Statistique et Probabilités
F-31062 Toulouse, France

Editor

Pierre Bernard
Université Blaise-Pascal
Clermont-Ferrand
Laboratoire de
Mathématiques Appliquées
URA CNRS 1501
F-63177 Aubière Cedex, France

Cataloging-in-Publication Data applied for

Die Deutsche Bibliothek - CIP-Einheitsaufnahme

Lectures on probability theory and statistics / Ecole d'Eté de Probabilités de Saint-Flour XXIV - 1994. R. Dobrushin ; P. Groeneboom ; M. Ledoux. Ed.: P. Bernard. - Berlin ; Heidelberg ; New York ; Barcelona ; Budapest ; Hong Kong ; London ; Milan ; Paris ; Santa Clara ; Singapore ; Tokyo : Springer, 1996
(Lecture notes in mathematics ; Vol. 1648)
ISBN 3-540-62055-9
NE: Dobrušin, Roland L.; Groeneboom, Piet; Ledoux, Michel; Bernard, Pierre [Hrsg.]; Ecole d'Eté de Probabilités <24, 1994, Saint-Flour>; GT

Mathematics Subject Classification (1991):
46N30, 47D07, 60-01, 60-06, 60B11, 60D05, 60G15, 60G60, 60K35, 60J60, 62-01, 62-06, 62G05, 82-01, 82B27, 82B44

ISSN 0075-8434
ISBN 3-540-62055-9 Springer-Verlag Berlin Heidelberg New York

Typesetting: Camera-ready TEX output by the authors
SPIN: 10520230 46/3142-543210 - Printed on acid-free paper

INTRODUCTION

This volume contains lectures given at the Saint-Flour Summer School of Probability Theory dur the period 7th - 23rd July, 1994.

The school brought together 74 participants, 40 of whom gave a lecture concerning their research work.

We thank the authors for all the hard work they accomplished. Their lectures are a work of reference in their domain.

Professor Roland DOBRUSHIN, one of the most brilliant representatives of the Russian School of Probability founded by Andrei KILMOGOROV, died in Moscow on the 12th November, 1995.

We thank his wife "Galina" and Professor Eugene DYNKIN who helped us to make possible the publication of these lecture notes, even if there are some imperfections resulting from these tragical circumstances.

At the end of this volume you will find the list of participants and their papers.

Finally, to facilitate research concerning previous schools we give here the number of the volume of "Lecture Notes" where they can be found :

Lecture Notes in Mathematics

1971 : n°307 - 1973 : n°390 - 1974 : n°480 - 1975 : n°539 - 1976 : n°598 -
1977 : n°678 - 1978 : n°774 - 1979 : n°876 - 1980 : n°929 - 1981 : n°976 -
1982 : n°1097 - 1983 : n°1117 - 1984 : n°1180 - 1985 - 1986 et 1987 : n°1362 -
1988 : n°1427 - 1989 : n°1464 - 1990 : n°1527 - 1991 : n°1541 - 1992 : n°1581
1993 : n°1608

Lecture Notes in Statistics

1986 : n°50

TABLE OF CONTENTS

PERTURBATION METHODS OF THE THEORY OF GIBBSIAN FIELDS

R.L. Dobrushin

§1. Introduction

The mathematical theory of Gibbsian fields arised in the sixtieth as an approach to the comprehension of the main notions of statistical mechanics at the mathematical level. Soon it became clear that in a natural way all the main problems of classical (i.e., non-quantum) statistical mechanics can be exposed in the language of probability theory. On the other hand, it turned out that the notion of the Gibbsian random field is almost identical to the notion of Markov random field and so the study of such fields have to be in the mainstream of probability theory. After the notion of Gibbsian distribution was released from its original physical accesories it found a lot of various non-physical applications (picture recognition, queuing systems, neuron networks and so on). Now more and more probabilists are beginning to work in this direction. Nevertheless the process of the merging of classical statistical mechanics and probability theory is far to be completed. This process slows down since the most part of specialists making ivestigations in this domain are physicists or mathematical physicist in their origin and education. It does not define the essence of their results but influences strongly on the style of exposition in their publications. They treat facts as well-known and use terminology which are not customary for probabilists and publish their results in journals which are not usually in the offices of probabilists. Some books on the theory of random fields written in the probabilistic style have appeared (see [Ge], [MM], [Pre], [Sin]) but they do not include all the richness of ideas developed in this domain.

There are two main classes of methods to study the structure of the set of Gibbs fields and the properties of these fields. They are the methods of correlation inequalities and the perturbation methods – the main of the perturbation methods is called the method of cluster expansions. The corelation inequalities methods are applicable only to the case of Gibbsian fields defined by attractive (= ferromagnetic) potentials, but for all possible values of parameters describing the field. The perturbation methods are applicable to a very wide class of potentials, but in a restricted domain of values of parameters, mainly to the cases, when the interaction is small enough (large temperatures, rarified gases) or large enough (small temperatures). Even in the domain of its applicability the correlation inequalities methods permit to obtain only some separate results (but very deep and important ones). The perturbation methods give a complete control of the situation in the domain of its applicability. The cluster expansions found also important applications in the quantum field theory which, in its so-called Euclidean variant, leads to the theory of distribution-valued Markov fields with continuous argument. (The last very interesting probabilistic object has not become to be popular enough among probabilists.) Sometimes it seems that the specialists in the probabilistic mathematical physics pronounce the words: "Now the cluster expansion can be applied" as some kind of a magic incarntation. They means that now we can be sure that all plausible facts can be rigorously proved.

Even the perturbation methods are intensively used by mathematical physicists, they are not so popular as correlation inequalities among the probabilists. Partly it is connected with their analytical and combinatorical complexity and the absence before the last time (but see [MM]) their systematical exposition oriented to the mathematicians. These lectures contain such expositon of the theme. It does not suppose preliminary knowledge of the theory of random fields and statistical mechanics. The emphasis is placed on probabilistic intrepretations and applications including such traditional themes as theory of inhomogeneous Markov chains, central limit theorems and large deviations theory. We use a new approach to the construction of cluster expansions introduced recently by the author ([Do94]). It is found on an application of elementary facts of theory of analytical functions and permits to avoid some tremendous combinatorical considerations.

Part 1. Kotezký-Preiss animal model and its applications

In this part of the review we introduce an variant of the cluster expansion method and illustrate its application to polymer systems, finite inhomogeneous Markov chains and the Ising model. Other more involved applications are discussed in the rest of the paper.

§2. Kotezký -Preiss animal model – definition

The literature on the method of cluster expansion and its application is enormous – see books and rewiev papers [GJ], [MM], [Sim], [Br], [Se], [Ma] and reference there. The history of the method traces to the deeps of theoretical physics. The introduction of cluster expansions as a way of rigorous mathematical investigation is due to Glimm, Jaffe, and Spencer [GJS]. Often the cluster expansion method is treated as a class of ideas and approaches which have to be additionally specialized and modified for application to any concrete situation. But there is also a tendency to find a unified approach. An essential contribution to it was made by Gruber and Kunz [GK] who introduced so-called polymer model (see $ 3). The method of cluster expansion can be applied to polymer models and a lot of various situations can be reduced to them. Even a more general and so more convenient for a reduction to it model was introduced by Kotezký and Preiss [KP]. This model described in this section seems very special from the first glance, but, as we explain below, it can be applied to a study of a surprisingly wide class of different probabilistic objects.

Describe the Kotezký-Preiss model. Let Θ be a finite set. Its elements will be called *animals* and so we call this model the *animal model.*[1] Assume that a subset $S \subseteq \Theta\times\Theta$ of the set of pairs (θ_1, θ_2) of animals is fixed such that it is symmetric, i.e.,

[1]In the previous papers the elements of this set were called polymers or alternatively contours. But in the terminology of statistical mechanics these words are attached to some concrete objects, even in applications these elements can have various and sometimes exotic nature. So we propose the term animal and even develop further this animal terminology.

a pair $(\theta_1,\theta_2) \in \mathcal{S}$ if and only if the pair $(\theta_2,\theta_1) \in \mathcal{S}$ and reflexive, i.e. the diagonal pairs $(\theta,\theta) \in \mathcal{S}$. About the pairs of animals $(\theta_1,\theta_2) \in \mathcal{S}$ we say that they are *compatible animals* and write $\theta_1 \leftrightarrow \theta_2$. About pairs of animals $(\theta_1,\theta_2) \in (\Theta\times\Theta)\setminus\mathcal{S}$ we say that they are *incompatible animals* and write $\theta_1 \not\leftrightarrow \theta_2$. The pair $(\Theta,\mathcal{S})$ is called the *animal model.* Sometimes it is convenient to consider the undirected graph without loops and multiple edges with the set Θ of its vertices such that the vertices $\theta_1,\theta_2 \in \Theta$ are connected by an edge of this graph if and only if $\theta_1 \not\leftrightarrow \theta_2$. It is evident that this graph describe the animal model $(\Theta,\mathcal{S})$ in a unique way.

Fix an animal model $(\Theta,\mathcal{S})$. A finite subset $\pi \subseteq \Theta$ is called *a herd*, if any two animals $\theta_1,\theta_2 \in \pi$ are compatible. For any finite $\Lambda \subseteq \Theta$ the set of all herds π such that $\pi \subseteq \Lambda$ is denoted $H(\Lambda)$ and is called the set of all *herds in* Λ. (The set $H(\Lambda)$ includes the empty herd which is denoted by $\emptyset$.)

Assume that a complex-valued function $w(\theta), \theta \in \Theta$, is given. The number $w(\theta)$ is called *the weight of the animal* θ. Let a finite set $\Lambda \subseteq \Theta$. The number

$$Z_w(\Lambda) = \sum_{\pi\in H(\Lambda)} \prod_{\theta\in\pi} w(\theta) \tag{2.1}$$

is called *the partition function in* Λ *defined by the weights* $w = (w(\theta),\theta \in \Theta)$. (In the case $\pi = \emptyset$ the product in (2.1) is interpreted as the number 1.[2] So if Λ is an empty set, the partition function $Z_w(\Lambda) = 1$.)

Assume now that the weighs $w(\theta) \geq 0$ are nonnegative and the function $w(\theta),\theta \in \Theta$, is not identically vanished. Consider the probabilty distribution

$$p_w(\pi) = \frac{\prod_{\theta\in\pi} w(\theta)}{Z_w(\Theta)}, \quad \pi \in H(\Theta). \tag{2.2}$$

This probability distribution is called *the probability distribution in the animal model* Θ *defined by the weights* w. The main objects connected with this probability distribution can be represented as ratios of some partition function. Indeed, fixing disjoint sets of animals $\Lambda_+,\Lambda_- \subseteq \Theta$, we consider the probability

$$P_w(\Lambda_+,\Lambda_-) = \sum_{\pi\in H(\Theta):\Lambda_+\subseteq\pi,\Lambda_-\cap\pi=\emptyset} p_w(\pi). \tag{2.3}$$

of the event that the herd π contains the set of animals Λ_+ and contains no animals from the set of animals Λ_-. It follows from the definitions (2.1) and (2.2) that

$$P_w(\Lambda_+,\Lambda_-) = \begin{cases} \frac{Z_w(\Theta\setminus(\Lambda_+\cup\Lambda_-))\prod_{\theta\in\Lambda_+} w(\theta)}{Z_w(\Theta)}, & \text{if } \Lambda_+ \in H(\Theta), \\ 0, & \text{if } \Lambda_+ \notin H(\Theta). \end{cases} \tag{2.4}$$

Further, consider a real-valued function $\alpha(\theta),\theta \in \Theta$, the random variable

$$S_\alpha(\pi) = \sum_{\theta\in\pi} \alpha(\theta), \quad \pi \in H(\Theta), \tag{2.5}$$

[2] We always suppose that the sum of an empty set of terms equals 0 and the product of the empty set of multipliers equals 1.

and its characteristic function

$$\phi_\alpha(z) = \sum_{\pi \in H(\Theta)} p_w(\pi) e^{zS_\alpha(\pi)}. \tag{2.6}$$

(Since we consider a random variable with a finite set of values, we can define the characteristic function for all complex $z \in \mathbb{C}$.) Then

$$\phi_\alpha(z) = \frac{Z_w(\Theta, z)}{Z_w(\Theta)}, \tag{2.7}$$

where $Z_w(\Theta, z)$ is the partition function defined as above, but for the new weight function

$$w^z(\theta) = w(\theta) e^{z\alpha(\theta)}, \quad \theta \in \Theta. \tag{2.8}$$

These examples demonstrate why a study of logarithms of partition functions is so important for a study of probabilistic properties of animal models and models which can be reduced to animal models.

A good control of logarithms of partition functions in the animal models is possible only if the absolute values $|w(\theta)|$ are small enough in some sence. It leads to the main restrictions on the domain of applicability of the discussed approach. To be more definite we describe a condition on the weights which is used below and was introduced by Kotezký and Preiss [KP]. Assume that a positive-valued function $b(\theta), \theta \in \Theta$, is given. The value $b(\theta)$ will be called *the might of the animal* θ. A selection of the function $b(\theta)$ for a concrete animal model with given weights is defined simply by a wish to satisfy the needed conditions. Roughly speaking, the might $b(\theta)$ has to be large if the animal θ is incompatible with a lot of other animals.

The Kotezký-Preiss Condition. *We say that a weight function* $w(\theta), \theta \in \Theta$, *satisfies the Kotezký-Preiss Condition if there exists an non-negative weight function* $w_0(\theta) \geq 0, \theta \in \Theta$, *such thar for any* $\theta \in \Theta$

$$\exp\left\{ \sum_{\tilde\theta \in \Theta: \tilde\theta \nleftrightarrow \theta} w_0(\tilde\theta) b(\tilde\theta) + w_0(\theta) b(\theta) \right\} \leq b(\theta). \tag{2.9}$$

and

$$|w(\theta)| \leq w_0(\theta), \quad \theta \in \Theta. \tag{2.10}$$

The Kotezký-Preiss condition is fulfilled for the weights $w^z(\theta), \theta \in \Theta$, defined in (2.8) and used in the animal representation (2.7) of the characteristic function, if the weights $w_0(\theta)$ satisfy to condition (2.9), $|w(\theta)| < w_0(\theta), \theta \in \Theta$, and $|z|$ is small enough. The requirement of smallness of $|z|$ is not very restrictive. We will see (in §5 and §?) that it does not disturb a possibility to derive the central limit theorem, to develop the theory of moderate deviations and so on and that in some cases a modification of the animal model permit to avoid this requirement (see §11).

§3. Reduction to the animal model – polymer models

In the following three sections we discuss some examples of probabilistic situations which can be reduced to a study of animal models. More involved examples will be discussed in the following parts of the paper.

Let $\mathbb{G}$ be a countable (or a finite) set and $\mathcal{G}$ be a fixed set of finite subsets $G \subseteq \mathbb{G}$. The elements $G \in \mathcal{G}$ are called *polymers.*[3] Let $V \subseteq \mathbb{G}$ be a finite set which is called a *volume* and $\mathcal{G}(V) \subseteq \mathcal{G}$ be a set of all polymers $G \in \mathcal{G}$ such that $G \subseteq V$. Any finite (including the empty one) subset $\gamma \subseteq \mathcal{G}(V)$ such that all the pairs of polymers $G_1, G_2 \in \gamma$ are disjoint is called a *polymer configuration in the volume* V. The set of all such polymer configuration is denoted by $R(V)$. Assume that a nonnegative weight function $w(G), G \in \mathcal{G}$, is given. The *polymer probability distribution* on the system of all polymer configurations $\gamma \in R(V)$ is defined by the relation

$$p_V(\gamma) = \frac{\prod_{G\in\gamma} w(G)}{Z(V)}, \quad \gamma \in R(V), \tag{3.1}$$

where the partition function

$$Z(V) = \sum_{\gamma\in R(V)} \prod_{G\in\gamma} w(G). \tag{3.2}$$

In the special case $w(G) \equiv 1$ all the possible polymer configurations are equiprobable.

It is evident that this polymer model is a special case of the animal model considered in $ 2. Indeed, it is possible to treat polymers as animals if we let $\Theta = \mathcal{G}(V)$ and say that polymers $G_1, G_2 \in \mathcal{G}(V), G_1 \neq G_2$ are compatible $G_1 \leftrightarrow G_2$ if $G_1 \cap G_2 = \emptyset$. Then the polymer probability distribution (3.1) coincides with the animal probability distribution (2.2).

It is possible to consider the random variable

$$S_V^{\alpha}(\gamma) = \sum_{G\in\gamma} \alpha(G), \quad \gamma \in R(V), \tag{3.3}$$

where $\alpha(G), G \in \mathcal{G}$, is a real-valued function, and its characteristic function (cf. (2.6))

$$\phi_V^{\alpha}(z) = \sum_{\gamma\in R(V)} p_V(\gamma) e^{zS_V^{\alpha}(\gamma)}. \tag{3.4}$$

It is clear that

$$\phi_V^{\alpha}(z) = \frac{Z(V,z)}{Z(V)}, \tag{3.5}$$

where

$$Z(V,z) = \sum_{\gamma\in R(V)} \prod_{G\in\gamma} w(G) e^{z\alpha(G)}. \tag{3.6}$$

---

[3] It would be naive to compare too seriously this mathematical definition of polymers with the corresponding physical object.

The partition function $Z(V, z)$ is again the partition function in an animal model with the same set of animals $\Theta = \mathcal{G}(V)$ and the same compatibility condition as above but with a weight function

$$w_z(G) = w(G)e^{z\alpha(G)}. \tag{3.7}$$

Consider now the situation, when $\mathbb{G}$ is the d-dimensional integer lattice $\mathbb{Z}^d$ with the points $t = (t_1, t_2, \ldots, t_d), t_i \in \mathbb{Z}^1, i = 1, 2, \ldots, d$. and the polymer model is translation invariant. The translation invariance means that, if a set G belongs to $\mathcal{G}$, then for any $a \in \mathbb{Z}^d$ the shift $G + a$ also belongs to $\mathcal{G}$ and

$$w(G) \equiv w(G + a), \quad G \in \mathcal{G}, a \in \mathbb{Z}^d. \tag{3.8}$$

In another words, the set of polymers $\mathcal{G}$ consists of disjoint classes of polymers $\overline{\mathcal{G}}$ invariant with respect to the shifts and for some $w(\overline{\mathcal{G}}) \geq 0$

$$w(G) = w(\overline{\mathcal{G}}), \quad \text{if } G \in \overline{\mathcal{G}}. \tag{3.9}$$

The classes $\overline{\mathcal{G}}$ are called the *types of polymers* and the numbers $w(\overline{\mathcal{G}})$ are called the *weights of the types of polymers*. These weights characterise (in some indirect way) the densities of polymers of different types. (If $w(\overline{\mathcal{G}}) \to 0$ for all types of polymers, the polymer probability distribution asymptotically concentrates on the empty polymer configuration.)

As an example we consider the *domino model*. It is a translation invariant polymer model with $d = 2$ in which there are only two types of polymers: the type of two-point sets $\{(t_1, t_2), (t_1, t_2 + 1)\}, (t_1, t_2) \in \mathbb{Z}^2$, (horizontally situated dominoes) and the type of two-point sets $\{(t_1, t_2), (t_1 + 1, t_2)\}, (t_1, t_2) \in \mathbb{Z}^2$, (vertically situated dominoes). See a configuration of the domino model at Fig 3.1.

Another example is a translation invariant polymer model in which there is only one type of polymers – the $(d+1)$-point sets $B_t, t = (t_1, t_2, \ldots, t_d) \in \mathbb{Z}^d$ such that $B_t = \{(t_1, t_2, \ldots, t_d), (t_1 + 1, t_2, \ldots, t_d), (t_1, t_2 + 1, \ldots, t_d), \ldots, (t_1, t_2, \ldots, t_d + 1)\}$ (see Fig 3.2). Let $V' \subset V$ be the set of all $t \in V$ such that $B_t \subseteq V$. It is easy to see that there is an one-to-one correspondence between the elements of the set $R(V)$ of configurations in a volume $V \subset \mathbb{Z}^d$ of this polymer model and the set $\mathcal{Q}(V')$ of all subsets $Q \subseteq V'$ such that the distances between any pair of points $t^1, t^2 \in Q$ are more then 1. In this case the polymer probability distribution induces the probability distribution on the set $\mathcal{Q}(V')$

$$p_{V'}(Q) = \frac{w^{|Q|}}{\sum_{Q \in \mathcal{Q}(V')} w^{|Q|}}, \quad Q \in \mathcal{Q}(V'), \tag{3.10}$$

where $|Q|$ is the number of elements in the set Q. This probability model is well known in theory of Gibbsian fields, where it is called the *hard core system* (see sect. ??). Of course, the hard core model can be described as an animal model in a direct way. For it we let $\Theta = V' \subset \mathbb{Z}^d$ and $t^1 \not\leftrightarrow t^2$ if dist $(t^1, t^2) > 1$.

For the polymer systems a reasonale choice of the mights $b(G), G \in \mathcal{G}$, is

$$b(G) = e^{\kappa|G|}, \tag{3.11}$$

where $\kappa > 0$ is an constant.[4]

Proposition 3.1. *The condition*

$$\sup_{t\in\mathbb{G}} \sum_{G\in\mathcal{G}:t\in G} w_0(G)e^{\kappa|G|} \le \kappa. \tag{3.12}$$

implies the fulfilment of Kotezký-Preiss conditions (2.9), (2.10) for $w(G) \le w_0(G), G \in \mathbb{G}$.

Proof. We can estimate the exponent in the Kotezky-Preiss condition (2.9) as

$$\begin{aligned}\sum_{\widetilde{G}\in\mathcal{G}(V):\widetilde{G}\nsim G} w_0(\widetilde{G})b(\widetilde{G}) + w_0(G)b(G) &\le \sum_{t\in G}\sum_{\widetilde{G}\in\mathcal{G}:t\in\widetilde{G}} w_0(\widetilde{G})e^{\kappa|\widetilde{G}|} \\ &\le |G| \sup_{t\in\mathbb{G}} \sum_{\widetilde{G}\in\mathcal{G}:t\in\widetilde{G}} w_0(\widetilde{G})e^{\kappa|\widetilde{G}|}.\end{aligned} \tag{3.13}$$

So the condition (3.12) implies the condition (2.9). □

For concrete models it is possible to optimize the choice of the parameter κ. For example, in the domino model the number of polymers which contain a fixed point $t \in \mathbb{Z}^2$ equals 4. So, if the weights of both horisontal and vertical dominoes are equal w_0, the condition (3.12) is valid if

$$w_0 \le \frac{1}{4}\kappa e^{-2\kappa}. \tag{3.14}$$

The minimum in the right part of (3.14) is achieved for $\kappa = 1/2$ and so we arrive to the condition

$$w_0 \le \frac{1}{8e}. \tag{3.15}$$

In a similar way for the case of d-dimensional hard core system we obtain the estimate

$$w_0 \le \frac{1}{3(d+1)e}. \tag{3.16}$$

The Kotezký-Preiss condition can be also checked for the weights (3.7) arising in the animal model representation of the partition function $Z(V, z)$ (see (3.6)).

Proposition 3.2. *Let the function α defining the random variable S_V^α be bounded: for some constant K*

$$|\alpha(G)| \le K, G \in \mathcal{G}, \tag{3.17}$$

and for some $\zeta > 0$ the variable z is such that

$$|z| \le \zeta. \tag{3.18}$$

[4] Here and in the following $|A|$ is the number of elements in a finite set A.

Let

$$w_0(G) = w(G)e^{K\varsigma}, \quad G \in \mathbb{G}, \tag{3.19}$$

and assume that for this function w_0 *the condition (3.12) is fulfilled. Then the Kotezký-Preiss condition (2.9) is valid and (cf. (2.10))*

$$|w_z(G)| \leq w_0(G), \quad ,G \in \mathbb{G}. \tag{3.20}$$

Proof. The statement of Proposition follows immediately from Proposition 3.1 and the definition (3.7). □

The restriction (3.18) do not obstruct a possibility to derive the Central Limit Theorem (see §?) and to develope the theory of modern deviations (see §?). From the other side it is impossible to expect that Proposition 3.2 is valid without some restriction of a similar type (see §?).

Of course, the previous estimates are only some sufficient conditions for a possibility to apply the animal model method and using another appropriate choice of mights and more involved considerations we can for some concrete models to improve these estimates, but the requirement that the weights of polymers are small enough is in the essence of the perturbation method. In some cases, for example for the hard core model (see $??), it is possible to use another method of reducing to animal models which permits to study the case of large enough weights $w(G)$. But the case of equiprobable polymer configurations, $w(G) \equiv 1$, belongs to the most difficult intermediate domain of values of weights and only in some cases can be studied by the help of some very involved methods (see again §??).

§4. Reduction to the animal model – inhomogeneous Markov chains

In this section we represent characteristic functions of sums of random variables connected in a time inhomogeneous Markov chain with a finite set of states as a partition function of an animal model. It seems that this reduction is new.

Consider a time inhomogeneous Markov chain with a finite set of states X. Such chain is defined by a sequence of stochastic transition matrices $P^i = (p^i(x,y), x, y \in X), i = 1, 2, \dots,$ and an initial probability distribution $\bar{p}^0 = (\bar{p}^0(x), x \in X)$. (A square matrix P is *stochastic*, if its matrix elements are nonnegative and their sums over any row equals 1.) The probability of a realization $x = (x_0, x_1, \dots, x_n) \in X^{[0,n]}$ on the interval $[0,n] = \{0, 1, \dots, n\} \subset \mathbb{Z}^1$ is equal to

$$\bar{p}^{[0,n]}(x) = \bar{p}^0(x_0)p^1(x_0,x_1)p^2(x_1,x_2)\cdots p^n(x_{n-1},x_n). \tag{4.1}$$

Consider a sequence $\alpha = (\alpha_i(x_i), x_i \in X, i = 0, 1, \dots)$ of real-valued function, the random variables

$$S^n_\alpha(x) = \sum_{i=0}^{n} \alpha_i(x_i), \quad x = (x_0, x_1, \dots, x_n) \in X^{[0,n]}, n = 0, 1 \dots. \tag{4.2}$$

and its characteristic functions

$$\phi^n_\alpha(z) = \sum_{x \in X^{[0,n]}} \bar{p}^{[0,n]}(x) e^{zS^n_\alpha(x)}. n = 0, 1, \dots \tag{4.3}$$

This characteristic function can be represented in a matrix form. Let $\Phi_\alpha^k(z) = (\varphi_{x,y}^k(z),$
$x, y \in X), k = 1, 2, \dots, n$, be the square matrices with the matrix elements

$$\varphi_{x,y}^k(z) = p^k(x, y)e^{z\alpha_k(y)}, \tag{4.4}$$

and

$$\Phi_\alpha^{[0,n]}(z) = \Phi_\alpha^1(z)\Phi_\alpha^2(z)\dots\Phi_\alpha^n(z) \tag{4.5}$$

be the product of these matrices. Let $\bar\varphi_\alpha^0(z) = (\bar\varphi_\alpha^0(z)_{x_0}, x_0 \in X)$ be the vector-row with the components

$$\bar\varphi_\alpha^0(z)_{x_0} = \bar p^0(x)e^{z\alpha_0(x_0)}. \tag{4.6}$$

Then the characteristic function (4.3)

$$\phi_\alpha^n(z) = \sum_{x_n \in X} (\bar\varphi_\alpha^0(z)\Phi_\alpha^{[0,n]}(z))_{x_n}, \tag{4.7}$$

where $(\bar\varphi_\alpha^0(z)\Phi_\alpha^{[0,n]}(z))_{x_n}$ is the component of the vector-row $\bar\varphi_\alpha^0(z)\Phi_\alpha^{[0,n]}(z)$ labeled x_n.

The Markov chain is time homogeneous if the transition matrices $P^k \equiv P, k = 1, 2, \dots$, do not depend on k. If also $\alpha_i(x_i) \equiv \alpha_0(x_i)$, then the matrices $\Phi_\alpha^k(z) \equiv \Phi_\alpha(z)$ coincide also and (recall (4.5))

$$\Phi_\alpha^{[0,n]}(z) = (\Phi_\alpha(z))^n. \tag{4.8}$$

There are well-known algebraic methods to study the asymptotic behaivour of powers of matrices. Applied to the characteristic function (4.7) these methods give a lot of results about the asymptotics of the probability distribution of random variables S_α^n (the central limit theorems, large and moderate deviations and so on).

In the general case of inhomogeneous chains the matrices $\Phi_\alpha^k(z)$ do not commutate and so the algebraic methods do not provide a help. Special ingenious probailistic and analytical methods are invented in theory of limit theorems for inhomogeneous Markov chains and a lot of papers is devoted to this problems (see the books of Ibragimov and Linnik [IL??] and Saulis and Statuleicius [SS]) and the references there. Below it is shown that in the general inhomogeneous case the characteristic function (4.3) can be represented as a partition function of an animal model and so the limit theorems for inhomogeneous Markov chains can be obtained in a more simple and general way.

In the derivation of the animal representation of the characteristic function (4.3) we will use the following notation. Let integers $0 \le a \le b \le n$ and the interval $[a, b] = \{a, a+1, \dots, b\}$. For any realization $x \in X^{[0,n]}$ let $x_{[a,b]}$ be its restriction $(x_a, x_{a+1}, \dots, x_b)$ to the interval $[a, b]$. In all the cases a sum over $x_{[a,b]}$ means the sum over all $x_a, x_{a+1}, \dots, x_b \in X$. For $a > 0$ the conditional probability of $x_{[a,b]}$ for a fixed x_{a-1} equals to

$$p^{[a,b]}(x_{a-1}, x_{[a,b]}) = p^{a-1}(x_{a-1}, x_a)p^a(x_a, x_{a+1}) \times \dots \times p^{b-1}(x_{b-1}, x_b). \tag{4.9}$$

Observe that

$$\sum_{x_{[a,b]}} p^{[a,b]}(x_{a-1}, x_{[a,b]}) \equiv 1. \tag{4.10}$$

Consider also the transition probabilities from the time $a-1$ to the time b

$$p^{a-1,b}(x_{a-1}, x_b) = \sum_{x_{[a,b-1]}} p^{[a,b]}(x_{a-1}, x_{[a,b]}). \tag{4.11}$$

which are the matrix elements of the matrix $P^a P^{a+1} \cdots P^b$. For any $a > 0$ consider unconditional probability distribution at the time a

$$\bar{p}^a(x_a) = \sum_{x_0} \bar{p}^0(x_0) p^{0,a}(x_0, x_a). \tag{4.12}$$

It is clear that for any $0 < a \leq b$

$$\bar{p}^b(x_b) = \sum_{x_{a-1}} \bar{p}^{a-1}(x_{a-1}) p^{a-1,b}(x_{a-1}, x_b). \tag{4.13}$$

At last consider unconditional probability distribution on the interval $[a,b]$

$$\bar{p}^{[a,b]}(x_{[a,b]}) = \bar{p}^a(x_a) p^{[a+1,b]}(x_a, x_{[a+1,b]}). \tag{4.14}$$

Removing the parentheses, we can rewrite the characteristic function (4.3) as

(4.15)

$$\begin{aligned}
\phi_\alpha^n(z) &= \sum_{x \in X^{[0,n]}} \bar{p}^{[0,n]}(x) \left(e^{z\alpha_0(x_0)} - 1 + 1\right) \left(e^{z\alpha_1(x_1)} - 1 + 1\right) \times \cdots \times \left(e^{z\alpha_n(x_n)} - 1 + 1\right) \\
&= \sum_{x \in X^{[0,n]}} \bar{p}^{[0,n]}(x) \left(\sum_{A \subseteq [0,n]} \prod_{a \in A} \left(e^{z\alpha_a(x_a)} - 1\right) \right) \\
&= \sum_{A \subseteq [0,n]} \left(\sum_{x \in X^{[0,n]}} \bar{p}^{[0,n]}(x) \prod_{a \in A} \left(e^{z\alpha_a(x_a)} - 1\right) \right).
\end{aligned}$$

(For the empty set $A = \emptyset$ the product is interpreted as 1). Now we fix a set $A \subseteq [0,n]$ and consider the corresponding term in the right part of (4.15). The set A can be represented in a unique way as a sum of some number k of intervals

(4.16)

$$A = \cup_{i=1}^k [a_i, b_i], \quad \text{where } a_i \leq b_i, i = 1, 2, \cdots, k \text{ and } b_i < a_{i+1} - 1, i = 1, 2, \ldots, k-1.$$

Using the notations (4.9) and (4.14) we can write that
(4.17)

$$\sum_{x\in X^{[0,n]}} \bar{p}^{[0,n]}(x)\prod_{a\in A}\left(e^{z\alpha_a(x_a)}-1\right)=\sum_{x\in X^{[0,n]}}\left(\bar{p}^{[0,a_1]}(x_{[0,a_1]})\right.$$

$$\times\, p^{[a_1+1,b_1]}(x_{a_1},x_{[a_1+1,b_1]})\prod_{u\in[a_1,b_1]}\left(e^{z\alpha_u(x_u)}-1\right)p^{[b_1+1,a_2]}(x_{b_1},x_{[b_1+1,a_2]})$$

$$\times\, p^{[a_2+1,b_2]}(x_{a_2},x_{[a_2+1,b_2]})\prod_{u\in[a_2,b_2]}\left(e^{z\alpha_u(x_u)}-1\right)\times\cdots\times p^{[a_k+1,b_k]}(x_{a_k},x_{[a_k+1,b_k]})$$

$$\left.\times\prod_{u\in[a_k,b_k]}\left(e^{z\alpha_u(x_u)}-1\right)p^{[b_k+1,n]}(x_{b_k},x_{[b_k+1,n]})\right).$$

(Here in a degenerate case $a_j=b_j$ the multiplier $p^{[a_j+1,b_j]}(x_{a_j},x_{[a_j+1,b_j]})$ is interpreted as the number 1. The same is assumed about the last multiplier $p^{[b_k+1,n]}(x_{b_k},x_{[b_k+1,n]})$, if $b_k=n$. In the case $k=0$ the right side of (4.17) is interpreted as 1). Let

$$\chi^{[a,b]}(x_a,x_b)=\sum_{x_{[a+1,b-1]}} p^{[a+1,b]}(x_a,x_{[a+1,b]})\prod_{u\in[a,b]}\left(e^{z\alpha_u(x_u)}-1\right). \tag{4.18}$$

(Here in a degenerate case $a=b$ the function $\chi^{[a,b]}(x_a,x_a)=e^{z\alpha_a(x_a)}-1$.) Now we sum the right side of (4.17) over all $x_t, t\in[0,n]$, except the variables $x_{a_j},x_{b_j}, j=1,2,\ldots,k$, corresponding to the ends of the intervals of the partition (4.16). Summing over $x_{[0,a_1-1]}$ we use the relation (4.11) and (4.13), summing over $x_{[a_j+1,b_j-1]}$ we use the definition (4.18), summing over $x_{[b_j+1,a_j-1]}$ we use the relation (4.11) and at last summing over $x_{[b_k+1,n]}$ we use the identity (4.10). In the result we find
(4.19)

$$\phi_\alpha^n(z)=\sum_{A subseteq[0,n]}\sum_{x\in X^{[0,n]}}\bar{p}^{[0,n]}(x)\prod_{a\in A}\left(e^{z\alpha_a(x_a)}-1\right)=1+$$

$$\sum_{A\subseteq[0,n],A\neq\emptyset}\ \sum_{x_{a_1},x_{b_1},\ldots,x_{a_k},x_{b_k}}\bar{p}^{a_1}(x_{a_1})\chi^{[a_1,b_1]}(x_{a_1},x_{b_1})p^{b_1,a_2}(x_{b_1},x_{a_2})\chi^{[a_2,b_2]}(x_{a_2},x_{b_2})$$

$$\times\, p^{b_2,a_3}(x_{b_2},x_{a_3})\times\cdots\times p^{b_{k-1},a_k}(x_{b_{k-1}},x_{a_k})\chi^{[a_k,b_k]}(x_{a_k},x_{b_k}).$$

It the concluding formula of the first stage of our calculations.

Going to the second stage of the calculations we consider the set $J(A)$ of all intervals $[b_1,a_2],[b_2,a_3],\ldots,[b_{k-1},a_k]$ connecting the intervals of the set A. We can rewrite the terms in (4.19) as
(4.20)

$$\sum_{x\in X^{[0,n]}}\bar{p}^{[0,n]}(x)\prod_{a\in A}\left(e^{z\alpha_a(x_a)}-1\right)=$$

$$\sum_{x_{a_1},x_{b_1},\ldots,x_{a_k},x_{b_k}}\bar{p}^{a_1}(x_{a_1})\chi^{[a_1,b_1]}(x_{a_1},x_{b_1})(p^{b_1,a_2}(x_{b_1},x_{a_2})-\bar{p}^{a_2}(x_{a_2})+\bar{p}^{a_2}(x_{a_2}))$$

$$\times\,\chi^{[a_2,b_2]}(x_{a_2},x_{b_2})(p^{b_2,a_3}(x_{b_2},x_{a_3})-\bar{p}^{a_3}(x_{a_3})+\bar{p}^{a_3}(x_{a_3}))$$

$$\times\cdots\times(p^{b_{k-1},a_k}(x_{b_{k-1}},x_{a_k})-\bar{p}^{a_k}(x_{a_k})+\bar{p}^{a_k}(x_{a_k}))\chi^{[a_k,b_k]}(x_{a_k},x_{b_k}).$$

Removing the brackets in the right side of (4.20), we find that

$$\sum_{x\in X^{[0,n]}} \bar{p}^{[0,n]}(x) \prod_{a\in A}\left(e^{z\alpha_a(x_a)}-1\right) = \sum_{B\subseteq J(A)} v(A,B), \tag{4.21}$$

where

(4.22)

$$v(A,B) = \sum_{x_{a_1},x_{b_1},\dots,x_{a_k},x_{b_k}} \bar{p}^{a_1}(x_{a_1})\chi^{[a_1,b_1]}(x_{a_1},x_{b_1})q^1_{A,B}(x_{b_1},x_{a_2})\chi^{[a_2,b_2]}(x_{a_2},x_{b_2})$$
$$q^2_{A,B}(x_{b_2},x_{a_3})\times\dots\times q^{k-1}_{A,B}(x_{b_{k-1}},x_{a_k})\chi^{[a_k,b_k]}(x_{a_k},x_{b_k})$$

and

(4.23)

$$q^j_{A,B}(x_{b_j},x_{a_{j+1}}) = \begin{cases} p^{b_j,a_{j+1}}(x_{b_j},x_{a_{j+1}}) - \bar{p}^{a_{j+1}}(x_{a_{j+1}}), & \text{if } [b_j,a_{j+1}]\in B,\\ \bar{p}^{a_{j+1}}(x_{a_{j+1}}), & \text{if } [b_j,a_{j+1}]\notin B.\end{cases}$$

Recalling (4.15), we find that the characteristic function

$$\phi^n_\alpha(z) = 1 + \sum_{A\subseteq[0,n],A\neq\emptyset,B\subseteq J(A)} v(A,B). \tag{4.24}$$

(The term 1 corresponds to the set $A=\emptyset$ in (4.25)).

Now we want to rewrite the quantities $v(A,B)$ in a product form. For fixed A,B let $j_1,j_2,\dots,j_s$ be the subsequence of the sequence of indexes $1,2,\dots,k-1$ containing all the indexes j such that the interval $[b_j,a_{j+1}]\notin B$. Let $j_0=0$. Using (4.23), we find that

(4.25)

$$\bar{p}^{a_1}(x_{a_1})\chi^{[a_1,b_1]}(x_{a_1},x_{b_1})q^1_{A,B}(x_{b_1},x_{a_2})\chi^{[a_2,b_2]}(x_{a_2},x_{b_2})q^2_{A,B}(x_{b_2},x_{a_3})$$
$$\times\dots\times q^{k-1}_{A,B}(x_{b_{k-1}},x_{a_k})\chi^{[a_k,b_k]}(x_{a_k},x_{b_k})$$
$$=\prod_{l=0}^{s}\Big(\bar{p}^{a_{j_l+1}}(x_{a_{j_l+1}})\chi^{[a_{j_l+1},b_{j_l+1}]}(x_{a_{j_l+1}},x_{b_{j_l+1}})$$
$$\times(p^{b_{j_l+1},a_{j_l+2}}(x_{b_{j_l+1}},x_{a_{j_l+2}})-\bar{p}^{a_{j_l+2}}(x_{a_{j_l+2}}))\chi^{[a_{j_l+2},b_{j_l+2}]}(x_{a_{j_l+2}},x_{b_{j_l+2}})$$
$$\times\dots\times(p^{b_{j_{l+1}-1},a_{j_{l+1}}}(x_{b_{j_{l+1}-1}},x_{a_{j_{l+1}}})-\bar{p}^{a_{j_{l+1}}}(x_{a_{j_{l+1}}}))\chi^{[a_{j_{l+1}},b_{j_{l+1}}]}(x_{a_{j_{l+1}}},x_{b_{j_{l+1}}})\Big).$$

Now we need a simple observation. Since we will use it also later, we formulated it in a general form. Let T and X be finite sets and configurations $x=(x_t,t\in T)\in X^T$. Let a partition $T=T_1\cup T_2\cup\ldots\cup T_k$ of the set T into mutually disjoint subsets $T_j, j=1,2,\dots,k$, be fixed and $q_j(x_i,i\in T_j), j=1,2,\dots,k$, be some functions of the components $x_i\in T_j$. Then

$$\sum_{x\in X^T}\left(\prod_{j=1}^k q_j(x_i,i\in T_j)\right) = \prod_{j=1}^k\left(\sum_{x_i\in X,i\in T_j} q_j(x_i,i\in T_j)\right). \tag{4.26}$$

Applying (4.26) for $T = \{a_1, b_1, a_2, b_2, \dots, a_k, b_k\}$ and $T_j = \{a_{j+1}, b_{j+1}, a_{j+2}, b_{j+2} \dots, a_{j_{l+1}}, b_{j_{l+1}}\}$ we derive from (4.22), (4.25) and (4.26) that
(4.27)

$$v(A, B) = \prod_{l=0}^{s} \Bigg(\sum_{x_{a_{j_l+1}}, x_{b_{j_l+1}}, x_{a_{j_l+2}}, x_{b_{j_l+2}}, \dots, x_{a_{j_{l+1}}}, x_{b_{j_{l+1}}}} \Big(\bar{p}^{a_{j_l+1}}(x_{a_{j_l+1}})\chi^{[a_{j_l+1}, b_{j_l+1}]}(x_{a_{j_l+1}}, x_{b_{j_l+1}})$$

$$\times (p^{b_{j_l+1}, a_{j_l+2}}(x_{b_{j_l+1}}, x_{a_{j_l+2}}) - \bar{p}^{a_{j_l+2}}(x_{a_{j_l+2}}))\chi^{[a_{j_l+2}, b_{j_l+2}]}(x_{a_{j_l+2}}, x_{b_{j_l+2}})$$

$$\times \cdots \times (p^{b_{j_{l+1}-1}, a_{j_{l+1}}}(x_{b_{j_{l+1}-1}}, x_{a_{j_{l+1}}}) - \bar{p}^{a_{j_{l+1}}}(x_{a_{j_{l+1}}}))\chi^{[a_{j_{l+1}}, b_{j_{l+1}}]}(x_{a_{j_{l+1}}}, x_{b_{j_{l+1}}})\Big)\Bigg).$$

The relations (4.24) and (4.27) can be interpreted as an animal model representation for the characteristic function $\phi_\alpha^n(z)$.

Proposition 4.1. *Let the set of animals Θ_n consists of all the sequences of integers $\theta = (s = s_1, t_1, s_2, t_2, \dots, s_m, t = t_m)$ such that $s_i \le t_i, i = 1, 2, \dots, m, t_i < s_{i+1} - 1, i = 1, 2, \dots, m-1, 0 \le s \le t \le n$ (for $m = 1$ the animal $\theta = (s, t)$.) The interval $[s, t]$ is called the support of the animal θ. Assume that a pair of animals is compatible, if the distance between their support is more 1. Let the weights*

$$\begin{aligned} w(\theta) = &\sum_{x_s, x_{t_1}, x_{s_2}, x_{t_2}, \dots, x_{s_m}, x_t} \Big(\bar{p}^s(x_s)\chi^{s,t_1}(x_s, x_{t_1}) \\ &\times (p^{t_1, s_2}(x_{t_1}, x_{s_2}) - \bar{p}^{s_2}(x_{s_2}))\chi^{s_2, t_2}(x_{s_2}, x_{t_2}) \times \dots \\ &\times (p^{t_{m-1}, s_m}(x_{t_{m-1}}, x_{s_m}) - \bar{p}^{s_m}(x_{s_m}))\chi^{s_m, t}(x_{s_m}, x_t)\Big), \quad \theta \in \Theta_n. \end{aligned} \tag{4.28}$$

Then the characteristic function

$$\phi_\alpha^n(z) = Z_w(\Theta_n), \tag{4.29}$$

where the partition function $Z_w(\Theta_n)$ is defined by the relation (2.1).

Proof. The right part of (4.28) differs from the multipliers in (4.27) by notations only. So recalling (4.24), it is enough to observe that when $A \subseteq [0, n], A \ne \emptyset, B \subseteq J(A)$ pass all possible values, any nonempty compatible sets of animals $\theta \in \Theta_n$ arise as sets of multipliers in (4.27). The term 1 in (4.24) corresponds to the empty set of animals. □

It is possible to construct an animal representtion for the characteristic function $\phi_\alpha^n(z)$ in a more simple way, but the advantage of the approach described above is a possibility to prove under some natural additional hypothesis that for small enough $|z|$ the Kotezk'y-Preiss condition is fulfilled for the considered construction. It seems that this approach is new.

We assume that the considered inhomogeneous Markov chain is *exponentially ergodic*. It means that for some constants $\delta > 0$ and $C < \infty$ and any $0 \le s \le t$ and $x_s \in X$

$$\sum_{x_t} |p^{s,t}(x_s, x_t) - \bar{p}^t(x_t)| \le C \exp\{-\delta |t - s|\}. \tag{4.30}$$

This restriction is weak enough in the case of a finite space X considered here. For example, it is certainly valid, if all the transiition probabilities are uniformly bounded away from zero:

$$p^j(x,y) \geq \mu > 0, \quad x, y \in X, j = 1, 2, \dots . \tag{4.31}$$

The problem of ergodicity of finite Markov chains is well studied in the probabilistic literature and we will not discuss it here. The question about applicability of animal model approach to Markov chains with an arbitrary space X is discussed in §?.

Proposition 4.2. **A)** *For some $\gamma > 0, \epsilon > 0$ and any $\theta = (s = s_1, t_1, s_2, t_2, \dots, s_m, t = t_m) \in \Theta_n$ we let*

$$w_0(\theta) = \epsilon^m \exp\{-\gamma(t-s)\}, \tag{4.32}$$

and consider the mights

$$b(\theta) = e^{\kappa(t-s)}, \tag{4.33}$$

where we assume that $0 < \kappa < \gamma$. Then there exists a small enough value $\epsilon > 0$ depending on the difference $\gamma - \kappa$ only such that for all n the Kotecký-Preiss condition (2.9) is valid for the weights (4.32) and the mights (4.33).

B) *Assume that the considered Markov chain is exponentially ergodic and there is a constant $A < \infty$ such that*

$$|\alpha_i(x_i)| \leq A, \quad x_i \in X, i = 0, 1, \dots . \tag{4.34}$$

For $\gamma = \delta$ and any $\epsilon > 0$ there exists a value $\zeta > 0$ so small that for

$$|z| \leq \zeta \tag{4.35}$$

the weights $w(\theta)$, defined by the relations (4.28) and (4.18), are such that

$$|w(\theta)| \leq w_0(\theta), \quad \theta \in \Theta_n, n = 1, 2, \dots . \tag{4.36}$$

Proof. **A)** Let $\Theta_n([s,t]), 0 \leq s \leq t \leq n$ be the set of all animals $\theta \in \Theta_n$ with a given support $[s,t]$ and $\Theta_n([s,t];m)$ be the subset of $\Theta_n([s,t])$ consisting of animals with a given value of the parameter m. It follows from the definition of animals that the set $\Theta_n([s,t];m)$ is empty, if $m > t-s+1$. For $2 \leq m \leq t-s+1$ the number of elements

$$|\Theta_n([s,t];m)| \leq \binom{t-s}{m-1}^2. \tag{4.37}$$

(We have estimated separately the numbers of ways to choose the $m-1$ points $s_i, i = 2, 3, \dots, m$ and the $m-1$ points $t_i, i = 1, 2, \dots, m-1$ by the binomial coefficients.) Observe also that $|\Theta_n([s,t];1)| = 1$. Recalling the definition (4.32), we find that

$$\sum_{\theta\in\Theta_n([s,t])} w_0(\theta) \le \exp\{-\gamma(t-s)\}\left(\epsilon+\left(\sum_{m=2}^{t-s+1}\binom{t-s}{m-1}^2\epsilon^m\right)\right)$$

(4.38)
$$\le \epsilon\exp\{-\gamma(t-s)\}\left(1+\left(\sum_{m=1}^{t-s}\binom{t-s}{m}\epsilon^{\frac{m}{2}}\right)^2\right)$$

$$\le \epsilon\exp\{-\gamma(t-s)\}\left(1+\sqrt{\epsilon})^{|t-s|}\right)^2.$$

It follows from (4.38) that for all small enough ϵ and some γ' such that $\kappa<\gamma'<\gamma$ and all intervals $[s,t]$

$$\sum_{\theta\in\Theta_n([s,t])} w_0(\theta) \le \epsilon\exp\{-\gamma'(t-s)\}. \tag{4.39}$$

Observe now that for any fixed interval $[s,t]$ and any integer k the number $N([s,t],k)$ of intervals $[\tilde{s},\tilde{t}]$ such that $\tilde{t}-\tilde{s}+1=k$ and the distance between $[\tilde{s},\tilde{t}]$ and $[s,t]$ do not exceed 1 can be estimated as

$$N([s,t],k)\le t-s+k+2. \tag{4.40}$$

Using (4.39), (4.40) and recalling the definition (4.33), we obtained that for any $\theta\in\Theta_n$ with the support $[s,t]$

(4.41)
$$\sum_{\tilde{\theta}\in\Theta_n:\tilde{\theta}\nleftrightarrow\theta} w_0(\tilde{\theta})b(\tilde{\theta})+w_0(\theta)b(\theta)$$

$$\le\sum_{k=1}^{n}N([s,t],k)\epsilon\exp\{-(\gamma'-\kappa)k\}\le\epsilon\sum_{k=1}^{\infty}(t-s+k+2)\exp\{-(\gamma'-\kappa)k\}.$$

Since $\gamma'-\kappa>0$, the series in the right side of (4.41) converges and, more of it, for small enough ϵ and all $t-s$ this right side do not exceed $\kappa(t-s)$. So for such ϵ

$$\sum_{\tilde{\theta}\in\Theta_n:\tilde{\theta}\nleftrightarrow\theta} w_0(\tilde{\theta})b(\tilde{\theta})+w_0(\theta)b(\theta)\le\ln b(\theta) \tag{4.42}$$

and it means that the Kotecký Preiss condition (2.9) is fulfilled.

B) Let

$$\mu=\left|e^{\zeta A}-1\right|. \tag{4.43}$$

It follows from the definition (4.18) and the conditions (4.34), (4.35) that for any $a\le b$ and any $x_a\in X$

$$\sum_{x_b}|\chi^{[a,b]}(x_a,x_b)|\le\mu^{b-a+1}. \tag{4.44}$$

In the first place we want to check that for $\theta = (s = s_1, t_1, s_2, t_2, \dots, s_m, t = t_m)$
(4.45)
$$|w(\theta)| \le \mu^{t_1 - s_1 + 1} C \exp\{-\delta|t_2 - s_2|\}\mu^{t_2 - s_2 + 1} \times \ldots \times C \exp\{-\delta|s_m - t_{m-1}|\}\mu^{t_m - s_m + 1}.$$

Really, recalling the definition (4.28) and applying the estimate (4.44) to the last multiplier in (4.28), we find that

$$(4.46)\quad \begin{aligned} |w(\theta)| \le \mu^{t_m - s_m + 1} & \sum_{x_s, x_{t_1}, x_{s_2}, x_{t_2}, \dots, x_{s_m}} |\bar{p}^s(x_s)\chi^{s,t_1}(x_s, x_{t_1}) \\ & \times (p^{t_1, s_2}(x_{t_1}, x_{s_2}) - p^{s_2}(x_{s_2}))\chi^{s_2, t_2}(x_{s_2}, x_{t_2}) \times \dots \\ & \times (p^{t_{m-1}, s_m}(x_{t_{m-1}}, x_{s_m}) - \bar{p}^{s_m}(x_{s_m}))|. \end{aligned}$$

Now we can apply the estimate (4.30) to the last multiplier in (4.46). We obtain that
(4.47)
$$\begin{aligned} |w(\theta)| \le C \exp\{-\delta|s_m - t_{m-1}|\}\mu^{t_m - s_m + 1} & \sum_{x_s, x_{t_1}, x_{s_2}, x_{t_2}, \dots, x_{t_{m-1}}} |\bar{p}^s(x_s)\chi^{s,t_1}(x_s, x_{t_1}) \\ \times (p^{t_1, s_2}(x_{t_1}, x_{s_2}) - \bar{p}^{s_2}(x_{s_2}))\chi^{s_2, t_2}(x_{s_2}, x_{t_2}) \times \dots \\ \times \chi^{s_{m-1}, t_{m-1}}(x_{s_{m-1}}, x_{t_{m-1}})|. \end{aligned}$$

Now we can estimate the last multiplier $\chi^{s_{m-1}, t_{m-1}}(x_{s_{m-1}}, x_{t_{m-1}})$ in (4.47) and so on. After $2m$ steps we arrive to the desired estimate (4.45).

Suppose now that the value ζ is so small that

$$(4.48)\qquad \sqrt{\mu} < e^{-\delta}.$$

and

$$(4.49)\qquad C\sqrt{\mu} < \epsilon.$$

Then the right side of (4.32) exceeds the right side of (4.45) and we obtained the statement (4.36). □

Observe for the following reference that in the conditions of Proposition 4.2 for small enough ϵ and any n

$$(4.50)\qquad \sum_{\theta \in \Theta_n} w_0(\theta) b(\theta) \le \kappa n.$$

The derivation of this estimate is completely similar to the derivation of the estimate (4.42).

§5. Reduction to the animal model – Ising model.

In this preliminary discussion we consider the simplest nontrivial case of Gibbsian field – the symmetric ferromagnet Ising model with plus-boundary conditions and reduce it to some animal models. It is a reexposition of constructions usually used in the application of cluster expansions to Gibbsian fields (see, for example the book [MM] and reference there. Applications of cluster expansions to wider classes of Gibbsian fields and boundary conditions are discussed in §?.

Let $V \subset \mathbb{Z}^d$ be a finite subset of d-dimensional integer lattice $\mathbb{Z}^d$, which is treated as a graph with edges connecting its points situated at the distance 1 and $X(V) = \{-1, 1\}^V$ be the set of configurations $x = (x_t, t \in V)$, where $x_t = \pm 1$. The (symmetric ferromagnet) *Ising distribution in the volume V with the plus-boundary condition and an inverse temperature* $\beta \geq 0$ is the probability distribution on the set $X(V)$ given by

$$(5.1) \qquad p_V^+(x) = p_V^{+,\beta}(x) = (Z^{+,\beta}(V))^{-1} \exp\{-\beta U_V^+(x)\}, \quad x \in X(V),$$

with the *partition function*

$$(5.2) \qquad Z^+(V) = Z^{+,\beta}(V) = \sum_{x \in X(V)} \exp\{-\beta U_V^+(x)\}$$

and the *energy*

$$(5.3) \qquad U_V^+(x) = - \sum_{\{s,t\} \subseteq V : |s-t|=1} x_s x_t - \sum_{t \in \partial^{\text{in}} V} x_t R_t(V), \quad x \in X(V).$$

Here and below we use the following definitions. The sets

$$(5.4) \qquad \partial V = \{t \in V^c : \text{dist}\,(t, V) = 1\}, \qquad \partial^{\text{in}} V = \{t \in V : \text{dist}\,(t, V^c) = 1\},$$

are called the *(outer) boundary of the set V* and the *inner boundary of the set V*. The numbers

$$(5.5) \qquad R_t(V) = |\{s \in \partial V : |s - t| = 1\}| \quad , t \in \partial^{\text{in}} V,$$

are the numbers of points of the set V^c neighboring to t. For any function $\phi(x), x \in X(V)$, its mean value

$$(5.6) \qquad \langle \phi \rangle_V^+ = \langle \phi \rangle_V^{+,\beta} = \sum_{x \in X(V)} p_V^{+,\beta}(x) \phi(x).$$

(The index β in the introduced notation is ommited when it can not generate misunderstanding).

Let a set $W \subset V$ and a configuration $\tilde{x} = (\tilde{x}_t, t \in W) \in X(W)$ be fixed and $W^+(\tilde{x}) = \{t \in W : \tilde{x}_t = 1\}$, $W^-(\tilde{x}) = \{t \in W : \tilde{x}_t = -1\}$. The probability that the restriction $x|_W$ of the configuration $x \in X(V)$ to W is equal to $\tilde{x}$

$$(5.7) \qquad p_{W,V}^+(\tilde{x}) = \sum_{x \in X(V) : x|_W = \tilde{x}} p_V^+(x) = \left\langle \prod_{t \in W^+(\tilde{x})} \chi_t \prod_{t \in W^-(\tilde{x})} (1 - \chi_t) \right\rangle_V^+,$$

where $\chi_t, t \in W$, are the indicator function of the sets of configuration $\{x = (x_s, s \in V) \in X(V) : x_t = 1\}$. If $\tilde{x} = x_W^+ = (x_t \equiv 1, t \in W)$ the probabilities (5.7) can be written as

$$p^+_{W,V}(x^+_W) = p^+_{W,V}$$

(5.8)

$$= \exp\left\{\beta\left(|\{\{s,t\} \subseteq W : |s-t| = 1\}| + \sum_{t\in W\cap\partial^{\text{in}} V} R_t(V)\right)\right\} \frac{Z^+(V\setminus W)}{Z^+(V)}.$$

The probabilities (5.8) are called the *correlation functions* in the literature on statistical mechanics. In the general case, removing the parenthesis in the right-hand part of (5.7), we find that

$$p^+_{W,V}(\tilde{x}) = \sum_{A:W^+(\tilde{x})\subseteq A\subseteq W} (-1)^{|A\setminus W^+(\tilde{x})|} \left\langle \prod_{t\in A} \chi_t \right\rangle^+_V$$

(5.9)

$$= \sum_{A:W^+(\tilde{x})\subseteq A\subseteq W} (-1)^{|A\setminus W^+(\tilde{x})|} p^+_{A,V}.$$

Relations (5.8) and (5.9) show that the finite-dimensional distributions of the Ising model can be expressed as the ratios of partition functions. So it is desirable to obtain representations of these partition functions in terms of animal models. We describe two such representations. Even formally the both representation are valid for all values of β, one of them is useful in the case of small enough β only (the high-temperature case) and the second one is useful in the case of large enough β only (the low-temperature case.) The intermediate interval of values of β is very interesting since it includes the critical value β_{cr}, where the phase transition in temperature occurs, but the known perturbations methods are not applicable to this situation.

5 A. The high-temperature case. In the previous probabilistic consideration we assume that $\beta > 0$, but the partition function $Z(V)$ is well defined for the comlex values of β also. This possibility is very useful (see §9) and so we assume below that β is a complex-valued parameter.

Let $\mathcal{Q}_V$ be the set

(5.10)
$$\mathcal{Q}_V = \{\{s,t\} \subseteq V : |s-t| = 1\}$$

of all pairs of neighboring points of V. Then the partition function (5.2) can be rewritten as

(5.11)
$$Z^+(V) = \sum_{x\in X(V)} \prod_{\{s,t\}\in\mathcal{Q}_V} (e^{-\beta x_s x_t} - 1 + 1) \prod_{t\in\partial^{\text{in}} V} \left(e^{-\beta x_t R_t(V)} - 1 + 1\right).$$

Removing the parenthesis in (5.11) and changing the order of summation, we find that

(5.12)

$$Z^+(V) = \sum_{A\subseteq\mathcal{Q}_V, B\subseteq\partial^{\text{in}} V} \left(\sum_{x\in X(V)} \prod_{\{s,t\}\in A} (e^{-\beta x_s x_t} - 1) \prod_{t\in B} \left(e^{-\beta x_t R_t(V)} - 1\right)\right).$$

The following transformation of the partition function uses a graph structure on the set of vertices

$$\widehat{\mathcal{Q}}_V = \mathcal{Q}_V \cup \partial^{\mathrm{in}} V. \tag{5.13}$$

A pair of distinct vertices $\{s_1, t_1\}, \{s_2, t_2\} \in \mathcal{Q}_V$ of this graph belongs to an edge, if the pair of sets $\{s_1, t_1\}, \{s_2, t_2\} \subseteq V$ has nonempty intersection. A pair of vertices $\{s_1, t_1\} \in \mathcal{Q}_V, t \in \partial^{\mathrm{in}} V$, belongs to an edge if $s_1 = t$ or $t_1 = t$. Pairs of vertices $t_1, t_2 \in \partial^{\mathrm{in}} V$ can not be vertices of the same edge. In a usual way any subset $(A, B) \subseteq \widehat{\mathcal{Q}}_V, A \subseteq \mathcal{Q}_V, B \subseteq \partial^{\mathrm{in}} V$, can be represented in a unique way as a sum of its maximal connected (with respect to the graph sructure in $\widehat{\mathcal{Q}}_V$) components:

$$(A, B) = (A_1, B_1) \cup (A_2, B_2) \cup \ldots \cup (A_m, B_m), \tag{5.14}$$

where $A_j \subseteq \mathcal{Q}_V, B_j \subseteq \partial^{\mathrm{in}} V, j = 1, 2, \ldots, m, m = m(A, B) = 0, 1, 2, \ldots$. For any subset $(A, B) \subseteq \widehat{\mathcal{Q}}_V$ we let

$$T(A, B) = \cup_{\{s,t\} \in A} \{s, t\} \cup_{t \in B} \{t\} \subseteq V. \tag{5.15}$$

The set $T(A, B) \subseteq V$ is called the *support of* (A, B). Let $\mathcal{S}_V$ be the set of all finite collections $(A_1, B_1), (A_2, B_2), \ldots, (A_m, B_m)$ of sets of vertices of the graph $\widehat{\mathcal{Q}}_V$ such that each of these sets is connected with respect to the graph structure and the supports $T_j = T(A_j, B_j)$ are mutually disjoint. It follows from the previous construction that relation (5.14) defines an one-to-one correspondence between the sets $(A, B) \in \widehat{\mathcal{Q}}_V$ and the collections $(A_1, B_1), (A_2, B_2), \ldots, (A_m, B_m) \in \mathcal{S}_V$. Using this correspondence and then the general relation (4.26) we rewrite (5.12) as

$$\begin{aligned} Z^+(V) = & \sum_{(A_1,B_1),(A_2,B_2),\ldots,(A_m,B_m) \in \mathcal{S}_V} \ \sum_{x \in X(V)} \prod_{j=1}^{m} \Bigg(\prod_{\{s,t\} \in A_j} \left(e^{-\beta x_s x_t} - 1\right) \\ & \times \prod_{t \in B_j} \left(e^{-\beta x_t R_t(V)} - 1\right)\Bigg) = \sum_{(A_1,B_1),(A_2,B_2),\ldots,(A_m,B_m) \in \mathcal{S}_V} \ \prod_{j=1}^{m} w_V(A_j, B_j), \end{aligned} \tag{5.16}$$

where

$$\begin{aligned} w_V(A, B) = & \sum_{x \in X(T(A,B))} \left(\prod_{\{s,t\} \in A} \left(e^{-\beta x_s x_t} - 1\right) \prod_{t \in B} \left(e^{-\beta x_t R_t(V)} - 1\right)\right), \\ & A \subseteq \mathcal{Q}_V, B \subseteq \partial^{\mathrm{in}} V. \end{aligned} \tag{5.17}$$

Proposition 5.1. *Consider the following animal model. The set of animals* $\Theta_V = \{(A, B) : A \subseteq \mathcal{Q}_V, B \subseteq \partial^{\mathrm{in}} V, (A, B) \text{ is connected in } \widehat{\mathcal{Q}}_V\}$*, two animals* $(A_1, B_1), (A_2, B_2) \in \Theta_V$ *are compatible, if their supports*

$$T(A_1, B_1) \cap T(A_2, B_2) = \emptyset \tag{5.18}$$

are mutually disjoint, the weights of animals are defined by relation (5.17). Then the partition function

$$Z^{+}(V) = Z_{w_V}(\Theta_V). \tag{5.19}$$

Proof. It follows from the definiton (2.1) and the previous construction. □

It is important to observe that the weights $w_V(A, B)$ do not depend on V if their supports are mutually disjoint with the inner boundary $\partial^{\text{in}} V$. If

$$T(A, B) \cap \partial^{\text{in}} V = \emptyset, \tag{5.20}$$

then $B = \emptyset$ and the weights

$$w_V(A, B) = w(A). \tag{5.21}$$

We consider also the random variable

$$S_V(x) = \sum_{t\in V} x_t, \quad x \in X(V), \tag{5.22}$$

and its characteristic function

$$\phi_V(z) = \langle e^{zS_V} \rangle_V^{+} = \frac{Z^{+}(V, z)}{Z^{+}(V)}, \quad z \in \mathbb{C}, \tag{5.23}$$

where

$$Z^{+}(V, z) = \sum_{x\in X(V)} \exp\{-\beta U_V^{+}(x) + zS_V(x)\} \tag{5.24}$$

Again it will usefull to treat the parameter β in (5.24) as a complex-valued one.

The derivation of the animal representation for the partition function $Z^{+}(V, z)$ is similar to its derivation for the partition function $Z^{+}(V)$ and we describe it in short only. Similarly to (5.11) and (5.12) we have

$$\begin{aligned} Z^{+}(V, z) = &\sum_{x\in X(V)} \Bigg(\prod_{\{s,t\}\in \mathcal{Q}_V} (e^{-\beta x_s x_t} - 1 + 1) \prod_{t\in\partial^{\text{in}} V} \left(e^{-\beta x_t R_t(V)} - 1 + 1\right) \\ &\times \prod_{t\in V} (e^{zx_t} - 1 + 1)\Bigg) = \sum_{A\subseteq \mathcal{Q}_V, B\subseteq \partial^{\text{in}} V, C\subseteq V} \Bigg(\sum_{x\in X(V)} \prod_{\{s,t\}\in A} (e^{-\beta x_s x_t} - 1) \\ &\times \prod_{t\in B} \left(e^{-\beta x_t R_t(V)} - 1\right) \left(\prod_{t\in C} (e^{zx_t} - 1)\right)\Bigg). \end{aligned} \tag{5.25}$$

Instead of the graph (5.13) we consider now the extended graph with the set of vertices

$$\tilde{\mathcal{Q}}_V = \mathcal{Q}_V \cup \partial^{\text{in}} V \cup V \tag{5.26}$$

and complement the definition of the graph structure by the conditions that a pair of vertices $\{s_1, t_1\} \in \mathcal{Q}_V, t \in V$ belongs to an edge if $s_1 = t$ or $t_1 = t$ and a pair of vertices $t_1 \in \partial^{\text{in}} V, t_2 \in V$ belongs to an edge if $t_1 = t_2$. The support

$$T(A, B, C) = (\cup_{\{s,t\} \in A} \{s, t\}) \bigcup (\cup_{t \in B} \{t\}) \bigcup (\cup_{t \in C} \{t\}) \subseteq V \tag{5.27}$$

Proposition 5.2. *Consider the following animal model. The set of animals* $\tilde{\Theta}_V = \{(A, B, C) : A \subseteq \mathcal{Q}_V, B \subseteq \partial^{\text{in}} V, C \subseteq V, (A, B, C) \text{ is connected in } \tilde{\mathcal{Q}}_V\}$. *Two animals* $(A_1, B_1, C_1) \in \Theta_V$ *and* $(A_2, B_2, C_2) \in \Theta_V$ *are compatible, if their supports are mutually disjoint. The weights*

$$\begin{aligned} \tilde{w}_V(A, B, C) = \sum_{x \in X(T(A,B))} \Bigg(\prod_{\{s,t\} \in A} (e^{-\beta x_s x_t} - 1) \prod_{t \in B} \left(e^{-\beta x_t R_t(V)} - 1\right) \\ \times \prod_{t \in C} \left(e^{z x_t} - 1\right)\Bigg), \quad A \subseteq \mathcal{Q}_V, B \subseteq \partial^{\text{in}} V, C \subseteq V. \end{aligned} \tag{5.28}$$

Then the partition function

$$Z^+(V, z) = Z_{\tilde{w}_V}(\Theta_V). \tag{5.29}$$

Proof. It follows from the previous construction. □

Proposition 5.3. *Assume that the mights*

$$b(A, B) = e^{T(A,B)}, \quad (A, B) \in \Theta_V, \qquad b(A, B, C) = e^{T(A,B,C)}, \quad (A, B, C) \in \tilde{\Theta}_V. \tag{5.30}$$

Then there exists a value of inverse temperature $\beta_0 = \beta_0(d) > 0$ *such that the Kotezký-Preiss conditions (2.9), (2.10) for the partition function* $Z^+(V)$ *holds for any complex* β *such that* $|\beta| < \beta_0$ *and any finite* $V \subset \mathbb{Z}^d$. *For the partition function* $Z^+(V, z)$ *the similar statement holds, if we asume additionally that* $|z|$ *is small enough.*

We emphasize that in the formulation of this Proposition, as in similar situations later, we do not set ourselves as an object to obtain an explicit exact estimate for β_0 and at all steps of the derivation prefer more simple estimates to more exact ones.

Before the proof of Proposition 5.3 we turn to a general lemma which will be used more than once below.

Lemma 5.4. *Let* $\mathbb{G}(r)$ *be the set of all finite graphs (undirected, without loops and multiple edges) such that for any* $G \in \mathbb{G}(r)$ *and any vertex* $t \in G$ *this vertex belongs to not more than* r *edges of the graph. Let* $\mathcal{C}(G)$ *be a set of all connected subsets* $C \subseteq G$ *of vertices of the graph* G. *There exists a constant* $R = R(r)$ *depending on* r *only such that for any graph* $G \in \mathbb{G}(r)$, *any integer* n, *and any vertex* $t_0 \in G$ *the number of connected sets* $C \in \mathcal{C}(G)$ *such that* $t_0 \in C$ *and* $|C| = n$

$$|\{C \in \mathcal{C}(G) : t_0 \in C, |C| = n\}| \le R^n. \tag{5.31}$$

Proof. Let $\mathbb{G}_N(r), N = 1, 2, \ldots$, be the set of all graphs $G \in \mathbb{G}(r)$ such that the number of vertices $|G| = N$. Let

$$K_N^n = \max_{G \in \mathbb{G}_N(r), t_0 \in G} |\{C \in \mathcal{C}(G) : t_0 \in C, |C| = n\}|. \tag{5.32}$$

Observe that for a fixed n the function K_N^n is monotone nondecreasing funcion of N and for any graph $G \in \mathbb{G}(r)$

$$|\{C \in \mathcal{C}(G) : t_0 \in C, |C| = n\}| \leq \lim_{N \to \infty} K_N^n. \tag{5.33}$$

Assume that a graph $G \in \mathbb{G}_{N+1}(r)$, a vertex $t_0 \in G$ and $(t_0, t_1), (t_0, t_2), \ldots, (t_0, t_s), s \leq r$ be the list of all edges connecting the vertex t_0 with the other vertices of the graph G. By $G \setminus t_0$ we denote the graph, obtained from G by deleting the vertex t_0 and the edges $(t_0, t_1), (t_0, t_2), \ldots, (t_0, t_s)$. Let a set $C \in \mathcal{C}(G)$ be such that $t_0 \in C$ and $|C| = n + 1, n \geq 1$. The set C can be represented (may be in many ways) as

$$C = \{t_0\} \cup C_{i_1} \cup C_{i_2} \cup \ldots \cup C_{i_k}, \tag{5.34}$$

where $1 \leq i_1 < i_2 < \ldots < i_k \leq s$, the connected sets $C_{i_1}, C_{i_2} \ldots C_{i_k} \in \mathcal{C}(G \setminus t_0)$, the vertices $t_{i_1} \in C_{i_1}, t_{i_2} \in C_{i_2}, \ldots, t_{i_k} \in C_{i_k}$ and

$$|C_{i_1}| + |C_{i_2}| + \ldots + |C_{i_k}| = n. \tag{5.35}$$

Since $G \setminus t_0 \in \mathbb{G}_N(r)$, it follows from (5.35) that

$$\begin{aligned} K_{N+1}^{n+1} &\leq \sum_{i_1, i_2, \ldots, i_k : 1 \leq i_1 < i_2 < \ldots < i_k \leq s} \ \sum_{j_1, j_2, \ldots, j_k : j_1 + j_2 + \ldots + j_k = n} K_N^{j_1} K_N^{j_2} \cdots K_N^{j_k} \\ &\leq \sum_{i_1, i_2, \ldots, i_k : 1 \leq i_1 < i_2 < \ldots < i_k \leq r} \ \sum_{j_1, j_2, \ldots, j_k : j_1 + j_2 + \ldots + j_k = n} K_N^{j_1} K_N^{j_2} \cdots K_N^{j_k}, \quad m = 1, 2, \ldots. \end{aligned} \tag{5.36}$$

Consider now the generating polynomials

$$f^N(\lambda) = \sum_{n=1}^{N} K_n^N \lambda^n, N = 1, 2, \ldots \tag{5.37}$$

and the polynomials

$$g^N(\lambda) = (1 + f^N(\lambda))^r. \tag{5.38}$$

It is easy to check that for $n \geq 1$ the n-th coefficiemt of the polynomial $g^N(\lambda)$ coincides with the right-hand part of (5.36). Now we assume that $\lambda \geq 0$, multiply the both parts of the inequality (5.36) on λ^{n+1} and sum the resulting inequalities over $n = 0, 1, \ldots, N$. Observing that $K_{N+1}^1 = 1$, we arrive to the inequality

$$f^{N+1}(\lambda) \leq \lambda(1 + f^N(\lambda))^r. \tag{5.39}$$

Since in the case $r = 1$ the statement of Lemma is trivial, we assume that $r \geq 2$. In this case it is easy to check that for small enough λ there exists a value $u_\lambda > 0$

such that

$$\lambda(1+u_\lambda)^r = u_\lambda \tag{5.40}$$

and $\lambda(1+u)^r \geq u$, if $0 \leq u \leq u_\lambda$. The value $u_\lambda \to 0$, as $\lambda \to 0$, and changing $\lambda(1+u_\lambda)^r$ on its dominant part $\lambda(1 + ru_\lambda + \frac{1}{2}r(r-1)(u_\lambda)^2)$, we can check that

$$\lambda < u_\lambda \tag{5.41}$$

for small enough λ. Below we consider such values of λ only. Observe that $f^1(\lambda) = \lambda$. So, using the monotonicity of the function $\lambda(1+u)^r$ in u, we derive from (5.39) and (5.40) that

$$f^2(\lambda) \leq \lambda(1+\lambda)^r \leq \lambda(1+u_\lambda)^r = u_\lambda. \tag{5.42}$$

Now we use the induction in N. If $f^{N-1}(\lambda) \leq u_\lambda$, it follows from (5.39) and (5.40) that

$$f^N(\lambda) \leq \lambda(1 + f^{N-1}(\lambda))^r \leq \lambda(1+u_\lambda)^r = u_\lambda \tag{5.43}$$

and so this estimate is valid for all N. Recalling the definition (5.37), we derive that for all N and n and all small enough λ

$$K_N^n \leq \lambda^{-n} f^N(\lambda) \leq u_\lambda \left(\frac{1}{\lambda}\right)^n \tag{5.44}$$

Recall that $u_\lambda < 1$ for small enough λ. So together with relation (5.33) the proved estimate (5.44) implies the desired relation (5.31). □

Proof of Proposition 5.3. Fix a value of $\beta_0 > 0$, which will be described explicitly below, and let

$$w_0(A,B) = \left(2(e^{2d\beta_0} - 1)\right)^{|A|+|B|}, \quad A \subseteq Q_V, B \subseteq \partial^{\text{in}} V. \tag{5.45}$$

Since $R_t(V) \leq 2d$, it follows from the definition of weights (5.17) that for $|\beta| \leq \beta_0$

$$|w_V(A,B)| \leq w_0(A,B), \quad A \subseteq Q_V, B \subseteq \partial^{\text{in}} V. \tag{5.46}$$

For any animal $\theta = (A,B) \in \Theta_V$ the set of animals

$$\{\tilde\theta \in \Theta_V : \tilde\theta \nleftrightarrow \theta\} \cup \{\theta\} \subseteq \bigcup_{t_0 \in T(A,B)} \{\tilde\theta = (\tilde A, \tilde B) \in \Theta_V : t_0 \in T(\tilde A, \tilde B)\}. \tag{5.47}$$

Observe that, if $t_0 \in T(\tilde A, \tilde B)$, then the set $(\tilde A, \tilde B)$ is such that either $\tilde A$ contains one of edges $(t_0, s) \in Q_V$ (the number of such edges is not greater than $2d$) or $t_0 \in \tilde B$. The animal $(\tilde A, \tilde B)$ is a connected subset of vertices of the graph $\widehat{Q}_V$ which belongs to the set of graphs $\mathbb{G}(r)$ with $r = 4d$ (see Lemma 5.4). So applying the estimate (5.31), we obtain that

$$|\{\tilde\theta = (\tilde A, \tilde B) \in \Theta_V : t_0 \in T(\tilde A, \tilde B), |\tilde A| + |\tilde B| = n\}| \leq (2d+1)R(4d)^n. \tag{5.48}$$

It follows from the definitions (5.30) and (5.15) that

$$b(\tilde{A},\tilde{B}) \le e^{2|\tilde{A}|+|\tilde{B}|}. \tag{5.49}$$

and so recalling the definition (5.45) we find that for any $t_0 \in V$

$$\sum_{\tilde{\theta}=(\tilde{A},\tilde{B})\in\Theta_V: t_0\in T(\tilde{A},\tilde{B})} w_0(\tilde{\theta})b(\tilde{\theta}) \le (2d+1)\sum_{n=1}^{\infty}\left(2e^2R(4d)\left(e^{2d\beta_0}-1\right)\right)^n. \tag{5.50}$$

The series in the right part of (5.50) tends to 0, as $\beta_0 \to 0$. So we derive from (5.47) and (5.50) that for a small enough β_0

$$\sum_{\tilde{\theta}\in\Theta_V:\tilde{\theta}\nrightarrow\theta} w_0(\tilde{\theta})b(\tilde{\theta}) + w_0(\theta)b(\theta) \le |T(A,B)|. \tag{5.51}$$

The fulfilment of the conditions (2.9) and (2.10) for the weights w_V follows from the estimate (5.51) and the definition (5.30).

In the case of the partition function $Z^+(V,z)$ defined in (5.24) the construction is completely similar to the previous one. The main distinction is that we fix also some $\zeta > 0$, instead of (5.45) write

(5.52)

$$w_0(A,B,C) = \left(2\max\{(e^{2d\beta_0}-1),(e^{\zeta}-1)\}\right)^{|A|+|B|+|C|}, \quad A\subseteq Q_V, B\subseteq\partial^{\text{in}}V, C\subseteq V,$$

instead of (5.46) find that for $|\beta|\le\beta_0, |z|\le\zeta$,

$$|w_V(A,B,C)| \le w_0(A,B,C) \quad A\subseteq Q_V, B\subseteq\partial^{\text{in}}V, C\subseteq V, \tag{5.53}$$

and, at last, instead of (5.50) obtain the estimate

(5.54)

$$\sum_{\tilde{\theta}=(\tilde{A},\tilde{B},\tilde{C})\in\tilde{\Theta}_V: t_0\in T(\tilde{A},\tilde{B},\tilde{C})} w_0(\tilde{\theta})b(\tilde{\theta}) \le (2d+1)\sum_{n=1}^{\infty}\left(2e^2R(4d+2)\left((e^{2d\beta_0}-1\right)\right)^n.$$

and it proves the statement of Propostion. □

5 B. The low-temperature case. The construction of an animal representation for the partition function $Z^+(V)$ adjusted for the low-temperature case is completely different and is founded on a use of a well-known contour method. This method was invented by Peierls [Pe] in 1936. After the papers [Gr], [Do65] this method was introduced in the mathematical literature and now is widely used in a study of phase transitions. Here we describe the contour method for the simplest case of two-dimensional Ising model, where it has the most clear geometrical interpretation. We consider the general situation in §?.

Assume that the lattice $\mathbb{Z}^2$ is embedded in the Euclidean space $\mathbb{R}^2$ in the natural way. Let $\mathbb{Z}^{2*}$ be the *conjugate lattice* with vertices $(n_1+\frac{1}{2}, n_2+\frac{1}{2}), n_1, n_2\in\mathbb{Z}^1$. Let $\mathbb{E}$ be the set of all *edges of the congugate lattice*, i.e. the set of all closed intervals of the length 1 connecting the adjacent points of this lattice. For each edge $e\in\mathbb{E}$ there are two vertices of the original lattice $\mathbb{Z}^2$ on the distance 1/2 from e. We say that they are *vertices adjacent to the edge* e. Let $V\subset\mathbb{Z}^2$ be a finite set. The set of edges $e \in \mathbb{E}$ such that at least one of two points to which the edge e is adjacent belongs to V is called the set of *edges in the volume* V and is denoted by $\mathbb{E}(V)$.

For each configuration $x \in X(V)$ of the Ising model we define its *boundary* $B(x)$ as the set of all edges $e \in \mathbb{E}(V)$ of the conjugate lattice such that, if t, t' are the points of the original lattice adjacent to e, the values $x_t \neq x_{t'}$. (We assume here that $x_t = 1$ for $t \in V^c = \mathbb{Z}^2 \setminus V$ and it corresponds to the considered case of the plus-boundary conditions.) The contour method of a study of the Ising model is founded on a possibility to represent the boundary $B(x)$ in a unique way as a sum of contours, i.e. non-selfintersecting closed lines consisting of edges $e \in E(V)$, but it is necessary to be careful in the definition of contours. (What is a unique natural way to divide to contours the boundary of the configuration x such that $x_t = (-1)^{t_1+t_2}, t = (t_1, t_2) \in V$?) So we introduce the following convention. Let $e_1, e_2 \in \mathbb{E}$ be two different edges containing a common vertex $t = (t_1, t_2) \in \mathbb{Z}^{2\star}$. We say that these edges make a *legitimate turn*, if either one of these edges connects the vertex t with the vertex $(t_1 + 1, t_2)$ and the other vertex connects it with the vertex $(t_1, t_2 + 1)$ or if one of these edges connects the vertex t with the vertex $(t_1 - 1, t_2)$ and the other vertex connects it with the vertex $(t_1, t_2 - 1)$. A contour is defined as a sequence $e_1, e_2, \ldots, e_k$ of mutually distinct edges such that the edges $e_i, e_{i+1}, i = 1, 2, \ldots, k$ (here $k + 1 = 1$) have a common vertex and, if there is another pair of edges $e_{i'}, e_{i'+1}$ of this contour having the same common vertex the edges e_i, e_{i+1} make, a legitimate turn. In other words, a contour is a closed broken line consisting of mutually different edges such that a small deformation conserving the legitimate turns transforms it to a non-selfintersecting closed curve. The set of all contours is denoted by $\mathbb{G}$. We say that a contour $\Gamma \in \mathbb{G}$ is a *contour in the volume* V, if all its edges belong to $\mathbb{E}(V)$. The set of all such contours is denoted by $\mathbb{G}(V)$. The number of edges in a contour $\Gamma \in \mathbb{G}$ is denoted $|\Gamma|$ and is called the *length* of this contour. The set of all points $t \in \mathbb{Z}^2$ such that there are no continuous curve in $\mathbb{R}^2$ which do not intersect a contour Γ and connect the point $t \in \mathbb{Z}^2 \subset \mathbb{R}^2$ with "infinity" will be called the *interior of the contour* Γ and will be denoted Int Γ. We say that the contours Γ_1 and Γ_2 are *compatible* if they have no common edges and at any vertex which is contained in both of contours these contours make legitimate turns.

Let $H(V)$ be the set of all sets $\pi \subseteq \mathbb{G}(V)$ of contours in the volume V such that any two different contours in π are compatible ($H(V)$ includes the empty set of contours). It is easy to understand that for any finite volume V and any configuration $x \in X(V)$ there exists a unique system of contours $\pi(x) \in H(V)$ such that

$$B(x) = \cup_{\Gamma \in \pi(x)} \Gamma. \tag{5.55}$$

Further for any system of contours $\pi \in H(V)$ there is a unique configuration $x(\pi) = (x_t(\pi), t \in V)$ such that

$$\pi(x(\pi)) = \pi. \tag{5.56}$$

This fact ,the proof of which is usually omitted as evident, is not so evident, as it seems at first sight. The configuration $x(\pi)$ can be defined by the following construction. For any point $t \in \mathbb{Z}^2$ denote by $O(t)$ the set of all contours $\Gamma \in \mathbb{G}$ such that $t \in \operatorname{Int}\Gamma$. Then we let

$$(5.57) \qquad x_t(\pi) = (-1)^{|\pi \cap O(t)|}, \quad t \in V, \pi \in H(V).$$

It is necessary to check that the boundary

$$(5.58) \qquad B(x(\pi)) = \sum_{\Gamma \in \pi} \Gamma.$$

Really, consider points $t \in V$ and $s \in \mathbb{Z}^2$ such that $|s-t| = 1$ and the edge $e \in \mathbb{E}(V)$ to which these points are adjacent. It is clear that, if a contour Γ is such that $e \notin \Gamma$, then either $\Gamma \in O(t)$ and $\Gamma \in O(s)$ or $\Gamma \notin O(t)$ and $\Gamma \notin O(s)$. So if $e \notin \sum_{\Gamma \in \pi} \Gamma$, then $x_t(\pi) = x_s(\pi)$ and $e \notin B(x(\pi))$. On the other hand, if $e \in \Gamma_0$ for some contour $\Gamma_0 \in \pi$, then either $\Gamma_0 \in O(t)$ but $\Gamma_0 \notin O(s)$ or $\Gamma_0 \in O(s)$ but $\Gamma_0 \notin O(t)$ and so $x_t(\pi) \neq x_s(\pi)$ and $e \in B(x(\pi))$.

The definition (5.3) implies that the energy

$$(5.59) \quad U_V^+(x) = \sum_{e \in B(x)} (+1) + \sum_{e \in \mathbb{E}(V) \setminus B(x)} (-1) = 2|B(x)| - |\mathbb{E}(V)|, \quad x \in X(V),$$

So recalling the definition (5.2), we find that

$$(5.60) \qquad Z^+(V) = \exp\{\beta|\mathbb{E}(V)|\} Z^{\text{cont}}(V),$$

where the *contour partition function*

$$(5.61) \qquad Z^{\text{cont}}(V) = \sum_{\pi \in H(V)} \exp\left\{-2\beta \sum_{\Gamma \in \pi} |\Gamma|\right\}.$$

The explicitly calculated factor $\exp\{\beta|\mathbb{E}(V)|\}$ in (5.60) is not essential for our aims, since we will mainly consider ratios of partition functions. So we describe an animal representation for the contour partition function $Z^{\text{cont}}(V)$.

Proposition 5.5. *Consider the following animal model. The set of animals $\Theta = \mathbb{G}(V)$. Two animals $\Gamma_1, \Gamma_2 \in \Theta$ are compatible if Γ_1, Γ_2 are compatible contours. The weight function*

$$(5.62) \qquad w(\Gamma) = e^{-2\beta|\Gamma|}, \quad \Gamma \in \Theta,$$

Then the partition function

$$(5.63) \qquad Z^{\text{cont}}(V) = Z_w(\mathbb{G}(V)).$$

Proof. It is evident that $H(V)$ is the set of all herds of animals in the considered animal model. So this Proposition is a simple reinterpretation of the definition (5.61). □

Proposition 5.6. *Assume that the mights*

$$b(\Gamma) = \epsilon^{|\Gamma|}, \quad \theta \in \Theta. \tag{5.64}$$

Then there exist a value of inverse temperature β_1 such that the Kotezký-Preiss condition (2.9) for the partition function (5.63) holds for any $\beta > \beta_1$ and finite $V \subset \mathbb{Z}^2$.

Proof. Observe that if a contour $\widetilde{\Gamma}$ is incompatible with a contour Γ, then there is a vertex t^* of the conjugate lattice $\mathbb{Z}^{2*}$ belonging to both contours. So

$$\sum_{\widetilde{\Gamma} \in \mathbb{G}(U):\widetilde{\Gamma} \nleftrightarrow \Gamma} w_0(\widetilde{\Gamma})b(\widetilde{\Gamma}) + w_0(\Gamma)b(\Gamma) \le \sum_{t^* \in \mathbb{Z}^{2*}:t^* \in \Gamma} \; \sum_{\widetilde{\Gamma} \in \mathbb{G}(U):t^* \in \widetilde{\Gamma}} w_0(\widetilde{\Gamma})b(\widetilde{\Gamma}). \tag{5.65}$$

The cardinality

$$|\{t^* \in \mathbb{Z}^{2*} : t^* \in \Gamma\}| \le |\Gamma|. \tag{5.66}$$

Observe also that the number of contours

$$|\{\widetilde{\Gamma} \in \mathbb{G}(U) : t^* \in \widetilde{\Gamma}, |\widetilde{\Gamma}| = r\}| \le 4 \cdot 3^r, \quad \text{if } r \ge 4 \text{ is even}, \tag{5.67}$$

and vanishes for other r. Really, there is a choice between four directions of the edge of the contour beginning in the vertex t^* and between three or less than three directions for each of the following edges of the contour. So it follows from (5.62), (5.64)–(5.67) that for any contour $\Gamma \in \mathbb{G}(V)$

$$\sum_{\widetilde{\Gamma} \in \mathbb{G}(U):\widetilde{\Gamma} \nleftrightarrow \Gamma} w_0(\widetilde{\Gamma})b(\widetilde{\Gamma}) + w_0(\Gamma)b(\Gamma) \le |\Gamma| \sum_{d \ge 4, d \text{ is even}} 4(3\epsilon^{-2\beta_0+1})^d. \tag{5.68}$$

The sum of the series in the right-hand side of (5.68) tends to 0, as $\beta_0 \to \infty$, and so this sum is less than 1 for large enough β_0. For such β_0 the condition (2.9) is fulfilled. □

A derivation of the animal representation for the partition function (5.24) adjusted for the case of large β is more difficult and we return to this problem in §?.

§6. Analiticity in the Kotezký-Preiss model

Now we return to the general situation of §2. Instead of the Kotezk'y-Preiss codition (2.9) we shall use the following milder restriction for a non-negative weight function $w_0(\theta), \theta \in \Theta$. For any animal $\theta \in \Theta$

$$1 - w_0(\theta) \exp\left\{ \sum_{\tilde{\theta} \in \Theta:\tilde{\theta} \nleftrightarrow \theta} w_0(\tilde{\theta})b(\tilde{\theta}) \right\} \ge \exp\{-w_0(\theta)b(\theta)\}. \tag{6.1}$$

We shall show that the condition (6.1) follows from the condition (2.9). Really, the condition (2.9) implies that

$$(6.2)\qquad w_0(\theta)\exp\left\{\sum_{\tilde\theta\in\Theta:\tilde\theta\nrightarrow\theta} w_0(\tilde\theta)b(\tilde\theta)\right\}\le w_0(\theta)b(\theta)\exp\{-w_0(\theta)b(\theta)\}.$$

We let $x = w_0(\theta)b(\theta)$ and see that the condition (6.1) follows from (2.9) and the general inequality

$$(6.3)\qquad 1-e^{-x}\ge xe^{-x}\quad \text{for } x\ge 0.$$

To derive this inequality it is enough to check that the function $f(x)=1-(x+1)e^{-x}$ has an nonnegative derivate for $x\ge 0$. In concrete applications of the type of applications of §3–5 the difference between the restrictions (6.1) and (2.9) is not substantial. The Kotezký-Preiss condition (2.9) is more convenient for a check, but a use of of the condition (6.1) facilates essentially the proof of the following Theorem 6.1, since this condition arises in a natural way in induction arguments.

At this point our exposition deviates from the standards of the theory of cluster expansions (see the references in §2.) Usually the exposition begins with a derivation of a cluster expansion for logarithms of partition functions and difficult combinatorial estimates for its terms. Estimates for the logarithms of partition functions in a complex domain which guarantees its analiticity after limit approaches arise as an implication of estimates for terms of the cluster expansion. We use an opposite order. First we obtain analyticity estimates for the logarithms of partition functions applying simple induction arguments. After it the cluster expansions turn out to be simple a Taylor expansion and the estimates for its terms can be obtained as usual Cauchy estimates for derivatives of analytical functions (see §10). Such approach was used earlier by the authors and his coauthors ([D87], [D88], [DM], [DW]) but for special situations and in a more complex variant. The variant used here appeared in the paper [Do 94].

Theorem 6.1. *Fix a positive weight function $w_0(\theta), \theta\in\Theta$, satisfying the condition (6.1). Let W_0 be the set of all (complex-valued) weight functions $w = w(\theta)$ of $\theta\in\Theta$ such that*

$$(6.4)\qquad |w(\theta)|\le w_0(\theta), \theta\in\Theta.$$

Consider a weight function $w\in W_0$. Then for any finite set $\Lambda\subseteq\Theta$ the partition function $Z_w(\Lambda)\ne 0$ and for any finite $\Lambda'\subseteq\Lambda$

$$(6.5)\qquad \left|\ln\left|\frac{Z_w(\Lambda)}{Z_w(\Lambda')}\right|\right|\le\sum_{\theta\in\Lambda\setminus\Lambda'} w_0(\theta)b(\theta).$$

In particular (for $\Lambda'=\emptyset$) we have

$$(6.6)\qquad |\ln|Z_w(\Lambda)||\le\sum_{\theta\in\Lambda} w_0(\theta)b(\theta).$$

Discuss the estimate (6.6) which, in the essence, is the main conclusion from the formulated result. In concrete situations like the cases of translation-invariant polymer model and the Ising model in a volume V (see §3 and §5), the sum in the right-nand part of (6.6) has an order $C|V|$, where C is a constant (see §7). In the cases, when the weight function is positive, estimates of such type can be obtained in a trivial way (see §?). But it is not so in the general complex case, since the contribution of different herds can mutually anihilate and the partition function can vanish or almost vanish. The main substance of the estimate (6.6) is to prevent such a possibility. It is well-known in statistical mechanics that an appearence of zeroes of partition functions for complex values of parameters, which tend to the real values, when $|V| \to \infty$, is closely connected with a possibility of phase transitions (see §).

Proof of Theorem 6.1. We shall use an induction in the number of elements $|\Lambda|$ in the set Λ. In the case $|\Lambda| = 0$, i.e. $\Lambda = \emptyset$, the statement of the Theorem is evident. So we fix a set Λ and suppose that the estimate (6.5) is valid for any sets with numbers of elements smaller than $|\Lambda|$. We fix also a subset $\Lambda' \subset \Lambda$ and an animal $\theta_0 \in \Lambda \setminus \Lambda'$. (The case of $\Lambda = \Lambda'$ is trivial.) Consider also the set $\widehat{\Lambda} = \Lambda \setminus \{\theta_0\}$. Then

$$\frac{Z_w(\Lambda)}{Z_w(\Lambda')} = \frac{Z_w(\Lambda)}{Z_w(\widehat{\Lambda})}\frac{Z_w(\widehat{\Lambda})}{Z_w(\Lambda')}. \tag{6.7}$$

It follows from the induction hypothesis that

$$\left|\ln\left|\frac{Z_w(\widehat{\Lambda})}{Z_w(\Lambda')}\right|\right| \leq \sum_{\theta\in\widehat{\Lambda}\setminus\Lambda'} w_0(\theta)b(\theta). \tag{6.8}$$

So we shall prove the statement of the Theorem, if we show that

$$\left|\ln\left|\frac{Z_w(\Lambda)}{Z_w(\widehat{\Lambda})}\right|\right| \leq w_0(\theta_0)b(\theta_0). \tag{6.9}$$

Let

$$\Lambda_0 = \{\theta \in \widehat{\Lambda} : \theta \not\leftrightarrow \theta_0\}. \tag{6.10}$$

It is clear from the definition (2.1) that

$$Z_w(\Lambda) = Z_w(\widehat{\Lambda}) + w(\theta_0)Z_w(\Lambda_0). \tag{6.11}$$

Since $\Lambda_0 \subseteq \widehat{\Lambda}$ and $|\widehat{\Lambda}| < |\Lambda|$ we find using the induction conjecture and the definition (6.4) that

$$\left|\frac{w(\theta_0)Z_w(\Lambda_0)}{Z_w(\widehat{\Lambda})}\right| \leq w_0(\theta_0)\exp\left\{\sum_{\tilde{\theta}:\tilde{\theta}\in\Lambda,\tilde{\theta}\leftrightarrow\theta_0} w_0(\tilde{\theta})b(\tilde{\theta})\right\}. \tag{6.12}$$

Observe now that for any complex z such that $|z| < 1$ we have $1-|z| \leq |1+z| \leq 1+|z|$ and so

$$(6.13) \qquad |\ln|1+z|| \leq \max(\ln(1+|z|), -\ln(1-|z|)) \leq -\ln(1-|z|).$$

It follows from the main condition (6.1) that

$$(6.14) \qquad w_0(\theta)\exp\left\{ \sum_{\tilde{\theta}\in\Theta:\tilde{\theta}\nrightarrow\theta} w_0(\tilde{\theta})b(\tilde{\theta}) \right\} < 1.$$

So the absolute value of the right part in (6.12) is smaller than 1. Now applying the relations (6.11), (6.12) and (6.13) we find that

$$(6.15) \qquad \begin{aligned} \left|\ln\left|\frac{Z_w(\Lambda)}{Z_w(\widehat{\Lambda})}\right|\right| &= \left|\ln\left|1+\frac{w(\theta_0)Z_w(\Lambda_0)}{Z_w(\widehat{\Lambda})}\right|\right| \\ &\leq -\ln\left(1-w_0(\theta_0)\exp\left\{\sum_{\tilde{\theta}:\tilde{\theta}\in\Lambda:\tilde{\theta}\nrightarrow\theta_0} w_0(\tilde{\theta})b(\tilde{\theta})\right\}\right). \end{aligned}$$

The estimate (6.15) together with the main condition (6.1) implies the desired estimate (6.9). □

§7. Central Limit Theorem

The main implication of Theorem 6.1 is a simple derivation of the cluster derivation (see §10). The various applications of the cluster expansion are discussed in the rest of the paper. However there are some important facts which can be obtained directly from the analyticity estimates of Theorem 6.1 avoiding applications of the cluster expansion. We illustrate such a possibility in this section and section 9.

We begin with a simple but very useful general lemma.

Lemma 7.1. *Let*

$$(7.1) \qquad \mathcal{O}_R = \{z \in \mathbb{C} : |z| \leq R\}$$

be a closed disk of a radius $R > 0$ in the complex plain $\mathbb{C}$. Let $f(z), z \in \mathcal{O}_R$, be a function analytical in this disk such that

$$(7.2) \qquad |f(z)| \leq C, \quad z \in \mathcal{O}_R.$$

Assume that for some integer k

$$(7.3) \qquad f^{(j)}(0) = 0, \quad j = 0, 1, \dots, k,$$

where $f^{(j)}$ is the j-th derivative of the function f (here $f^{(0)} = f$). Then

$$(7.4) \qquad |f(z)| \leq C\left(\frac{|z|}{R}\right)^{k+1}, \quad z \in \mathcal{O}_R.$$

Proof. Consider the function

$$\tilde{f}(z) = z^{-(k+1)} f(z), \quad z \in \mathcal{O}_R. \tag{7.5}$$

The condition (7.3) implies that this function is analytical in $\mathcal{O}_R$. It follows from the maximum principle that the maximum of $|\tilde{f}(z)|$ attained for $|z| = R$. So

$$|\tilde{f}(z)| \leq CR^{-(k+1)}, \quad z \in \mathcal{O}_R. \tag{7.6}$$

The desired estimate (7.4) follows from (7.5) and (7.6). □

Let $S_n, n = 1, 2, \dots$, be a sequence of real-valued random variables having finite mean values E_n and finite positive variances D_n[5]

$$E_n = \langle S_n \rangle, \quad D_n = \text{Var } S_n = \langle (S_n)^2 \rangle - (\langle S_n \rangle)^2 \tag{7.7}$$

We say that the *central limit theorem* is valid for the sequence S_n, if for any real x the limit of probabilities

$$\lim_{n \to \infty} P \left\{ \frac{S_n - E_n}{\sqrt{D_n}} < x \right\} = \frac{1}{\sqrt{2\pi}} \int_{-\infty}^{x} e^{-\frac{t^2}{2}} \, dt. \tag{7.8}$$

Lemma 7.2. *Assume that for some $R > 0$ the characteristic functions of the variables S_n (the Laplace transformations of their probability distributions)*

$$\phi_n(z) = \langle e^{zS_n} \rangle \tag{7.9}$$

are finite and do not vanish for all $z \in \mathcal{O}_R$ (recall (7.1)). Moreover, we suppose that for some constant $K > 0$

$$|\ln |\phi_n(z)|| \leq Kn, \quad z \in \mathcal{O}_R, n = 1, 2, \dots . \tag{7.10}$$

Assume additionally that there exists a constant $d > 0$ such that the variances

$$D_n \geq dn, \quad n = 1, 2, \dots . \tag{7.11}$$

Then for the sequence of random variables S_n the central limit theorem is valid.

Proof. Recall that the mean values

$$E_n = \left. \frac{d \ln \phi_n(z)}{dz} \right|_{z=0} \tag{7.12}$$

and the variances

$$D_n = \left. \frac{d^2 \ln \phi_n(z)}{dz^2} \right|_{z=0} . \tag{7.13}$$

Consider the characteristic functions of the normalized variables

$$\hat{\phi}_n(\hat{z}) = \left\langle \exp \left\{ \frac{\hat{z}(S_n - E_n)}{\sqrt{D_n}} \right\} \right\rangle = \phi_n \left(\frac{\hat{z}}{\sqrt{D_n}} \right) \exp \left\{ -\frac{E_n \hat{z}}{\sqrt{D_n}} \right\} . \tag{7.14}$$

[5] Here and in the following $\langle \xi \rangle$ denotes the mean value of a random variable ξ.

It is sufficient to prove that for any $r > 0$ uniformly in $\hat{z} \in \mathcal{O}_r$

$$\lim_{n\to\infty} \ln \hat{\phi}_n(\hat{z}) = -\frac{\hat{z}^2}{2}. \tag{7.15}$$

It follows from (7.14) that

$$\ln \hat{\phi}_n(\hat{z}) = \ln \phi_n(z) - E_n z, \tag{7.16}$$

where

$$z = \frac{\hat{z}}{\sqrt{D_n}}. \tag{7.17}$$

Consider the function

$$f_n(\hat{z}) = \ln \hat{\phi}_n(\hat{z}) + \frac{\hat{z}^2}{2}. \tag{7.18}$$

Observe that it follows from (7.11) that $z \in O_R$, if $\hat{z} \in O_{\widehat{R}_n}$, where

$$\widehat{R}_n = R\sqrt{dn}. \tag{7.19}$$

So it follows from the condition (7.10) that

$$|\ln \phi_n(z)| \le Kn, \quad \text{if } \hat{z} \in \mathcal{O}_{\widehat{R}_n}. \tag{7.20}$$

Since E_n is the derivative of the function $\ln \phi_n(z)$ (see (7.12)), it follows from the conditions (7.10) and the integral Cauchy formula that

$$|E_n z| \le KnR^{-1}|z| \le Kn, \quad \text{if } \hat{z} \in \mathcal{O}_{\widehat{R}_n} \tag{7.21}$$

Now we obtain from (7.16)–(7.21) that

$$|f_n(\hat{z})| \le 2Kn + \frac{R^2 dn}{2}, \quad \text{if } \hat{z} \in \mathcal{O}_{\widehat{R}_n}. \tag{7.22}$$

It follows from relations (7.12) and (7.13) that the value of the function $f_n(\hat{z})$ and the values of its two first derivatives vanish for $\hat{z} = 0$. So we can apply Lemma 7.1 for $k = 2$ and R changed to $\widehat{R}_n$. We find that

$$|f_n(\hat{z})| \le \left(2Kn + \frac{R^2 dn}{2}\right)\left(\frac{r}{R\sqrt{dn}}\right)^3, \quad \text{if } \hat{z} \in \mathcal{O}_r, \text{ where } r \le R\sqrt{dn}. \tag{7.23}$$

For any fixed r and $\hat{z} \in \mathcal{O}_r$ the right-hand part of this estimate tends to 0 as $n \to \infty$. It implies the desired relation (7.15). □

Note 7.1. The following implication of the estimate (7.23) will be used in §?. In the conditions of Lemma 7.2

$$\lim_{n\to\infty} \left(\ln \hat{\phi}_n(\hat{z}_n) + \frac{\hat{z}_n^2}{2}\right) = 0, \quad \text{if } |\hat{z}_n|\sqrt{n} \to 0 \quad \text{as } n \to \infty. \tag{7.24}$$

In applications of Lemma 7.2 to concrete situations we have to check the bound (7.10) for characteristic functions and the lower estimate (7.11) for variances. We shall see that the bound (7.10) can be obtained as an implication of Theorem 7.1. Now we explain a simple approach which is very useful for obtaining lower estimates of variance.

Lemma 7.3. *Let Ω be a finite set and $\{p(\omega), \omega \in \Omega\}$ be a probability distribution. Let mutually disjoint sets $A_1, A_2, \dots, A_k \subseteq \Omega$ are such that $\cup_{j=1}^k A_j = \Omega$ and their probabilities*

$$p_i = \sum_{\omega \in A_i} p(\omega), \quad i = 1, 2, \dots, k, \tag{7.25}$$

do not vanish. By

$$p_i(\omega) = \frac{p(\omega)}{p_i}, \quad \omega \in A_i, \tag{7.26}$$

we denote the conditional probability distribution under the condition A_i, $i = 1, 2, \dots, k$. Let $S(\omega), \omega \in \Omega$, be a random variable and

$$\mathrm{Var}\, S = \sum_{\omega \in \Omega} S(\omega)^2 p(\omega) - \left(\sum_{\omega \in \Omega} S(\omega) p(\omega) \right)^2 \tag{7.27}$$

be its variance. At last let

$$\mathrm{Var}\,(S/A_i) = \sum_{\omega \in A_i} S(\omega)^2 p_i(\omega) - \left(\sum_{\omega \in A_i} S(\omega) p_i(\omega) \right)^2, \quad i = 1, 2, \dots, k, \tag{7.28}$$

be its conditional variances under the conditions A_i. Then

$$\mathrm{Var}\, S \geq \sum_{i=1}^k p_i \,\mathrm{Var}\,(S/A_i). \tag{7.29}$$

Proof. Let the mean values be

$$E(S) = \sum_{\omega \in \Omega} S(\omega) p(\omega), \quad E(S/A_i) = \sum_{\omega \in A_i} S(\omega) p_i(\omega). \tag{7.30}$$

The estimate (7.29) follows from the identity

$$\mathrm{Var}\, S = \sum_{i=1}^k p_i \mathrm{Var}\,(S/A_i) + \sum_{i=1}^k p_i (E(S/A_i) - E(S))^2, \tag{7.31}$$

since the second sum in (7.31) is non-negative. It is easily to check the identity (7.31) by a direct calculation. This identity is well-known in mathematical statistics. It means that the variance is equal to the sum of the mean value of conditional variances and the variance of the conditional mean value. □

The problem of the low estimate of the variance for Markov chains is a classical problem of the theory of probability. Bernstein (see [Be]) introduced the method described above in application to Markov chains. Thiis method was applied to Gibbsian fields in the paper [DN]. The skill of application of estimate (7.29) consists in a choice of a system of events A_i for which it is easy to estimate the conditional variances. In all the application we use the windows-wall construction (see fig.?) , in which A_i mean fixing of a configuration on an appropriate part of the volume called a wall.

Theorem 7.4. *Consider a translation-invariant polymer model with a finite set of types $\overline{\mathcal{G}}$ of polymers (see (3.8), (3.9)) and assume that the function $\alpha(G)$ defining the random variables S_V^α (see (3.3)) is also translation-invariant*

$$\alpha(G) = \alpha(\overline{\mathcal{G}}), \quad \text{if } G \in \overline{\mathcal{G}}. \tag{7.32}$$

Let $V_n = \{t = (t_1, t_2, \dots, t_d) : |t_i| \le n\} \subseteq \mathbb{Z}^d$ be a sequence of cubes and $S_n = S_{V_n}^\alpha$. Assume that the weights $w(G)$ defining polymer probability distribution (3.10) are such that (cf. (3.13))

$$\sum_{G \in \mathcal{G}: 0 \in G} w(G) e^{\kappa |G|} < \kappa \quad \text{for some } \kappa > 0. \tag{7.33}$$

Then for the sequence of random variables S_n the central limit theorem is valid.

Proof. We need to check the conditions (7.10), (7.11) with n changed to $|V_n|$. The condition (7.33) implies Proposition 3.1 and so Theorem 4.1 is applicable to the partition functions $Z(V_n)$ (see (3.2)). So the estimates (6.6) together with (7.33) implies that

$$|\ln |Z(V_n)|| \le \sum_{G \in \mathcal{G}(V_n)} w(G) e^{\kappa |G|} < \kappa |V_n|. \tag{7.34}$$

It follows from the condition (7.33) that the main conditions of Proposition 3.2 are fulfilled. So Theorem 4.1 is applicable to the partition functions $Z(V_n, z)$, if $|z| \le \zeta$, and the estimates (6.6) and (3.12) imply that

$$|\ln |Z(V_n, z)|| \le e^{K\zeta} \sum_{G \in \mathcal{G}(V)} w(G) e^{\kappa |G|} < e^{K\zeta} \kappa |V_n|, \tag{7.35}$$

where the constant K is such that the condition of boundness (3.17) is fulfilled. Recalling (3.5) we find that

$$|\ln |\phi_{V_n}^\alpha(z)|| \le |\ln |Z(V_n, z)|| + |\ln |Z(V_n)|| \le (1 + e^{K\zeta}) \kappa |V_n| \tag{7.36}$$

and we have checked the condition (7.10).

For obtaining an estimate from below for variances we develop the following construction. Fix an integer r such that $\operatorname{diam} G < \frac{r}{3}$ for all $G \in \mathcal{G}$. Consider a partition of the lattice $\mathbb{Z}^d$ to the cubes

$$\begin{aligned} W_u = &\{t = (t_1, t_2, \dots, t_d) \in \mathbb{Z}^d : r u_i \le t_i < r(u_i + 1), i = 1, 2 \dots, d\}, \\ &\text{where } u = (u_1, u_2, \dots, u_d) \in \mathbb{Z}^d. \end{aligned} \tag{7.37}$$

Let $\mathbb{Z}_2^d \subset \mathbb{Z}^d$ be the sublattice consisting of all u such that u_i is even for all $i = 1, 2, \ldots, d$. We will use the terminology suggested by the two-dimensional fig. 7.1. So the sets $W_{u,n} = W_u \cap V_n, u \in \mathbb{Z}_2^d$, will be called the *windows* and the set

$$\overline{W}_n = V \setminus \left(\bigcup_{u \in \mathbb{Z}_2^d} W_{u,n} \right) \tag{7.38}$$

will be called the *wall*. Let $\mathbb{Z}_{2,n}^d$ be the set of all $u \in \mathbb{Z}_2^d$ such that the window $W_{u,n}$ are non-emtpty. The random variable

$$S_{V_n}^{\alpha}(\gamma) = S_n^{\text{wall}}(\gamma) + \sum_{u \in \mathbb{Z}_{2,n}^d} S_{u,n}(\gamma), \quad \gamma \in R(V_n), \tag{7.39}$$

where

$$S_n^{\text{wall}}(\gamma) = \sum_{G \in \gamma : G \cap \overline{W}_n \neq \emptyset} \alpha(G), \qquad S_{u,n}(\gamma) = \sum_{G \in \gamma : G \subseteq W_{u,n}} \alpha(G), \quad u \in \mathbb{Z}_{2,n}^d. \tag{7.40}$$

Now we want to apply Lemma 7.3 for $\Omega = R(V_n)$. The main point is to define the sets A_i used in this Lemma. They are the sets of polymer configurations $\gamma \in R(V_n)$ such that the set of all polymers $G \in \gamma$ having non-empty intersections with the wall: $G \cap \overline{W}_n \neq \emptyset$ is fixed in a way. Observe that for each of events A_i the conditional distribution of the variable $S_n^{\text{wall}}(\gamma)$ is concentrated in a point and so this term in (7.39) gives no contribution in the conditional variance (7.28). For each of events A_i the random variables $S_{u,n}(\gamma), u \in \mathbb{Z}_{2,n}^d$ are conditionally independent (the knowledge of a configuration in some of windows creates no restrictions on configurations in other windows, since the each of piers is wider then the diameters of all polymers). So for any i

$$\text{Var}\,(S_{V_n}^{\alpha}/A_i) = \sum_{u \in \mathbb{Z}_{2,n}^d} \text{Var}\,(S_{u,n}/A_i). \tag{7.41}$$

It is easy to understand that the conditional distributions of $S_{V_n}^{\alpha}$ do not degenerate to probability distributions concentrated at a point uniformly in u, i and n and so for some constant $\delta > 0$ and all u, i and n

$$\text{Var}\,(S_{u,n}/A_i) \geq \delta > 0. \tag{7.42}$$

The number of windows in the volume V_n is asymptotically proportional to $|V_n|$. So for some constant $d > 0$ and all i and n

$$\text{Var}\,(S_{\alpha}^{V_n}/A_i) \geq d|V_n| \tag{7.43}$$

and so the desired estimate (7.11) follows from the inequality (7.29). □

Note 7.2. Observe that the condition of Theorem 7.4 about the finiteness of the number of types of polymers was used in the derivation of the low estimate of the variances only. The generalizations are possible, but requires a more complex construction (cf. §?).

Consider (recall (4.2)) a time inhomogeneous Markov chain with a finite set of states X, transition matrices $(p^i(x,y), x, y \in X), i = 1, 2, \ldots$, and an initial probability distribution $\bar{p}^0 = (\bar{p}^0(x), x \in X)$.

Theorem 7.5. *Assume that for some constant $\mu > 0$ and all $x_i, x_{i-1} \in X$ and all $i = 1, 2, \ldots$ the transition probabilities*

$$p^i(x_{i-1}, x_i) \geq \mu \tag{7.44}$$

Then for the sequence of random variables S_α^n defined in (4.2) the central limit theorem is valid.

Proof. Again we need to check the conditions (7.10) and (7.11) of Lemma 7.2. Recall that the condition (7.44) implies that the considered Markov chain is exponentially ergodic (see (4.30)). Thus, it follows from Proposition 4.2 that it is possible to apply Theorem 6.1 for the values $b([s,u])$ and $w_0([s,u])$ defined in (4.33) and (4.32) to the estimate of characteristic function (4.3). The estimates (6.6) and (4.50) imply the desired estimate (7.10).

The derivation of low estimate for variances is similar to its derivation in the previous theorem and even simpler. So we describe it briefly only. Again we define windows and a wall. The windows are now one-point subsets $t \in [0,n]$ such that t is even. The wall is the set of all odd $t \in [0,n]$. The events A_i again fix the values of components x_s, where s belongs to the wall. If follows from the definition of Markov chain that under any of the conditions A_i the values $\{x_s, s \text{ is even}\}$, are conditionally independent. The condition (7.44) of Theorem guarantees that the conditional probabilities distributions of variables x_s are bounded away from zero uniformly in s and A_i. So the same is valid for the conditional variances and now the estimate (7.11) follows from the estimate (7.29). □

Turn now to consideration of the Ising model (see §5).

Theorem 7.6. *Let $V_n \subseteq \mathbb{Z}^d$ be a sequence of cubes defined in the formulation of Theorem 7.4 and $S_n = S_{V_n}$ be the sequence of random variables defined by the relation (5.22) for the Ising model. There is a $\beta_0 > 0$ such that for all values $\beta \leq \beta_0$ of the inverse temperature the central limit theorem is valid for the sequence S_n.*

Proof. Again we will check the conditions (7.10), (7.11) with n changed to $|V_n|$. We obtain from formula (5.23) that

$$|\ln|\phi_{V_n}(z)|| \leq |\ln|Z^+(V_n, z)|| + |\ln|Z^+(V_n)||. \tag{7.45}$$

It was shown in Proposition 5.3 that for the both partition functions $Z^+(V_n)$ and $Z^+(V_n, z)$ (if $|z| \leq \zeta$ and ζ is small enough) the Kotezký-Preiss condition is fulfilled and so we can apply Theorem 6.1 for an estimate of this partition functions. It follows from the estimates (6.6) and (5.50) that for small enough β

(7.46)

$$|\ln|Z^+(V_n)|| \leq \sum_{\theta=(A,B)\in\Theta_{V_n}} w_0(\theta)b(\theta) \leq \sum_{t_0 \in V_n} \sum_{\theta\in\Theta_{V_n}: t_0 \in T(A,B)} w_0(\theta)b(\theta) \leq K|V_n|,$$

where K is a constant which do not depend on n. In a similar way, using (5.54) instead of (5.50) we obtain the estimate

$$|\ln|Z^+(V_n, z)|| \leq K|V_n|, \quad \text{if } z \leq \zeta. \tag{7.47}$$

The desired estimate (7.10) follows from (7.45)-(7.47).

The derivation of the low estimate (7.11) for the variance is again use the windows-wall approach similar to the approach used in the proof of the previous two theorems. The windows are now the points $t = (t_1, t_2, \dots, t_d) \in V_n$ such that all all the components $t_i, i = 1, 2, \dots, d$ are even. The wall is the set of points $t \in V_n$ such that at least one of components $t_i, i = 1, 2, \dots, d$, is odd. From the definition (5.1) of the Ising distribution it follows immediately that, under a condition that the the restriction of the configuration to the wall is fixed, the values x_t in the windows are conditionally independent. Also it is clear that their conditional variances are uniformly bounded away from zero. So the desired estimate (7.11) follows from the estimate (7.29). □

Note 7.3. Observe that the described above dervation of a low estimate for the variances is equally applicable for all β. But the derivation of the estimate for characteristic function use the smallness of β in a very essential way. In §? we prove the central limit theorem for large enough β. It is impossible to expect that it is valid for all β. The physical common sense states that it is not so for $\beta = \beta_{cr}$ and $d = 2$ (but it seems that it is not proved).

§8. Semiinvariants

We need to introduce general probabilistic objects which have many names. They are called *semiinvariants* in the literature on the probability theory, *cumulants* in the literature on mathematical statistics, *truncated correlation functions* and *Ursell functions* in the literature on mathematical physics. This discord reflects the naturalness of this notion which was invented independently in different domain of science. We will use the probabilistic terminology here.

The semiinvariants are a very useful tool in probabilistic investigations. However this notion is not discussed in the standard manuals on probability theory and stochastic processes. In this section we give a definition of semiinvariants and derive their property used below.

Let $\xi = (\xi_1, \xi_2, \dots, \xi_m)$ be a m-dimensional random vector. Assume that its *moment generating functions*

$$\Phi_\xi(z) = \left\langle \exp\left\{\sum_{i=1}^{m} z_i \xi_i\right\}\right\rangle, \quad z = (z_1, z_2, \dots, z_m) \in \mathbb{C}^m, \tag{8.1}$$

is finite for for all complex $z_i, i = 1, 2, \dots, m$. Recall that for any integers $k_1, \dots, k_m$ the *mutual moment of the order* $k_1, k_2, \dots, k_m$

$$\langle (\xi_1)^{k_1} (\xi_2)^{k_2} \cdots (\xi_m)^{k_m} \rangle = \left. \frac{\partial^{k_1+k_2+\cdots+k_m}}{\partial z_1^{k_1} \partial z_2^{k_2} \dots \partial z_m^{k_m}} \Phi_\xi(z) \right|_{z=0} \tag{8.2}$$

The Taylor expansion of the moment generating function is
(8.3)

$$\Phi_\xi(z) = 1 + \sum_{s=1,2,\dots,m, k_1\geq 1, k_2\geq 1,\dots,k_s\geq 1} \frac{1}{k_1!k_2!\cdots k_s!}\langle(\xi_1)^{k_1}(\xi_2)^{k_2}\cdots(\xi_s)^{k_s}\rangle z_1^{k_1}z_2^{k_2}\cdots z_s^{k_s}.$$

The *logarithmic moment generating functions*

$$\Psi_\xi(z) = \ln\left\langle \exp\left\{\sum_{i=1}^{m} z_i\xi_i\right\}\right\rangle, \quad z=(z_1,z_2,\dots,z_m)\in\mathbb{C}^m. \tag{8.4}$$

Its derivate

$$\langle \xi_1^{,k_1}, \xi_2^{,k_2}, \dots, \xi_m^{,k_m}\rangle = \frac{\partial^{k_1+k_2+\cdots+k_m}}{\partial z_1^{k_1}\partial z_2^{k_2}\dots\partial z_m^{k_m}}\Psi_\xi(z)\bigg|_{z=0}. \tag{8.5}$$

is called the *semiinvariant of the order* $k_1, k_2, \dots, k_m$ of the random variable $\xi = (\xi_1, \xi_2, \dots, \xi_m)$. The comma notation for semiinvariants introduced in (8.5) is used sometimes in the literature on mathematical physics and turns out convenient. We will use also its abridged variants. In the case $k_1 = k_2 = \dots = k_m = 1$ of semiinvariants of order 1

$$\langle\xi_1,\xi_2,\dots,\xi_m\rangle = \langle\xi_1^{,1},\xi_2^{,1},\dots,\xi_m^{,1}\rangle = \frac{\partial^m}{\partial z_1\partial z_2\dots\partial z_m}\Psi_\xi(z)\bigg|_{z=0}. \tag{8.6}$$

If a subset of indexes $D = \{i_1, i_2, \dots, i_s\} \subseteq \{1, 2, \dots, m\}$ we let

$$\langle\xi_D'\rangle = \langle\xi_{i_1},\xi_{i_2},\dots,\xi_{i_s}\rangle \tag{8.7}$$

The Taylor expansion of the logarithmic moment generating function is

$$\Psi_\xi(z) = \sum_{s=1,2,\dots,m,k_1\geq 1,\dots,k_s\geq 1}\frac{1}{k_1!\cdots k_s!}\langle(\xi_1)^{,k_1},\cdots(\xi_s)^{,k_s}\rangle z_1^{k_1}\cdots z_s^{k_s}. \tag{8.8}$$

For $m = 1$ the semiinvariant of the order 1 is the mean value, the semiinvariant of the order 2 is the variance; for $m = 2$ the semiinvariant of the order $(1,1)$ is called *covariance*.

Obtain some properties of semiinvariants used below.

I)For any integer $k_1, k_2, \dots, k_m$

$$\langle\xi_1^{,k_1},\xi_2^{,k_2},\dots,\xi_m^{,k_m}\rangle = \langle\overbrace{\xi_1,\xi_1,\dots,\xi_1}^{k_1 \text{ times}},\overbrace{\xi_2,\xi_2,\dots,\xi_2}^{k_2 \text{ times}},\dots,\overbrace{\xi_m,\xi_m,\dots,\xi_m}^{k_m \text{ times}}\rangle. \tag{8.9}$$

To simplify the notation we explain the derivation of this identity for the case $m = 1$ only. The general case is similar. In the case $\xi = (\overbrace{\xi_1,\xi_1,\dots,\xi_1}^{k \text{ times}})$ the logarithmic generating function

$$\Psi_\xi(z) = \Psi_{\xi_1}(z_1 + z_2 + \cdots + z_k). \tag{8.10}$$

Now applying the definition (8.5), we derive the identity (8.9) by the direct differentiation.

The identity (8.9) permits to reduce the properties of multiple semiinvariants to the properties of semiinvariants of orders 1. Since in this special case formulations of these properties are simpler, we described below the properties of semiinvariants of order 1 only.

II) The following identity connecting the moments with semiinvariants holds

$$(8.11)\qquad \langle \xi_1, \xi_2, \dots, \xi_m \rangle = \sum_{\mathcal{D}\in\mathbb{D}_m} \prod_{D\in\mathcal{D}} \langle \xi_D \rangle,$$

where $\mathbb{D}_m$ is the set of all partitions $\mathcal{D}$ of the set of indexes $\{1, 2, \dots, m\}$ into mutually disjoint subsets $D \in \mathcal{D}$. The identity (8.11) follows from the equality

$$(8.12)\qquad \Phi_\xi(z) = e^{\Psi_\xi(z)},$$

in which we substitute the Taylor expansions (8.3) and (8.8) and then with a help of the usual Taylor expansion for the exponential function compare Taylor coeficients in two sides.

III) The semiinvariant

$$(8.13)\qquad \langle \xi_1, \xi_2, \dots, \xi_m \rangle = \sum_{\mathcal{D}\in\mathbb{D}_m} (-1)^{|\mathcal{D}|} \prod_{D\in\mathcal{D}} \langle \xi_D \rangle,$$

where $|\mathcal{D}|$ is the number of subsets $D \in \mathcal{D}$. The identity (8.13) follows from the equality

$$(8.14)\qquad \Psi_\xi(z) = \ln \Phi_\xi(z),$$

where for comparision of Taylor coefficients in the left and right parts we use the usual Taylor expansion of logarithmic function.

IV) *Property of symmetry.* For any permutation $i_1, i_2, \dots, i_m$ of indexes $1, 2, \dots, m$

$$(8.15)\qquad \langle \xi_{i_1}, \xi_{i_2}, \dots, \xi_{i_m} \rangle = \langle \xi_1, \xi_2, \dots, \xi_m \rangle.$$

This equality follows immediately from the definition (8.6).

V) *Property of semilinearity.* Let ξ_1', ξ_1'' be random variables and c', c'' be some constants. Then the semiminvariant

$$(8.16)\qquad \langle c'\xi_1' + c''\xi_1'', \xi_2, \dots, \xi_m \rangle = c'\langle \xi_1', \xi_2, \dots, \xi_m \rangle + c''\langle \xi_1'', \xi_2, \dots, \xi_m \rangle.$$

Really, consider the logarithmic generating function

(8.17)

$$\Psi_{\xi'}(z') = \ln \left\langle \exp\left\{ z_1'\xi_1' + z_1''\xi_1'' + \sum_{i=2}^{m} z_i\xi_i \right\} \right\rangle, \qquad z = (z_1', z_1'', z_2, \dots, z_m) \in \mathbb{C}^{m+1},$$

of the $(m+1)$-dimensional random variable $\xi' = (\xi_1', \xi_1'', \xi_2, \dots, \xi_m)$. Then for $\xi_1 = c_1\xi_1' + c_2\xi_1''$ and $\xi = (\xi_1, \xi_2, \dots, \xi_m)$ the logarithmic generating function

$$(8.18)\qquad \Psi_\xi(z_1, z_2, \dots, z_m) = \Psi_{\xi'}(c'z_1, c''z_1, z_2, z_3, \dots, z_m).$$

Differentiating both the left and the right part of (8.18) in $z_1, z_2, \dots, z_m$ and using the definition (8.6) we arrive to the relation (8.16).

VI) *Property of vanishing.* Let the random vector $\xi = (\xi_1, \xi_2, \dots, \xi_m)$ be such that for some $l, 1 \le l < m$ the random vectors $\xi' = (\xi_1, \xi_2, \dots, \xi_l)$ and $\xi'' = (\xi_{l+1}, \xi_{l+2}, \dots, \xi_m)$ are mutually independent. Then the semiinvariant

$$(8.19) \qquad \langle \xi_1, \xi_2, \dots, \xi_m \rangle = 0.$$

Really, it follows from the independence that the logarithmic generating function

$$(8.20) \qquad \Psi_\xi(z_1, z_2, \dots, z_m) = \Psi_{\xi'}(z_1, z_2, \dots, z_l) + \Psi_{\xi''}(z_{l+1}, z_{l+2}, \dots, z_m)$$

Differentiating the both part of (8.20) in $z_1, z_2, \dots, z_m$, we obtain (8.19).

Note 8.1. All the previous considerations are applicable, if we suppose only that the moment generating function (8.1) is defined for small enough z_i. More of it, if we suppose only that for some integer n the mutual moments $\langle (\xi_1)^{k_1} (\xi_2)^{k_2} \cdots (\xi_m)^{k_m} \rangle$ are finite, if $k_1 + k_2 + \ldots + k_m \le n$, then it is possible to define the seminvariants of the orders $k_1, k_2, \dots, k_m$, where $k_1 + k_2 + \ldots + k_m \le n$, by the help of the formulas (8.13) and (8.9). The properties of semiinvariants described above are valid in this case also. It is easy to check them, if we apply the previous construction to the cut-off random vectors $\xi^C = (\xi_1^C, \xi_2^C, \dots, \xi_m^C)$, where $C > 0$ and

$$(8.21) \qquad \xi_i^C = \begin{cases} \xi_i & \text{if } |\xi_i| \le C \\ 0 & \text{if } |\xi_i| > C, \end{cases}$$

and then make the limit approach $C \to \infty$.

§9. Virial expansion and infinite-volume probabilities for high-temperature Ising model.

We consider in this section the Ising distribution defined by the relation (5.1) for real $\beta \ge 0$. This definition can be extended to complex β, if the partition function $Z^{+,\beta}(V)$ (see (5.2)) included in (5.1) as a denominator, does not vanish. A priori it is possible only to state that $Z^{+,\beta}(V) \ne 0$ in a small enough complex neighborhood of the real axis and the wide of this neihborhood can tend to 0, as $|V| \to \infty$. (This possibility corresponds physically to a possibility of phase transitions; see the discussioin in sect. §?).

But it was checked in Proposition 5.3 that for some $\beta_0 > 0$ and all finite volumes V and all complex $\beta, |\beta| \le |\beta_0|$, the partition function $Z^{+,\beta}(V)$ can be represented as a partition function in an animal model satisfying the Kotezký-Preiss condition. So it follows from Theorem 6.1 that

$$(9.1) \qquad Z^{+,\beta}(V) \ne 0, \quad \text{for } |\beta| \le \beta_0 \text{ and any finite } V \subset \mathbb{Z}^d.$$

The functions (5.1) and (5.6) are well defined for all complex $\beta, |\beta| \le |\beta_0|$. (Of course, for complex β they can not more to be interpret as probabilities and mean values).

Fix a finite volume $W \subset \mathbb{Z}^d$ and a function $\phi(x_W), x_W \in X(W)$. With a slight abuse of notation we let for any finite $V \supseteq W$

$$(9.2) \qquad \langle \phi \rangle_V^{+,\beta} = \langle \phi_V \rangle_V^{+,\beta},$$

where ϕ_V is a function of configuration $x_V \in X(V)$ such that $\phi_V(x_V) = \phi(x_W)$, if the restriction $x_V|_W$ of the configuration x_V to W is equal x_W.

Theorem 9.1. *There exist $\beta_0 > 0$ such that for any finite $V \supseteq W$ the function*

$$f_V(\beta) = \langle \phi \rangle_V^{+,\beta} \tag{9.3}$$

is an analytical function of β for $\beta \in \mathcal{O}_{\beta_0}$, where the disk

$$\mathcal{O}_{\beta_0} = \{\beta \in \mathbb{C} : |\beta| \le \beta_0\}. \tag{9.4}$$

More of it, there exists a constant $K = K_\phi < \infty$ such that for all $V \supseteq W$

$$|f_V(\beta)| \le K, \quad \beta \in \mathcal{O}_{\beta_0}. \tag{9.5}$$

Proof. The analiticity of the functions $f_V(\beta)$ in the disk $\mathcal{O}_{\beta_0}$ follows immediately from the definition of this function. To prove its uniform boundness we observe that it is followed from the relations (5.6) and (5.9) that

$$f_V(\beta) = \sum_{S \subseteq W} q_S p_{S,V}^{+,\beta}, \tag{9.6}$$

where $q_S = q_{S,\phi}, S \subseteq W$ are some constants. It would be not difficult to write the coefficients q_S explicitly, but it is only essential for us that they do not depend on V. It follows from the representtion (9.6) that instead of (9.5) it is enough to prove that for any fixed $S \subseteq V$ the estimate

$$|p_{S,V}^{+,\beta}| \le K, \quad \beta \in \mathcal{O}_{\beta_0} \tag{9.7}$$

holds uniformly in V. We will use for this estimate the relation (5.8). The first exponent in the right part of (5.8) do not depend of V and so

$$|p_{S,V}^{+,\beta}| \le K' \left| \frac{Z^{+,\beta}(V \setminus S)}{Z^{+,\beta}(V)} \right|, \tag{9.8}$$

where K' is a constant. We will apply the estimate (6.5) of Theorem 6.1 to estimate the ratio in (9.8). It is necessary to be a little careful, since even it follows from the construction of §5 that

$$Z^{+,\beta}(V \setminus S) = Z_{w_{V\setminus S}}(\Theta_{V\setminus S}) \quad Z^{+,\beta}(V \setminus S) = Z_{w_V}(\Theta_V), \tag{9.9}$$

where the weights functions $w_{V\setminus S}$ and w_V are defined in (5.17), it is not true that $\Theta_{V\setminus S}$ is a subset of Θ_V. So we consider the set of animals $\Theta' = \{(A, B) : A \subseteq V \setminus S, B \subseteq \partial^{\text{in}} V \cap \partial^{\text{in}}(V \setminus S)\} \subseteq \Theta_V \cap \Theta_{V\setminus S}$ (recall the notation (5.4)). Observe that $w_{V\setminus S}(A, B) = w_V(A, B)$ for $(A, B) \in \Theta'$. So using (9.9) we obtain that

$$\left| \ln \left| \frac{Z^{+,\beta}(V \setminus S)}{Z^{+,\beta}(V)} \right| \right| \le \left| \ln \left| \frac{Z_{w_{V\setminus S}}(\Theta_{V\setminus S})}{Z_{w_{V\setminus S}}(\Theta')} \right| \right| + \left| \ln \left| \frac{Z_{w_V}(\Theta_V)}{Z_{w_V}(\Theta')} \right| \right|. \tag{9.10}$$

To each of the logarithms in the right part of (9.10) we can apply the estimate (6.5). Observe that each of pairs $(A, B) \in \Theta_V \setminus \Theta'$ is such that the support $T(A, B) \cap S \neq \emptyset$.

So using the estimates (6.5) and (5.50) we obtain that for some constant K_1 which do not depend on V

$$\left|\ln\left|\frac{Z_{w_V}(\Theta_V)}{Z_{w_V}(\Theta')}\right|\right| \leq \sum_{(A,B)\in\Theta_V\setminus\Theta'} w_0(A,B)b(A,B) \leq K_1|S|. \tag{9.11}$$

Similarly, each of pairs $(A, B) \in \Theta_{V\setminus S}\setminus\Theta'$ is such that the support $T(A, B)\cap\partial S \neq \emptyset$ (recall again the notation (5.4)) and so

$$\left|\ln\left|\frac{Z_{w_{V\setminus S}}(\Theta_{V\setminus S})}{Z_{w_{V\setminus S}}(\Theta')}\right|\right| \leq K_2|S|. \tag{9.12}$$

So we have proved that

$$|\ln|p_{S,V}^{+,\beta}|| \leq (K_1 + K_2)|S|. \tag{9.13}$$

The desired estimate (9.7) follows from the estimate (9.13). □

Since $\langle\phi\rangle_V^{+,\beta}$ is an analytical function of $\beta, \in \mathcal{O}_{\beta_0}$ it can be expanded to the Taylor series

$$\langle\phi\rangle_V^{+,\beta} = \sum_{k=0}^{\infty} \frac{q_k^V}{k!}\beta^k, \quad \beta \in \mathcal{O}_{\beta_0}. \tag{9.14}$$

where

$$q_k^V = \left.\frac{\partial^k\langle\phi\rangle_V^{+,\beta}}{\partial\beta^k}\right|_{\beta=0}. \tag{9.15}$$

The expansion (9.14) is called the *virial expansion* in statistical physics and the coefficients q_k^V are called the *virial coefficients*.

We discuss now a possibilities to obtain more explicit formulas for virial coefficients. For it we consider a partition function (cf. (5.2))

$$F_V(h,\beta) = \sum_{x\in X(V)} \exp\{h\phi(x_W) - \beta U_V^+(x)\}, \tag{9.16}$$

where $h \in \mathbb{C}$ is a new complex parameter, the energy function U_V^+ is defined by the relation (5.3) and x_W is a restriction of the configuration $x \in X(V)$ to W. It is easy to calculate recalling the definitions (5.6) and (5.1) that the derivate

$$\left.\frac{\partial\ln F_V(h,\beta)}{\partial h}\right|_{h=0} = \langle\phi\rangle_V^{+,\beta}. \tag{9.17}$$

So (recall (9.15))

$$q_k^V = \left.\frac{\partial^{k+1}\ln F_V(h,\beta)}{\partial h\partial\beta^k}\right|_{h=0,\beta=0}. \tag{9.18}$$

Another interpretation of the function $F_V(h,\beta)$ is useful. Consider the system of independent random variables $\{\xi_t, t\in V\}$ taking the values $+1$ and -1 with probabilities $\frac{1}{2}$ and the random variables

$$\phi_\xi = \phi(\xi_t, t\in W) \tag{9.19}$$

and (recall (5.3))

$$U^+_{V,\xi} = U^+_V(\xi_t, t\in V) = -\sum_{\{s,t\}\subseteq V:|s-t|=1} \xi_s\xi_t - \sum_{t\in\partial^{\text{in}}V} \xi_t R_t(V) \tag{9.20}$$

which are functions of the random variables $\{\xi_t, t\in V\}$. Consider also the moment generating function (recall (8.1)) of the pair of random variables $\phi_\xi, U^+_{V,\xi}$

$$\Phi_V(h,\beta) = \langle \exp\{h\phi_\xi + \beta U^+_{V,\xi}\}\rangle. \tag{9.21}$$

Comparing (9.16) and (9.21), it is easy to check that

$$F_V(h,\beta) = 2^{|V|}\Phi_V(h,-\beta). \tag{9.22}$$

So comparing (9.18) and (9.5), we find that

$$q_k^V = (-1)^k\langle \phi_\xi, (U^+_{V,\xi})^{,k}\rangle \tag{9.23}$$

and we have find a representation of the viral coefficients as semiinvariants. Using the properties of semiinvariants descrbed in §8, we can simplify this answer. First of all it follows from the property I) that

$$q_k^V = (-1)^k\langle \phi_\xi, \overbrace{U^+_{V,\xi}, U^+_{V,\xi}, \dots, U^+_{V,\xi}}^{k \text{ times}}\rangle. \tag{9.24}$$

Recall now that $U^+_{V,\xi}$ is the sum (9.20) of other random variables and so we can apply property V) of semilinearity of seminvariants. Using it together with Property IV) of symmetry we find that

(9.25)

$$q_k^V = \sum_{A,B}\langle \phi_\xi, (\xi_{s_1}\xi_{t_1})^{,k_1}, \dots,$$

$$(\xi_{s_m}\xi_{t_m})^{,k_m}, (\xi_{u_1}R_{u_1}(V))^{,l_1}, \dots, (\xi_{u_r}R_{u_r}(V))^{,l_r}\rangle,$$

where the sum is taken over all the pairs of sets $A = \{\{s_1,t_1\},\{s_2,t_2\},\dots,\{s_m,t_m\}\subseteq \mathcal{Q}_V$ (recall the notation (5.10)) and $B = \{u_1,u_2,\dots,u_r\}\subseteq\partial^{\text{in}}V$ such that $m+r\le k$ and all sets of indexes $k_1,k_2,\dots,k_m,l_1,l_2,\dots,l_r$ such that $k_1+k_2+\dots+k_m+l_1+l_2+\dots+l_r = k$. In a happy way, the most part of terms in (9.25) vanishes. Recall the graph structure on the set $\widehat{\mathcal{Q}}_V = \mathcal{Q}_V\cup\partial^{\text{in}}V$ introduced in §5 A. Any pair A,B used in the sum (9.25) is a subset of the set of vertices of this graph. We say that a pair $(A,B)\subseteq\widehat{\mathcal{Q}}_V$ is a *cluster connected with the set* W,if this pair is a connected subset of this graph and its support $T(A,B)$ has an non-empty intersection with the set W. If a pair $(A,B)\subseteq\widehat{\mathcal{Q}}_V$ is not a cluster connected with W, then the set of random variables considered in the corresponding term in (9.25) can be divided into two groups such that the variables of the first group are functions of the random variables $\xi_t, t \in V_1$, the variables of the second group are functions of the random variables $\xi_t, t \in V_2$, and the intersection $V_1 \cap V_2 = \emptyset$. Since the random variables $\xi_t, t \in V$, are mutually independent, it follows from the property VI) of vanishing of semiinvariants that the considered term in (9.25) vanishes. So if the subset of all clusters connected with the set W is denoted by $\widehat{\mathcal{Q}}^{\rm cl}_{V,W}$, we can rewrite (9.25) as

(9.26)
$$q_k^V = \sum_{(A,B)\in\widehat{\mathcal{Q}}^{\rm cl}_{V,W}} \langle \phi_\xi, (\xi_{s_1}\xi_{t_1})^{,k_1}, (\xi_{s_2}\xi_{t_2})^{,k_2}, \dots,$$
$$(\xi_{s_m}\xi_{t_m})^{,k_m}, (\xi_{u_1}R_{u_1}(V))^{,l_1}, (\xi_{u_2}R_{u_2}(V))^{,l_2}, \dots, (\xi_{u_r}R_{u_r}(V))^{,l_r}\rangle,$$

where again the sum is taken over the pairs of sets $A = \{\{s_1, t_1\}, \{s_2, t_2\}, \dots, \{s_m, t_m\} \subseteq \Omega_V$ and $B = \{u_1, u_2, \dots, u_r\} \subseteq \partial^{\rm in} V$ such that $m + r \le k$ and all sets of indexes $k_1, k_2, \dots, k_m, l_1, l_2, \dots, l_r$ such that $k_1 + k_2 + \dots + k_m + l_1 + l_2 + \dots + l_r = k$. It is the final formula for the virial coefficients.

The following implication of the formula (9.26) is used below. The virial coefficients

$$q_k^{V_1} = q_k^{V_2}, \quad \text{if dist } (W, V_i^c) > k, i = 1, 2. \tag{9.27}$$

Really, it is clear that the support $T(A, B)$ of any cluster (A, B) connected with the set W belongs to the k-neighborhood of the set W and so in the case, when dist $(W, V_i^c) > k, i = 1, 2$, the sets $\widehat{\mathcal{Q}}^{\rm cl}_{V_1,W}$ and $\widehat{\mathcal{Q}}^{\rm cl}_{V_2,W}$ coincide.

Now we are ready to discuss the limit behaivour of the mean values $\langle\phi\rangle_V^{+,\beta}$, when the volume V tends to infinity.

Theorem 9.2. *Let $V_n \subset \mathbb{Z}^d$ be a sequence of finite volumes such that*

$$\lim_{n\to\infty} dist(W, V_n) = \infty. \tag{9.28}$$

Assume that the inverse temperature β is such that $0 \le \beta \le \beta_1$, where β_1 is a a small enough constant, Then the limit

$$\langle\phi\rangle_\infty^{+,\beta} = \lim_{n\to\infty} \langle\phi\rangle_{V_n}^{+,\beta} \tag{9.29}$$

exists. There exist a constant $\gamma > 0$ (which do not depend on ϕ and β) and constants C_ϕ such that for any finite $V \subset \mathbb{Z}^d$

$$|\langle\phi\rangle_V^{+,\beta} - \langle\phi\rangle_\infty^{+,\beta}| \le C_\phi \exp\{-\gamma dist(W, V)\}. \tag{9.30}$$

The function $\langle\phi\rangle_\infty^{+,\beta}$ is an analytical function of β in the disk $\beta \in \mathcal{O}_{\beta_1}$ recall (9.4)).

Observe that the statement of this Theorem about analiticity reflects the absence of phase transitions in the Ising model for small β. It is not expected that it is valid in the point $\beta = \beta_{\rm cr}$ (see a discussion in §?).

Proof. The relation (9.6) used in the proof of Theorem 9.1 shows that it is enough to prove the convergence (9.29) for the case, when ϕ is the indicator function of the event $\{x_t = 1, t \in S\}$, where the set $S \subseteq V$, i.e., to prove the existence of the limit

$$\lim_{n\to\infty} p^{+,\beta}_{S,V_n} = p^{+,\beta}_{S,\infty}. \tag{9.31}$$

For any finite subvolumes $V_1, V_2 \subset \mathbb{Z}^d$ consider the function

$$g_{S,V_1,V_2}(\beta) = p^{+,\beta}_{S,V_1} - p^{+,\beta}_{S,V_2} \quad ,\beta \in \mathcal{O}_{\beta_0}. \tag{9.32}$$

The function $g_{S,V_1,V_2}(\beta)$ is an analytical function of β and it follows from the estimate (9.7) that for a constant K which do not depend on V_1, V_2, S

$$|g_{S,V_1,V_2}(\beta)| \leq 2K. \tag{9.33}$$

Assume now that for some integer k

$$\text{dist}\,(W, V_1^c) > k, \quad , \text{ and dist}\,(W, V_2^c) > k. \tag{9.34}$$

Then it follows from the equality (9.27) that the derivates

$$\left.\frac{d^j g_{S,V_1,V_2}(\beta)}{d\beta^j}\right|_{\beta=0} = 0, \quad \text{if } j = 0, 1, \dots, k. \tag{9.35}$$

In this way we have checked that the conditions (7.2) and (7.3) of Lemma 7.1 are fulfilled for the function $g_{S,V_1,V_2}(\beta)$ of β and $R = \beta_0$. So the estimate (7.4) implies that

$$|g_{S,V_1,V_2}(\beta)| \leq 2K\left(\frac{|\beta|}{\beta_0}\right)^{k+1}, \quad ,\text{if } |\beta| \leq \beta_0. \tag{9.36}$$

Assuming that $\beta_1 < \beta_0$ and recalling (9.34) we find that for an appropriate choice of constants $\gamma > 0$ and $C = C_S$

$$|p^{+,\beta}_{S,V_1} - p^{+,\beta}_{S,V_2}| \leq C \exp\{-\gamma \min\{\text{dist}\,(W, V_1^c), \text{dist}\,(W, V_2^c)\}\} \tag{9.37}$$

The existence of the limit (9.31) follows immediately from the estimate (9.37). Also it implies the estimate

$$|p^{+,\beta}_{S,V} - p^{+,\beta}_{S,\infty}| \leq C_\phi \exp\{-\gamma \text{dist}(S, V)\}. \tag{9.38}$$

The more general estimate (9.30) follows from (9.38) and the representation (9.6). The function $\langle\phi\rangle^{+,\beta}_\infty$ is analytical, since it a uniform limit of a sequence of analytical function. □

In the limit case of infinite volume the virial coeffients are a little simpler than in the finite volume case. We have

$$\langle\phi\rangle^{+,\beta}_\infty = \sum_{k=0}^{\infty} \frac{q_k}{k!}\beta^k, \quad \beta \in \mathcal{O}_{\beta_1}, \tag{9.39}$$

where

$$q_k = \sum_{A\in Q^{\rm cl}_W} \langle (\phi_\xi, \xi_{s_1}\xi_{t_1})^{,k_1}, (\xi_{s_2}\xi_{t_2})^{,k_2}, \dots, (\xi_{s_m}\xi_{t_m})^{,k_m}\rangle \tag{9.40}$$

Here $Q^{\rm cl}_W$ is the set of all subsets $A = \{\{s_1, t_1\}, \{s_2, t_2\}, \dots, \{s_m, t_m\}\} \subset \mathbb{Q}_{\mathbb{Z}^d}$ such that they are connected with respect to the graph structure in $\mathbb{Q}_{\mathbb{Z}^d}$ described in §5 A and the support $T(A)$ has non-empty intersection with W, and the sum is taken over all such sets A with $|A| \le k$ and all system of indexes $k_1, k_2, \dots, k_m$ such that $k_1 + k_2 + \dots + k_m = k$. Really, it is clear from (9.27) that there exists the limit

$$q_k = \lim_{n\to\infty} q_k^{V_n} \tag{9.41}$$

and for large enough V_n formula (9.26) is reduced to (9.40).(Recall that, as we observed in the derivation of (9.27), the support $T(A, B)$ belongs to the k-neiborhood of the set W and so for a fixed k and large enough n the set $B = \emptyset$.) The expansion (9.39) follows from the limit relation (9.40), since for a unformly convergent sequence of analytical functions the limits of its derivates equals to the derivates of the limit.

Even in the form (9.40) the expression for the virial coefficients are cumbersome enough. The number of terms in (9.40) grows exponentially in k. It is not so easy to calculate the separate terms also. Since the variables ξ_t are independent and uniformly distributed, the moments are simply the sums

$$\begin{aligned}&\langle (\xi_{s_1}\xi_{t_1})^{k_1}(\xi_{s_2}\xi_{t_2})^{k_2}\cdots(\xi_{s_m}\xi_{t_m})^{k_m}\rangle \\ &= 2^{-|V|} \sum_{x_{s_1}, x_{t_1}, x_{s_2} x_{t_2}, \dots, x_{s_m}, x_{t_m}} (x_{s_1}x_{t_1})^{k_1}(x_{s_2}x_{t_2})^{k_2}\cdots(x_{s_m}x_{t_m})^{k_m},\end{aligned} \tag{9.42}$$

where $V \subseteq \mathbb{Z}^d$ is the set, the elements of which are the points $s_i, t_i, i = 1, 2, \dots, m$. After the calculation of such sums it is necessary to calculate semiinvariants by the help of moments using the formula (8.13). Nevertheless, similar approaches are applied widely for numerical computation of physical quantities.

§10. Cluster expansion in animal models.

Now we return to the general situation of the animal model studied in §2 and§6 and construct a cluster expansion for logarithms of partition functions. For any finite $\Lambda \subseteq \Theta$ the pairs $\rho = (\bar\rho, \alpha)$ such that $\bar\rho \subseteq \Lambda$ is a set and $\alpha = \alpha(\theta) \ge 1, \theta \in \bar\rho$, is an integer-valued function of $\theta \in \bar\rho$ and will be called a *group of animals in* Λ. The set $\bar\rho$ will be called the *support of the group* and the value $\alpha(\theta)$ will be interpreted as the multiplicity of animals of the kind θ in the group ρ. The set of all groups of animals in Λ will be denoted by $D(\Lambda)$. We say that a group $\rho = (\bar\rho, \alpha)$ is a *sum of groups* $\rho_i = (\bar\rho_i, \alpha_i), i = 1, 2, \dots, k$, if $\bar\rho_i \subseteq \bar\rho, i = 1, 2, \dots, k$, and

$$\alpha(\theta) = \sum_{i=1,2,\dots,k:\theta\in\bar\rho_i} \alpha_i(\theta), \quad \theta \in \bar\rho. \tag{10.1}$$

A *gang of animals in* Λ is a non-empty group of animals $\rho = (\bar\rho, \alpha) \in D(\Lambda)$ such that for any two animals $\theta, \theta' \in \bar\rho$ there is a sequence $\theta = \theta_1, \theta_2, \dots, \theta_n = \theta'$ of animals in $\bar{\rho}$ such that the animals θ_i and θ_{i+1} are incompatible for all $i = 1, 2 \dots, n-1$, i.e. $\bar{\rho}$ is a connected subset of the graph Θ. The set of all gangs in Λ will be denoted by $G(\Lambda)$.

Theorem 10.1. *Let the conditions of Theorem 6.1 are fulfilled and a finite set $\Lambda \subseteq \Theta$ be fixed. Consider a polydisk $W_0(\Lambda) = \{w = (w(\theta), \theta \in \Lambda) : |w(\theta)| \le w_0(\theta), \theta \in \Lambda\} \subset \mathbb{C}^\Lambda$ and the set $W_0^{\text{in}}(\Lambda)$ of all inner points of the polydisk $W_0(\Lambda)$. The partition function $Z_w(\Lambda)$ will be treated as a function of $w \in \mathbb{C}^\Lambda$. For any $w \in W_0^{\text{in}}(\Lambda)$ a convergent expansion*

$$\ln Z_w(\Lambda) = \sum_{\rho \in G(\Lambda)} q_w(\rho) = \sum_{\rho \in G(\Lambda)} r(\rho) \prod_{\theta \in \bar{\rho}} w(\theta)^{\alpha(\theta)} \tag{10.2}$$

holds. The coefficients $r(\rho)$ are real numbers depending on the restriction of the graph structure on Θ to $\bar{\rho}$ only. For any gang $\rho = (\bar{\rho}, \alpha)$

$$|q_w(\rho)| = |r(\rho) \prod_{\theta \in \bar{\rho}} w(\theta)^{\alpha(\theta)}| \le \left(\sum_{\theta \in \bar{\rho}} w_0(\theta) b(\theta) \right) \left(\prod_{\theta \in \bar{\rho}} \left(\frac{|w(\theta)|}{w_0(\theta)} \right)^{\alpha(\theta)} \right). \tag{10.3}$$

Proof. For any $w \in W_0(\Lambda)$ let

$$F_\Lambda(w) = \ln Z_w(\Lambda). \tag{10.4}$$

It follows from Theorem 6.1 that $Z_w(\Lambda)$ do not vanish and so the function $F_\Lambda(w)$ is a holomorphic function of $w \in W_0^{\text{in}}(\Lambda)$. The Taylor expansion of this function at the point $w = 0$ can be written as

$$F_\Lambda(w) = \sum_{\rho \in D(\Lambda)} r_\Lambda(\rho) \prod_{\theta \in \bar{\rho}} w(\theta)^{\alpha(\theta)}, \tag{10.5}$$

For $\bar{\rho} = (\theta_1, \theta_2, \dots, \theta_n)$ the coefficients

$$r_\Lambda(\rho) = (\alpha(\theta_1)!\alpha(\theta_2)! \dots \alpha(\theta_n)!)^{-1} \left. \frac{\partial^{\alpha(\theta_1)+\alpha(\theta_2)+\dots+\alpha(\theta_n)} F_\Lambda(w)}{\partial^{\alpha(\theta_1)} w(\theta_1) \partial^{\alpha(\theta_2)} w(\theta_2) \dots \partial^{\alpha(\theta_n)} w(\theta_n)} \right|_{w=0}. \tag{10.6}$$

For any set $\bar{\rho} \subseteq \Lambda$ and any function $w = (w(\theta), \theta \in \bar{\rho}) \in W_0(\bar{\rho})$ consider its continuation $w_\Lambda = (w_\Lambda(\theta), \theta \in \Lambda)$ to Λ such that $w_\Lambda(\theta) = 0$ for $\theta \in \Lambda \setminus \bar{\rho}$. It follows from the definitions (2.1) and (10.4) that

$$F_\Lambda(w_\Lambda) = F_{\bar{\rho}}(w) \tag{10.7}$$

and so we see from the relation (10.6) that

$$r_\Lambda(\rho) = r_{\bar{\rho}}(\rho). \tag{10.8}$$

Now letting $r(\rho) = r_{\bar{\rho}}(\rho)$ we can rewrite the expansion (10.5) as

$$F_\Lambda(w) = \sum_{\rho \in D(\Lambda)} r(\rho) \prod_{\theta \in \bar{\rho}} w(\theta)^{\alpha(\theta)}. \tag{10.9}$$

The coefficients $r(\rho)$ are real, since the function $F_{\bar\rho}(w)$ takes real values for real w.

Since $Z_w(\Lambda) = 1$ for $w = 0$ the coefficient $r(\rho) = 0$ for the empty group of animals $\rho = \emptyset$. Let now an non-empty group of animals $\rho \in D(\Lambda) \setminus G(\Lambda)$. There exists a representation $\bar\rho = \bar\rho_1 \cup \bar\rho_2$, where the sets $\bar\rho_1, \bar\rho_2$ are non-empty, their intersection $\bar\rho_1 \cap \bar\rho_2 = \emptyset$ and for any $\theta_1 \in \bar\rho_1, \theta_2 \in \bar\rho_2$ the animals θ_1, θ_2 are compatible. Then it follows from the definitions (2.1) and (10.4) that

$$F_{\bar\rho}(w) = F_{\bar\rho_1}(w) + F_{\bar\rho_2}(w). \tag{10.10}$$

Differentiating we obtain from (10.6) and (10.10) that

$$r(\rho) = r_{\bar\rho}(\rho) = 0, \quad \text{if } \rho \in D(\Lambda) \setminus G(\Lambda), \tag{10.11}$$

and so the expansion (10.9) is reduced to the desired expansion (10.2).

It follows from the relation (10.6) and the Cauchy formula that

$$r(\rho) = (2\pi i)^{-|\bar\rho|} \oint_{w:|w(\theta)|=w_0(\theta),\theta\in\bar\rho} \frac{F_{\bar\rho}(w)}{\prod_{\theta\in\bar\rho}(w(\theta))^{\alpha(\theta)+1}}\, dw. \tag{10.12}$$

The inequality (6.6) applied for $\Lambda' = \emptyset$ gives a priori estimate

$$|F_{\bar\rho}(w)| \le \sum_{\theta\in\bar\rho} w_0(\theta) b(\theta), \quad w \in W_0(\Lambda), \tag{10.13}$$

The desired estimate (10.3) follows from the relations (10.12) and (10.13). □

The previous simplified derivation of an cluster expasion is taken from the paper [Do 94].

Part 2. Probability models reduced to the Kotezký-Preiss animal model

In Part 1 we show how some probabilistic models can be reduced to the animal models for which the K-P condition are fulfilled. In this part we extend the list of such situations. Applications of these results discussed in Parts 2 and 3.

§11. Extension of the domain of analiticity for Markov chains

The construction of §4 and §6 permits to check the analiticity of characteristic functions of sums of random variables connected in an inhomogeneous Markov chain in a neighborhood of the origin. It is enough for a derivation of the central limit theorem (see Theorem 7.5) but will be not enough for the development of the complete theory of large deviations (see §?). Here we show that an additional modification of the construction permits to prove the analiticity of the characteristic functions in a neighborhood of the real axis.

Theorem 11.1. *Let a Markov chain be defined by a sequence of stochastic transition matrices $P^i = (p^i(x,y), x, y \in X), i = 1, 2, \dots$, and an initial probability distribution $\bar p^0 = (\bar p^0(x), x \in X)$ (see (4.1)) and the condition of uniform positivity (7.44) is fulfilled. Let a sequence of real-valued functions $\alpha_i(x_i), i = 0, 1, \dots$, satisfy to the condition of uniform boundness (4.34) and $\phi_\alpha^n(z), z \in \mathbb{C}$, be the characteristic functions of their sums defined by the relation (4.3). Then for any real $\lambda \in \mathbb{R}$ there* *exist values $R(\lambda) > 0$ and $C(\lambda) > 0$ such that the function $\ln \phi_\alpha^n(z)$ is holomorphic in the disk*

$$O_{R(\lambda)} = \{z \in \mathbb{C} : |z - \lambda| \le R(\lambda)\} \tag{11.1}$$

and the estimate

$$|\ln \phi_\alpha^n(z)| \le C(\lambda)(n+1), \quad z \in O_{R(\lambda)}, n = 0, 1, \dots \tag{11.2}$$

holds.

Proof. Consider a family of "tilted" probability distributions on the space of realizations $x = (x_0, x_1, \dots, x_n) \in X^{[0,n]}$ depending on the real parameter $\lambda \in \mathbb{R}$ defined by the relation

$$\bar{p}_\lambda^{[0,n]}(x) = \frac{e^{\sum_{i=0}^n \lambda \alpha_i(x_i)} \bar{p}^{[0,n]}(x)}{Z_\lambda^{[0,n]}}, \quad x \in X^{[0,n]}, \tag{11.3}$$

where the normalizing multiplier

$$Z_\lambda^{[0,n]} = \sum_{x \in X^{[0,n]}} e^{\sum_{i=0}^n \lambda \alpha_i(x_i)} \bar{p}^{[0,n]}(x) \tag{11.4}$$

and $\bar{p}^{[0,n]}(x), x \in X^{[0,n]}$, is the probability distribution of the original Markov chain (see (4.1)). (The probability distribution (11.3) is a Gibbsian distribution and $Z_\lambda^{[0,n]}$ is its partition function; see (??)).

We want to check that the probability distribution (11.3) is the probability distribution of an inhomogeneous Markov chain and transition probabilities of this "tilted" Markov chain

$$p_\lambda^i(\tilde{x}_{i-1}, \tilde{x}_i) = \frac{\sum_{x \in X^{[0,n]}: x_i = \tilde{x}_{i-1}, x_i = \tilde{x}_i} \bar{p}_\lambda^{[0,n]}(x)}{\sum_{x \in X^{[0,n]}: x_i = \tilde{x}_i} \bar{p}_\lambda^{[0,n]}(x)} \tag{11.5}$$

satisfy the following condition of uniform positivity. For some constant $\mu(\lambda) > 0$ and all $x_{i-1}, x_i \in X$ and $i = 1, 2, \dots, n, n = 1, 2, \dots$

$$p_\lambda^i(x_{i-1}, x_i) \ge \mu(\lambda). \tag{11.6}$$

To prove these facts we consider the conditional proablity of a value x_i, when all the previous values $x_j, j = 1, 2, \dots, i-1$ are fixed, induced by the probability distribution (11.3),

$$\bar{p}_\lambda^{[0,n]}(x_i / x_j, j = 1, 2, \dots, i-1) = \frac{\sum_{x_{i+1}, x_{i+2}, \dots, x_n \in X} \bar{p}_\lambda^{[0,n]}(x)}{\sum_{x_i, x_{i+1}, \dots, x_n \in X} \bar{p}_\lambda^{[0,n]}(x)}. \tag{11.7}$$

If we substitute the formulas (11.3) and (4.1) into the definition (11.7), we obtain after canceling a common multiplier that

(11.8)

$$\bar{p}_\lambda^{[0,n]}(x_i/x_j, j = 1,2,\dots,i-1)$$
$$= \frac{\sum_{x_{i+1},\dots,x_n\in X} \bar{p}^0(x_0)e^{\lambda\alpha(x_0)}p^1(x_0,x_1)e^{\lambda\alpha(x_1)}p^2(x_1,x_2)\cdots p^n(x_{n-1},x_n)e^{\lambda\alpha(x_n)}}{\sum_{x_i,x_{i+1},\dots,x_n\in X} \bar{p}^0(x_0)e^{\lambda\alpha(x_0)}p^1(x_0,x_1)e^{\lambda\alpha(x_1)}p^2(x_1,x_2)\cdots p^n(x_{n-1},x_n)e^{\lambda\alpha(x_n)}}$$
$$= \frac{\sum_{x_{i+1},x_{i+2},\dots,x_n\in X} p^i(x_{i-1},x_i)e^{\lambda\alpha(x_i)}p^{i+1}(x_i,x_{i+1})\cdots p^n(x_{n-1},x_n)e^{\lambda\alpha(x_n)}}{\sum_{x_i,x_{i+1},x_{i+2},\dots,x_n\in X} p^i(x_{i-1},x_i)e^{\lambda\alpha(x_i)}p^{i+1}(x_i,x_{i+1})\cdots p^n(x_{n-1},x_n)e^{\lambda\alpha(x_n)}}.$$

The right part of (11.8) do not depend on $x_j, j = 1,2,\dots,i-2$ and it proves the Markov property of the probability distribution (11.3).

Now we need to check the estimate (11.6) for the transition probbilities. Observe that for any $i = 1,2,\dots,n-1$ and $x_1,x_2,\dots,x_n,x_i' \in X$ the ratio

(11.9)

$$\frac{\bar{p}_\lambda^{[0,n]}((x_1,x_2,\dots,x_{i-1}x_i',x_{i+1}x_{i+2},\dots,x_n))}{\bar{p}_\lambda^{[0,n]}((x_1,x_2,\dots,x_n))} = \frac{p^i(x_{i-1},x_i')e^{\lambda\alpha_i(x_i')}p^{i+1}(x_i',x_{i+1})}{p^i(x_{i-1},x_i)e^{\lambda\alpha_i(x_i)}p^{i+1}(x_i,x_{i+1})}.$$

and so the conditional probability of a value x_i', when all the other values $x_j, j = 1,2,\dots,n$,
$j \neq i$, are fixed, induced by the probability distribution (11.3),

(11.10)

$$\bar{p}_\lambda^{[0,n]}(x_i'/x_j, j = 1,2,\dots,n, j\neq i) = \frac{\bar{p}_\lambda^{[0,n]}((x_1,x_2,\dots,x_{i-1},x_i',x_{i+1},x_{i+2},\dots,x_n))}{\sum_{x_i\in X}\bar{p}_\lambda^{[0,n]}((x_1,x_2,\dots,x_n))}$$
$$= \frac{p^i(x_{i_1},x_i')e^{\lambda\alpha_i(x_i')}p^{i+1}(x_i',x_{i+1})}{\sum_{x_i\in X} p^i(x_{i_1},x_i)e^{\lambda\alpha_i(x_i)}p^{i+1}(x_i,x_{i+1})}.$$

Let $\bar{p}_\lambda^{[0,n]}(x_i'/x_{i-1}),x_{i+1})$ be the conditional probability of a value x_i', when the previous value x_{i-1} and the next value x_{i+1} are fixed. Using the conditions (4.34) and (7.44), we derive from (11.10) that for all integers $n, i = 1,2,\dots,n-1$ and all $x_i', x_{i-1}, x_{i+1} \in X$

$$\bar{p}_\lambda^{[0,n]}(x_i'/x_{i-1}),x_{i+1})$$
$$= \frac{p^i(x_{i-1},x_i')e^{\lambda\alpha_i(x_i')}p^{i+1}(x_i',x_{i+1})}{\sum_{x_i\in X} p^i(x_{i-1},x_i)e^{\lambda\alpha_i(x_i)}p^{i+1}(x_i,x_{i+1})} \geq |X|^{-1}e^{-2A|\lambda|}\mu^2. \tag{11.11}$$

Let

$$\mu(\lambda) = |X|^{-1}e^{-2A|\lambda|}\mu^2. \tag{11.12}$$

and $\bar{p}_\lambda^{[0,n]}(x_{i+1}/x_{i-1}))$ be the conditional probability of a value x_{i+1}, when the value x_{i-1} is fixed. It follows from (11.11) that for all integers $n, i = 1,2,\dots,n-1$ and all $x_{i-1}, x_i \in X$ the transition probability (recall (11.5))

$$(11.13)\qquad \begin{aligned} p^i_\lambda(x_{i-1}, x_i) &= \sum_{x_{i+1}\in X} \bar{p}^{[0,n]}_\lambda(x'_i/x_{i-1}, x_{i+1})\bar{p}^{[0,n]}_\lambda(x_{i+1}/x_{i-1})) \\ &\geq \min_{x_{i+1}\in X} \bar{p}^{[0,n]}_\lambda(x_i/x_{i-1}, x_{i+1}) \geq \mu(\lambda). \end{aligned}$$

An simplified variant of the previous calculations shows that for all integer n and all $x_n, x_{n-1} \in X$

$$(11.14)\qquad p^n_\lambda(x_{n-1}, x_n) = \frac{p^n(x_{n-1}, x_n)e^{\lambda\alpha_n(x_n)}}{\sum_{\tilde{x}_n\in X} p^n(x_{n-1}, \tilde{x}_n)e^{\lambda\alpha_n(\tilde{x}_n)}} \geq |X|^{-1}e^{-2A|\lambda|}\mu \geq \mu(\lambda).$$

We have derived the desired estimate (11.6).

Now we can apply proposition 4.1 and 4.2 to the tilted Markov chains constructed above. It follows from the estimate (11.6) that these chains are exponentially ergodic with constants $\delta = \delta(\lambda)$ and $C = C(\lambda)$ (see (4.30), which do not depend on n. So the estimate (4.36) is fulfilled under the condition (4.35) with $\zeta = \zeta(\lambda)$, which do not depend on n also. Let the characteristic functions (cf. (4.3))

$$(11.15)\qquad \phi^n_{\alpha,\lambda}(z) = \sum_{x\in X^{[0,n]}} \bar{p}^{[0,n]}_\lambda(x)e^{zS^n_\alpha(x)}.$$

Applying Theorem 6.1 together with the estimate (4.50), we find that for $|z| \leq \zeta(\lambda)$ and some constant $\overline{C}(\lambda)$

$$(11.16)\qquad |\ln\phi^n_{\alpha,\lambda}(z)| \leq \overline{C}(\lambda)(n+1), \quad n = 0, 1, \ldots$$

Comparing the definitions (4.3) and (11.3) and (11.15), we see that

$$(11.17)\qquad \phi^n_{\alpha,\lambda}(z) = \frac{\phi^n_\alpha(z+\lambda)}{Z^{[0,n]}_\lambda}.$$

So

$$(11.18)\qquad \ln\phi^n_\alpha(z+\lambda) = \ln\phi^n_{\alpha,\lambda}(z) + \ln Z^{[0,n]}_\lambda.$$

It follows from the definition (11.4) that

$$(11.19)\qquad |X|^{n+1}e^{-(n+1)|\lambda|A} \leq Z^{[0,n]}_\lambda \leq |X|^{n+1}e^{(n+1)|\lambda|A}.$$

The relation (11.18) and the estimates (11.16) and (11.19) imply the desired estimate (11.2) with $R(\lambda) = \zeta(\lambda)$ and $C(\lambda) = \overline{C}(\lambda) + \ln|X| + |\lambda|A$. □

Note 11.1. In typical situations the radius of the disk of analiticity $R(\lambda) \to 0$, as $|\lambda| \to \infty$.

Note 11.2. A possibility to extend the domain of analiticity of characteristic function to a neighborhood of the all real axes is typical for one-dimensional situations only, when the the phase transitions are not typical.

§12. Additive functionals

In §4 we considered sums of random variables depending on the values of a Markov chain in the separate times. Here we study a more general case of sums of random variables depending on the values of a Markov chain in intervals of time.

We use here all the notations introduced in §4 but instead of sums (4.2) we introduce a system $\alpha = (\alpha_{[a,b]}(x_{[a,b]}), 0 \le a \le b < \infty$, of real-valued functions depending on the values $x_{[a,b]} \in X^{[a,b]}$ of the Markov chain on intervals $[a, b]$ and let

$$S_\alpha^n(x) = \sum_{[a,b]\subseteq[0,n]} \alpha_{[a,b]}(x_{[a,b]}) \quad x \in X^{[0,n]}, \tag{12.1}$$

where $x_{[a,b]}$ are the restrictions of the realization x to the subintervals $[a, b]$. Let again (cf. (4.3)) characteristic functions

$$\phi_\alpha^n(z) = \sum_{x\in X^{[0,n]}} \bar{p}^{[0,n]}(x) e^{zS_\alpha^n(x)}, n = 0, 1, \dots . \tag{12.2}$$

We want to modify the derivation of the cluster expansion of §4 for an application to characteristic functions (12.2).

Instead of the transformation (4.14) we write

$$\begin{aligned} \phi_\alpha^n(z) &= \sum_{x\in X^{[0,n]}} \bar{p}^{[0,n]}(x) \prod_{[a,b]\subseteq[0,n]} \left(e^{z\alpha_{[a,b]}(x_{[a,b]})} - 1 + 1\right) \\ &= \sum_{\mathbb{B}\in\mathcal{B}_{[0,n]}} \left(\sum_{x\in X^{[0,n]}} \bar{p}^{[0,n]}(x) \prod_{[a,b]\in\mathbb{B}} \left(e^{z\alpha_{[a,b]}(x_{[a,b]})} - 1\right)\right), \end{aligned} \tag{12.3}$$

where we sum over all systems $\mathbb{B} \in \mathcal{B}_{[0,n]}$ of subintervals $[a, b] \subseteq [0, n]$. For any $\mathbb{B} \in \mathcal{B}_{[0,n]}$ we let

$$\text{supp } \mathbb{B} = \bigcup_{[a,b]\in\mathbb{B}} [a, b]. \tag{12.4}$$

For any interval $[a, b]$ the set of all systems $\mathbb{B} \in \mathcal{B}_{[0,n]}$ such that supp $\mathbb{B} = [a, b]$ is denoted $\mathcal{C}([a, b])$. For any system $\mathbb{B} \in \mathcal{B}_{[0,n]]}$ the set $A =$ supp $\mathbb{B}$ can be represented (cf. (4.16)) in a unique way as a sum of some number k of intervals

(12.5)

$$A = \bigcup_{i=1}^{k} [a_i, b_i], \quad \text{where } a_i \le b_i, i = 1, 2, \cdots, k \text{ and } b_i < a_{i+1} - 1, i = 1, 2, \dots, k - 1,$$

and the system $\mathbb{B}$ can be represented in a unique way as the sum

$$\mathbb{B} = \bigcup_{i=1}^{k} \mathbb{B}_i, \tag{12.6}$$

where the systems $\mathbb{B}_i \in \mathcal{C}([a_i, b_i]), i = 1, 2, \ldots, k$. So for any fixed set $A \subseteq [0, n]$
(12.7)

$$\sum_{\mathbb{B}\in\mathcal{B}_{[0,n]}:\text{supp }\mathbb{B}=A} \prod_{[a,b]\in\mathbb{B}} \left(e^{z\alpha_{[a,b]}(x_{[a,b]})} - 1\right) = \prod_{i=1}^{k} \left(\sum_{\mathbb{B}\in\mathcal{C}([a_i,b_i])} \prod_{[c,d]\in\mathbb{B}} \left(e^{z\alpha_{[c,d]}(x_{[c,d]})} - 1\right)\right).$$

Let (cf. (4.18))
(12.8)

$$\chi^{[a,b]}(x_a, x_b) = \sum_{x_{[a+1,b-1]}} p^{[a+1,b]}(x_a, x_{[a+1,b]}) \sum_{\mathbb{B}\in\mathcal{C}([a,b])} \prod_{[c,d]\in\mathbb{B}} \left(e^{z\alpha_{[c,d]}(x_{[c,d]})} - 1\right).$$

With this modification of the functions $\chi^{[a,b]}(x_a, x_b)$ we can, using the identity (12.7), repeat all the calculations of §4.

Proposition 12.1. *All the statements of Proposition 4.1 are valid for the characteristic function $\phi_\alpha^n(z)$ defined in (12.2) and the function $\chi^{[a,b]}(x_a, x_b)$ defined in (12.8).*

Proof. This proof coincides with the proof of Proposition 4.1. □

To check the K-P condition we need some additional restrictions. Let

$$A_m = \sup_{k=0,1,\ldots} \sup_{x_{[k,k+m]}\in X^{[k,k+m]}} |\alpha^n_{[k,k+m]}(x_{[k,k+m]})|. \tag{12.9}$$

Proposition 12.2. *All the statements of Proposition 4.2 are valid, if instead of the condition (4.34) we assume that*

$$\sum_{m=0}^{\infty} (m+1)(A_m)^{\frac{1}{m+1}} < \infty. \tag{12.10}$$

Proof. We need to explain the proof of the statement **B)** only. It follows from the definition (12.9) and the condition (4.35) that for some constant $K < \infty$ and all intervals $[a, b]$ and realizations $x_{[a,b]} \in X^{[a,b]}$

$$|e^{z\alpha_{[a,b]}(x_{[a,b]})} - 1| \leq K\zeta A_{b-a}. \tag{12.11}$$

So it follows from the definition (12.8) that
(12.12)

$$\sum_{x_b} |\chi^{[a,b]}(x_a, x_b)| \leq \sum_{\mathbb{B}\in\mathcal{C}([a,b])} \prod_{[c,d]\in\mathbb{B}} (K\zeta A_{d-c}) = \sum_{\mathbb{B}\in\mathcal{C}([a,b])} \prod_{[c,d]\in\mathbb{B}} \prod_{t\in[c,d]} (K\zeta A_{d-c})^{\frac{1}{d-c+1}}$$
$$\leq \prod_{t\in[a,b]} \sum_{[c,d]:t\in[c,d]} (K\zeta A_{d-c})^{\frac{1}{d-c+1}} \leq \mu^{b-a+1},$$

where

$$\mu = \mu(\zeta) = \sum_{m=0}^{\infty} (m+1)(K\zeta A_m)^{\frac{1}{m+1}}. \tag{12.13}$$

It follows from the condition (12.10) that $\mu(\zeta) \to 0$ as $\zeta \to 0$. Using the estimate (12.12) instead of the estimate (4.44) we can repeat the construction used in the proof of the statement **B** of Proposition 4.2. □

Theorem 12.3. *Assume that the considered Markov chain is exponentially ergodic and the condition (12.10) and the lower estimate (7.11) on the variances are fulfilled. Then for the sequence of random variables S^n_α defined in (12.1) the central limit theorem is valid.*

Proof. The proof coinsides with the proof of Theorem 7.5. Of course, it is necessary to use Proposition 12.2 instead of Proposition 4.2. □

Note 12.1. In the difference with the situation of Theorem 7.5 for the considered here more general situations the lower estimate (7.11) on variances can be not valid. The simplest example is the case, when

$$\alpha_{[t,t]}(x_t) + \alpha_{[t+1,t+1]}(x_{t+1}) \equiv -\alpha_{[t,t+1]}(x_{[t,t+1]}), \quad t = 0, 1, \ldots \tag{12.14}$$

and $\alpha_{[a,b]}(x_{[a,b})$ vanishes, if $b - a > 1$. The estimate (7.11) can checked under some additional constrains with a help of the method discussed in §7.

Note 12.2. There is a situation more general than considered above. It is the case, when (cf. (12.1)) the functions $\alpha_A(x_A)$ of $x_A \in X^A$ are defined for all finite A and

$$S^n_\alpha(x) = \sum_{A \subseteq [0,n]} \alpha_A(x_A) \quad x \in X^{[0,n]}. \tag{12.15}$$

This situation can be reduced to the situation studied in this section, if we consider the functions

$$\alpha'_{[a,b]}(x_{[a,b]}) = \sum_{A \subseteq [a,b], \text{diam } A = b-a} \alpha_A(x_A), \tag{12.16}$$

where x_A is the restriction of the realization $x_{[a,b]}$ to A.

§13. Markov chains with a general space of states

In §4, 7, and 11, 12 we consider Markov chains with a finite space of states only. Here we extend the previous constructions to the general case.

Let $(X, \mathfrak{B})$ be an arbitrary measurable space. A *Markov transition funcion* is defined as a function $P = (P(x, B), x \in X, B \in \mathfrak{B})$ such that for any fixed $B \in \mathfrak{B}$ the function $P^i(x, B)$ is a measurable function of $x \in X$ and for any fixed x the functions $P(x, B)$ defines a probability measures on $\mathfrak{B}$. A time inhomogeneous Markov chain with the state space X is defined by a sequence of Markov transition functions $P^i, i = 1, 2, \ldots$, and an initial probability distribution $\overline{P}^0 = \overline{P}^0(B)$, which is a probability measure on $\mathfrak{B}$. The probability distribution $\overline{P}^{[0,n]}$ of the Markov chain on the interval $[0, n] \subset \mathbb{Z}^1$ is defined as the probability measure on

the measurale space $(X^{[0,n]}, \mathfrak{B}^{[0,n]})$ such that (cf. (4.1)) for any measurable sets $B_1, B_2, \dots, B_n \in \mathfrak{B}$ the probability of their product
(13.1)

$$\overline{P}^{[0,n]}(B_0 \times B_1 \times B_2 \times \cdots B_n)$$

$$= \int_{B_0} \overline{P}^0(dx_0) \int_{B_1} P^1(x_0, dx_1) \dots \int_{B_{n-1}} P^{n-1}(x_{n-2}, dx_{n-1}) P^n(dx_{n-1}, B_n).$$

For any $0 < a \le b$ and any fixed state $x_{a-1} \in X$ of the chain in the time $a-1$ the conditional distribution $P^{[a,b]}(x_{a-1}, B)$ for the realization on the interval $[a, b]$ (cf. (4.9)) is the probability measure on the space $(X^{[a,b]}, \mathfrak{B}^{[a,b]})$ such that for any measurable sets $B_a, B_{a+1}, \dots, B_b \in \mathfrak{B}$ the probability of their product
(13.2)

$$P^{[a,b]}(x_{a-1}, B_a \times B_{a+1} \times \cdots B_b)$$

$$= \int_{B_a} P^a(x_{a-1}, dx_a) \int_{B_{a+1}} P^a(x_a, dx_{a+1}) \dots \int_{B_{b-1}} P^{b-1}(x_{b-2}, dx_{b-1}) P^b(dx_{b-1}, B_b).$$

The transition probabities from the time $a-1$ to the time b (cf. (4.11)) is the probability measure $P^{a-1,b}(x_{a-1}, B)$ on $\mathfrak{B}$ such that

$$P^{a-1,b}(x_{a-1}, B) = P^{[a,b]}(x_{a-1}, X \times X \dots \times X \times B). \tag{13.3}$$

For any $a > 0$ the unconditional probabiity in the moment a is the probability measure $\overline{P}^a(B)$ on $\mathfrak{B}$ such that

$$\overline{P}^a(B) = \int_X \overline{P}^0(dx_0) P^{0,a}(x_0, B), B \in \mathfrak{B}. \tag{13.4}$$

Consider a sequence $\alpha = (\alpha_i(x_i), x_i \in X, i = 0, 1, \dots)$ of measurable real-valued functions and the random variables

$$S_\alpha^n(x) = \sum_{i=0}^n \alpha_i(x_i), \quad x = (x_0, x_1, \dots, x_n) \in X^{[0,n]}, n = 0, 1, \dots. \tag{13.5}$$

Its characteristic functions (cf. (4.3))

$$\phi_\alpha^n(z) = \int_{X^{[0,n]}} \bar{P}^{[0,n]}(dx) e^{zS_\alpha^n(x)}, n = 0, 1, \dots. \tag{13.6}$$

For any $0 \le a \le b$ and $x_a \in X$ consider the complex-valued measure (cf. (4.18)) $\chi^{[a,b]}(x_a, B)$ on $\mathfrak{B}$

$$\chi^{[a,b]}(x_a, B) = \int_{X^{[a+1,b-1]} \times B} \prod_{u \in [a,b]} \left(e^{z\alpha_u(x_u)} - 1\right) P^{a+1,b}(x_a, dx_{[a+1,b]}). \tag{13.7}$$

The following analogue of Proposition 4.1 holds in the considered case.

Proposition 13.1. *Let Θ_n be the animal model described in Proposition 4.1 and the weights (cf. (4.28))*

(13.8)

$$w(\theta) =$$

$$\int_X \overline{P}^s(dx_s) \int_X \chi^{s,t_1}(x_s, dx_{t_1}) \int_X (P^{t_1,s_2}(x_{t_1}, dx_{s_2}) - \overline{P}^{s_2}(dx_{s_2})) \int_X \chi^{s_2,t_2}(x_{s_2}, x_{dt_2})$$

$$\times \dots \times \int_X (P^{t_{m-1},s_m}(x_{t_{m-1}}, dx_{s_m}) - \overline{P}^{s_m}(dx_{s_m})) \int_X \chi^{s_m,t}(x_{s_m}, dx_t), \quad \theta \in \Theta_n.$$

Then the characteristic function

$$\phi_\alpha^n(z) = Z_w(\Theta_n), \tag{13.9}$$

where the partition function $Z_w(\Theta_n)$ is defined by the relation (2.1).

Proof. Callculations leading to the representation (13.9) literally repeat the calculations used for the derivation of the representation (4.29). It is necessary only to change all the sums over X to the integrals over X. □

Recall that for any (in general, complex-valued) measure λ on an measurale space $(X, \mathfrak{B})$ its *variation* is

$$\text{Var}\ \lambda = \sup_{\gamma: |\gamma| \le 1} \left| \int_X \gamma(x)\lambda(dx), \right| \tag{13.10}$$

where the upper bound is taken over all measurable complex-valued functions $\gamma(x), x \in X$, such that $|\gamma(x)| \le 1$ for all $x \in X$ and for any pair of probability measures P_1, P_2 on a measurable space $(X, \mathfrak{B})$ the *variation distance* between these measures

$$\text{Var}\ (P_1, P_2) = \frac{1}{2}\text{Var}\ (P_1 - P_2) = \sup_{B \in \mathfrak{B}} |P_1(B) - P_2(B)|. \tag{13.11}$$

It follows from this definition that for any measurable bounded function $\gamma(x), x \in X$

$$\left| \int_X \gamma(x)(P_1(dx) - P_2(dx)) \right| \le 2 \sup_{x \in X} |\gamma(x)| \text{Var}\ (P_1, P_2). \tag{13.12}$$

We say that the considered Markov chain is *exponentially ergodic with respect to the variation distance*, if for some constants $\delta > 0$ and $C < \infty$ and any $0 \le s \le t$ and $x_s \in X$

$$\text{Var}\ (P^{s,t}(x_s, \cdot), \overline{P}^t) \le C \exp\{-\delta |t - s|\}. \tag{13.13}$$

It is well-known (see, for example, [Do 56] or Note 14.1) that the chain is exponentially ergodic, if for some $\epsilon > 0$ and all $s > 0$ and $x_s, x'_s \in X$

$$\text{Var}\ (P^s(x_s, \cdot), P^s(x'_s, \cdot)) \le 1 - \epsilon. \tag{13.14}$$

Proposition 13.2. *All the statement of Proposition 4.2 are valid for Markov chain which are exponentially ergodic with respect to the variation distance, if the weights $w(\theta)$ are defined by the relation (3.8).*

Proof. The proof is completely similar to the proof of Proposition 4.2. We mention only that instead of the estimate (4.44) we have now for any $x_a \in X$ and $a \leq b$

$$\text{Var}\ \chi^{[a,b]}(x_a,\cdot) \leq \mu^{b-a+1} \tag{13.15}$$

and need to use the relations (13.10) and (13.12). □

Theorem 13.3. A) *Assume that the considered Markov chain is exponentially ergodic with respect to the variation distance and the condition of boundness (4.34) and the lower estimate (7.11) on the variances are fulfilled. Then for the sequence of random variables S^n_α defined in (13.5) the central limit theorem is valid.*

B) *If all the transition functions $P^i(x, B)$ are given by their densities $p^i(x,x')$, $x, x' \in X$, i.e., for some probability measure m on $(X, \mathfrak{B})$*

$$P^i(x,B) = \int_B p^i(x,x')m(dx') \quad x \in X, B \in \mathfrak{B}, i = 1,2,\dots, \tag{13.16}$$

and these densities uniformly in $x, x' \in X$ and i bounded away from zero, then lower estimate (7.11) on the variances is valid.

Proof. The proof coinsides with the proof of Theorem 7.5. Of course, it is necessary to use Proposition 13.2 instead of Proposition 4.2. and in the lower etimates of the variances to use this conditional densities instead of conditional probilities. □

In §14 we extend the results of this section changing the condition of exponential ergodicity with respect to the variation distance to a more general condition. This extension use an important notion of Kantorovich distance discussed below.

§14. Kantorovich distance and its applications

Let $(X, \mathfrak{B})$ be a measurable space and $\rho(x_1,x_2), (x_1,x_2) \in X \times X$, be a $\mathfrak{B} \times \mathfrak{B}$–measurable function such that (X, ρ) is a metric space. Let $\mathcal{P}(X)$ be the space of all probability measures on $(X, \mathfrak{B})$. The *Kantorovich distance* between probability measures $P_1, P_2 \in \mathcal{P}(X)$ is defined by the relation

$$K(P_1,P_2) = \inf_{Q \in \mathcal{Q}(P_1,P_2)} \int_{X \times X} \rho(x_1,x_2)Q(dx_1,dx_2), \tag{14.1}$$

where $\mathcal{Q}(P_1,P_2)$ is the set of all probability measures Q on the measurable space $(X \times X, \mathfrak{B} \times \mathfrak{B})$ such that their marginal distrbutions coincide with P_1 and P_2, i.e.

$$Q(X \times B) \equiv P_1(B), \quad Q(B \times X) \equiv P_2(B), \quad B \in \mathfrak{B}. \tag{14.2}$$

This notion was introduced by Kantorovich [Ka] in the following important interpretation, which makes more clear the essence of this notion. Let P_1 be a departure distributon of a load and P_2 be its destination distribution and $\rho(x_1,x_2)$ be the cost of transportation from the point x_1 to the point x_2. Then $K(P_1,P_2)$ is the optimal cost of transportation. The discussion of this transportation problem goes back to Monge. The notion of the Kantorovich distance is so natural that it was rediscovered by many authors in different situations. For example, in applications to Markov interaction processes the Kantorovich distance is often called the Vasserstein distance.

Below we discuss the main properties of the Kantorovich distance, but omit nontrivial proves of the facts which will be not used later. The literature on the Kantorovich distance and its applications is very extensive. We can recommend a review paper [Ra], which contains 155 referenes.

The cases, when the minimum in (14.1) can be founded more explicitly, are rare and discussed below. Nevertheless, the Kantorovich distance is very useful, since in applications we typically need only good upper estimates of this distance and for this aim it is enough to construct a mutual distribution $Q \in \mathcal{Q}(P_1, P_2)$ in a reasonable way.

Let $\mathcal{P}_o(X) \subseteq \mathcal{P}(X)$ be the space of all probability measures on $(X, \mathfrak{B})$ such that for some (equivalently any) point $x_0 \in X$

$$\int_X \rho(x, x_0) P(dx) < \infty. \tag{14.3}$$

Observe that for any $P_1, P_2 \in \mathcal{P}_o(X)$ the Kantorovich distance $K(P_1, P_2)$ is finite. Really, for any $Q \in \mathcal{Q}(P_1, P_2)$
(14.4)

$$\int_{X\times X} \rho(x_1, x_2) Q(dx_1, dx_2) \leq \int_{X\times X} \rho(x_1, x_0) Q(dx_1, dx_2) + \int_{X\times X} \rho(x_0, x_2) Q(dx_1, dx_2)$$
$$= \int_X \rho(x_1, x_0) P_1(dx) + \int_X \rho(x_0, x_2) P_2(dx) < \infty.$$

Assume now that the metric space (X, ρ) is separable and $\mathfrak{B}$ is the Borel σ-algebra with respect to the metric ρ. It is useful to compare the Kantorovich distance with the well-known Prokhorov distance (see [Pro]) which metrizes the weak convergence in the space $\mathcal{P}(X)$). The *Prokhorov distance* $\pi(P_1, P_2)$ between probaility measures $P_1, P_2 \in \mathcal{P}(X)$ is defined as the infimum of numbers π such that for all closed subsets $F \subseteq X$

$$P_2(F) - P_1(F_\pi) \leq \pi \quad \text{and} \quad P_1(F) - P_2(F_\pi) \leq \pi, \tag{14.5}$$

where F_π is π-neighborhood of the set F. Prove that for any $P_1, P_2 \in \mathcal{P}(X)$

$$(\pi(P_1, P_2))^2 \leq K(P_1, P_2). \tag{14.6}$$

Really, recall the following evident inequality: for any probability measure P and any events A and B

$$P(A \cap B) \geq P(A) + P(B) - 1. \tag{14.7}$$

Using this inequality, we find that for any $Q \in \mathcal{Q}(P_1, P_2)$ and closed F
(14.8)

$$\int_{X\times X} \rho(x_1, x_2) Q(dx_1, dx_2) \geq \int_{(X\setminus F_\pi)\times F} \rho(x_1, x_2) Q(dx_1, dx_2) \geq \pi(Q((X \setminus F_\pi) \times F)$$
$$\geq \pi(Q(\{(x_1, x_2) : x_1 \in X \setminus F_\pi\}) + Q(\{(x_1, x_2) : x_2 \in F\}) - 1) = \pi(P_2(F) - P_1(F_\pi)).$$

Similarly we check that

$$\int_{X\times X} \rho(x_1,x_2)Q(dx_1,dx_2) \geq \pi(P_1(F) - P_2(F_\pi)). \tag{14.9}$$

The estimate (14.6) follows from the estimates (14.7), (14.8) and the definition of the Prokhorov distance.

Proposition 14.1. *Assume now that the metric space (X,ρ) is separable and $\mathfrak{B}$ is its Borel σ-algebra. Then the distance $K(P_1,P_2)$ defines a metric on the space $\mathcal{P}_o(X)$. The metric space $(\mathcal{P}_o(X),K)$ is a complete metric space. On the space $\mathcal{P}_o(X)$ the Kantorovich distance metrizes the topology of the weak convergence.*

Proof. We check the axioms of metric space. It is evident that if the Prochorov distance $\pi(P_1,P_2)$ vanishes, then $P_1(F) = P_2(F)$ for any closed F and so $P_1 = P_2$. It follows from the estimate (14.6) that the same holds for the Kantorovich distance. The symmetry property $K(P_1,P_2) = K(P_2,P_1)$ is evident from the definition (14.1). Now we show that the triangle inequality for the metric ρ implies the triangle inequality for the Kantorovich distance. Fix probability distributions $P_1,P_2,P_3 \in \mathcal{P}(X)$ and mutual distributions $Q_{1,2} \in \mathcal{Q}(P_1,P_2)$ and $Q_{2,3} \in \mathcal{Q}(P_2,P_3)$. Consider also a probability measure $Q_{1,2,3}$ on the space $(X\times X\times X, \mathfrak{B}\times\mathfrak{B}\times\mathfrak{B})$ such that its projections coincides with $Q_{1,2}$ and $Q_{2,3}$, i.e.,

$$\begin{aligned} Q_{1,2,3}(B_1\times B_2\times X) &= Q_{1,2}(B_1\times B_2), \quad B_1,B_2\in\mathfrak{B},\\ Q_{1,2,3}(X\times B_2\times B_3) &= Q_{1,2}(B_2\times B_3), \quad B_2,B_3\in\mathfrak{B}. \end{aligned} \tag{14.10}$$

Then the projection $Q_{1,3}$ of the measure $Q_{1,2,3}$, defined by the relation

$$Q_{1,3}(B_1\times B_3) = Q_{1,2,3}(B_1\times X\times B_3), \quad B_1,B_3\in\mathfrak{B}, \tag{14.11}$$

belongs to $\mathcal{Q}(P_1,P_3)$. So
(14.12)

$$\begin{aligned} K(P_1,P_3) &\leq \int_{X\times X} \rho(x_1,x_3)Q_{1,3}(dx_1,dx_3) = \int_{X\times X\times X} \rho(x_1,x_3)Q_{1,2,3}(dx_1,dx_2,dx_3)\\ &\leq \int_{X\times X\times X} \rho(x_1,x_2)Q_{1,2,3}(dx_1,dx_2,dx_3) + \int_{X\times X\times X} \rho(x_2,x_3)Q_{1,2,3}(dx_1,dx_2,dx_3)\\ &= \int_{X\times X} \rho(x_1,x_2)Q_{1,2}(dx_1,dx_2) + \int_{X\times X} \rho(x_2,x_3)Q_{2,3}(dx_2,dx_3). \end{aligned}$$

Minimizing the right part of (14.12) over $Q_{1,2} \in \mathcal{Q}(P_1,P_2)$ and $Q_{2,3} \in \mathcal{Q}(P_2,P_3)$, we arrive to the desired inequality

$$K(P_1,P_3) \leq K(P_1,P_2) + K(P_2,P_3) \tag{14.13}$$

However, we need to emphasize that the existence of the measure $Q_{1,2,3}$, satisfying the relation (14.10), is not so evident, as it can seem at first sight. The measure $Q_{1,2,3}$ can be constructed, for example, in such a way that the components x_1, x_2, x_3 forms the Markov chain such that the transition probabilities from x_1 to x_2 are defined by the probability distribution $Q_{1,2}$ and the transition probabilities from x_2 to x_3 are defined by the probability distribution $Q_{2,3}$.

It follows immediately from the estimate (14.6) and the well-known properties of the Prokhorov distance (see. for example, [Du], Th. 11.3.3) that the convergence $K(P_n, P) \to 0$ implies that the sequence of measures P_n weakly converges to the measure P.

We omit the proof of other statements of Proposition (see. for example, [Do70] for a proof.) □

The following fact will be useful for us.

Proposition 14.2. *For any function $f(x), x \in X$, satisfying the Lipschitz condition*

$$|f(x_1) - f(x_2)| \leq \rho(x_1, x_2), \quad x_1, x_2 \in X. \tag{14.14}$$

and any probability distributions $P_1, P_2 \in \mathcal{P}(X)$

$$\left| \int_X f(x)(P_1(dx) - P_2(dx)) \right| \leq K(P_1, P_2). \tag{14.15}$$

Proof. It follows from the definition (14.1) that for any $\epsilon > 0$ there exists a probability distribution $Q^\epsilon \in \mathcal{Q}(P_1, P_2)$

$$\int_{X \times X} \rho(x_1, x_2) Q^\epsilon(dx_1, dx_2) \leq (1 + \epsilon) K(P_1, P_2). \tag{14.16}$$

Then it follows from the estimates (14.14), (14.15) that

$$\begin{aligned} \left| \int_X f(x)(P_1(dx) - P_2(dx)) \right| &= \left| \int_{X \times X} (f(x_1) - f(x_2)) Q^\epsilon(dx_1, dx_2) \right| \\ &\leq \int_{X \times X} \rho(x_1, x_2) Q(dx_1, dx_2). \end{aligned} \tag{14.17}$$

and, since $\epsilon > 0$ can be arbitrary small, it implies the desired estimate (14.15). □

Really, in the situation of Proposition 4.1 the following more deep fact holds

$$K(P_1, P_2) = \sup_f \left| \int_X f(x)(P_1(dx) - P_2(dx)) \right|, \tag{14.18}$$

where the supremum is taken over all the function $f(x), x \in X$, with the Lipschitz condition (14.14) (for a proof see [Ra] and references there). Some analytically inclined authors takes the relation (14.18) as the dual definition of the distance $K(P_1, P_2)$, but we do not use this relation. We prefer the definition (14.1) which is well adjusted for coupling constructions popular in the modern probability theory.

Now we want to discuss the cases, when the Kantorovich distance can be calculated in a more explicit way.

Case 1. Let X is the Euclidean space $\mathbb{R}^n$ with a usual Euclidean metric

$$\rho(x_1, x_2) = |x_1 - x_2|. \tag{14.19}$$

Consider probabiity measures $P_1, P_2 \in \mathcal{P}_o(\mathbb{R}^n)$ which can be obtained one from another by a shift, i.e. for some vector $a \in \mathbb{R}^n$

$$P_2(B) = P_1(B - a), \quad B \in \mathfrak{B}. \tag{14.20}$$

Then it is easy to check that

$$K(P_1, P_2) = |a|. \tag{14.21}$$

Really, if the measure $Q_a \in \mathcal{P}(X)$ is concentrated on the diagonal

$$Q_a(\{(x_1, x_2) : x_2 = x_1 + a\}) = 1, \tag{14.22}$$

then

$$\int_{X \times X} |x_1 - x_2| Q_a(dx_1, dx_2) = |a|. \tag{14.23}$$

On the other hand, for any measure $Q \in \mathcal{Q}(P_1, P_2)$
(14.24)

$$\int_{X \times X} |x_1 - x_2| Q(dx_1, dx_2) \geq \left| \int_{X \times X} x_1 Q(dx_1, dx_2) - \int_{X \times X} x_2 Q(dx_1, dx_2) \right| =$$
$$\left| \int_X x_1 P_1(dx_1) - \int_X x_2 P_2(dx_2) \right| = |a|.$$

The formula (14.21) follows immediately from the relations (14.23), (14.24) and the definition (14.1).

Case 2. Let $X = \mathbb{R}^1$ be a real line, the distance is defined again by (14.19) and F_1, F_2 are the distribution functions of probability distributions $P_1, P_2 \in \mathcal{P}(\mathbb{R}^1)$:
(14.25)

$$F_1(x_0) = P_1(\{x \in \mathbb{R}^1 : x < x_0\}), \quad F_2(x_0) = P_2(\{x \in \mathbb{R}^1 : x < x_0\}), \quad x_0 \in \mathbb{R}^1.$$

Then for any $P_1, P_2 \in \mathcal{P}(\mathbb{R}^1)$

$$K(P_1, P_2) = \int_{-\infty}^{\infty} |F_1(x) - F_2(x)| dx = \int_0^1 |F_1^{-1}(u) - F_2^{-1}(u)| du, \tag{14.26}$$

where F_1^{-1}, F_2^{-1} are the functions inverse to the distribution functions F_1, F_2 (see [Ra] and references there).

Case 3. Let the metric

$$\rho(x_1, x_2) = \begin{cases} 1, & \text{if } x_1 \neq x_2 \\ 0, & \text{if } x_1 = x_2, \end{cases} \tag{14.27}$$

Then the Kantorovich distance coincides with the variation distance (see (13.11))

$$K(P_1, P_2) = \mathrm{Var}\,(P_1, P_2). \tag{14.28}$$

Really, using a well-known Hahn-Jordan decomposition (see, for example, [Du], Th. 5.6.1), we see that

$$P_1(B) = \mu(B) + \lambda_1(B), \quad P_2(B) = \mu(B) + \lambda_2(B), \quad B \in \mathfrak{B}, \tag{14.29}$$

where μ, λ_1, λ_2 are some finite measures and the measures λ_1 and λ_2 are concentrated on some mutually disjoint sets A and $X \setminus A$. It is clear from the definition (13.11) that

$$\text{Var}\,(P_1, P_2) = 1 - \mu(X) = \lambda_1(A) = \lambda_2(X \setminus A). \tag{14.30}$$

Now using the inequality (14.7) and the formula (14.30), we find that for any measure $Q \in \mathcal{Q}(P_1, P_2)$
(14.31)

$$\int_{X\times X} \rho(x_1, x_2) Q(dx_1, dx_2) = Q(\{(x_1, x_2) : x_1 \neq x_2)\}) \geq Q(A \times (X \setminus A))$$
$$\geq Q(\{(x_1, x_2) : x_1 \in A\}) + Q(\{(x_1, x_2) : x_2 \in X \setminus A\}) - 1 = P_1(A) + P_2(X \setminus A) - 1$$
$$= \mu(X) + \lambda_1(A) + \lambda_2(X \setminus A) - 1 = \text{Var}\,(P_1, P_2).$$

On the other hand, we can explicily present a measure $Q_o \in \mathcal{Q}(P_1, P_2)$ such that

$$\int_{X\times X} \rho(x_1, x_2) Q_o(dx_1, dx_2) = \text{Var}\,(P_1, P_2). \tag{14.32}$$

The measure Q_o is defined by the relation

$$Q_o(B_1 \times B_2) = \mu(B_1 \cap B_2) + \frac{\lambda_1(B_1)\lambda_2(B_2)}{1 - \mu(X)}, \quad B_1, B_2 \in \mathfrak{B}. \tag{14.33}$$

(We exluded the trivial case $\mu(X) = 1$).

It turned out that the Kantorovich distance is well adjusted for applications to Markov chains. Let $P = P(x, B), x \in X, B \in \mathfrak{B}$, be a Markov transition function (see §13). Its *Kantorovich norm* is defined as

$$K(P) = \sup_{x_1, x_2 \in X : x_1 \neq x_2} \frac{K(P(x_1, \cdot), P(x_2, \cdot))}{\rho(x_1, x_2)}. \tag{14.34}$$

Let T_P be the *Markov operator* corresponding to the transition function P. It is an operator transforming probability measures P^0 on the space $(X, \mathfrak{B})$ into the probability measures on the same space and is defined by the relation

$$T_P P^0(B) = \int_X P(x, B) P^0(dx), \quad B \in \mathfrak{B}. \tag{14.35}$$

Proposition 14.3. *Let P be a Markov transition function such that its Kantorovich norm $K(P)$ is finite. Let P_1^0, P_2^0 be probability measures on $(X, \mathfrak{B})$ such that the Kantorovich distance between them $K(P_1^0, P_2^0)$ is finite. Then the Kantorovich distance*

$$K(T_P P_1^0, T_P P_2^0) \leq K(P) K(P_1^0, P_2^0). \tag{14.36}$$

Proof. It follows from the definitions (14.1) and (14.34) that for any $x_1^0, x_2^0 \in X, x_1^0 \neq x_2^0$, and any $\epsilon > 0$ there exists a probability measure $Q^{\epsilon}_{x_1^0, x_2^0} \in \mathfrak{Q}(P_1(x_1^0, \cdot), P_2(x_2^0, \cdot))$ such that

$$\int_{X\times X} \rho(\tilde{x}_1, \tilde{x}_2) Q^{\epsilon}_{x_1^0, x_2^0}(d\tilde{x}_1, d\tilde{x}_2) \leq (1+\epsilon)\rho(x_1^0, x_2^0) K(P). \tag{14.37}$$

Similarly there exists a probability measure $Q^{0,\epsilon} \in \mathfrak{Q}(P_1^0, P_2^0)$ such that

$$\int_{X\times X} \rho(x_1^0, x_2^0) Q^{0,\epsilon}(dx_1^0, dx_2^0) \leq (1+\epsilon) K(P_1^0, P_2^0). \tag{14.38}$$

Consider the probability measure $\widehat{Q}^{\epsilon}$ on the product space $(X \times X \times X \times X, \mathfrak{B} \times \mathfrak{B} \times \mathfrak{B} \times \mathfrak{B})$ defined by the relation: for any measurable sets $B^0, \widetilde{B} \in X \times X$

$$\widehat{Q}^{\epsilon}(B^0 \times \widetilde{B}) = \int_{B^0} Q^{\epsilon}_{x_1^0, x_2^0}(B) \widetilde{Q}^{0,\epsilon}(dx_1^0, dx_2^0). \tag{14.39}$$

Really, to make this construction correct we need to know that we can choose the measures $Q^{0,\epsilon}_{x_1^0, x_2^0}$ in such a way that they depend on x_1^0, x_2^0 in a measurable way. We omit a discussion of this question. Let $\widetilde{Q}^{\epsilon}$ be the projection of the measure $\widehat{Q}^{\epsilon}$ to the third and fourth components

$$\widetilde{Q}^{\epsilon}(\widetilde{B}) = \widehat{Q}^{\epsilon}(X \times X \times \widetilde{B}), \quad \widetilde{B} \in \mathfrak{B} \times \mathfrak{B}. \tag{14.40}$$

It follows from the condition $Q^{0,\epsilon} \in \mathfrak{Q}(P_1^0, P_2^0)$ and the definition of the Markov operator (14.35) that the probability measure $\widetilde{Q}^{\epsilon} \in \mathfrak{Q}(T_P P_1^0, T_P P_2^0)$. However it follows from the definitions (14.39) and (14.40) that

(14.41)

$$\begin{aligned}
\int_{X\times X} \rho(\tilde{x}_1, \tilde{x}_2) \widetilde{Q}^{\epsilon}(d\tilde{x}_1, d\tilde{x}_2) &= \int_{X\times X\times X\times X} \rho(\tilde{x}_1, \tilde{x}_2) \widehat{Q}^{\epsilon}(dx_1^0, dx_2^0, d\tilde{x}_1, d\tilde{x}_2) \\
&= \int_{X\times X} \left(\int_{X\times X} \rho(\tilde{x}_1, \tilde{x}_2) Q^{\epsilon}_{x_1^0, x_2^0}(d\tilde{x}_1, d\tilde{x}_2) \right) Q^{0,\epsilon}(dx_1^0, dx_2^0).
\end{aligned}$$

We can use the estimate (14.37) for the estimate of the inner integral in (14.41) and so obtain that

$$\int_{X\times X} \rho(\tilde{x}_1, \tilde{x}_2) \widetilde{Q}^{\epsilon}(d\tilde{x}_1, d\tilde{x}_2) \leq (1+\epsilon) K(P) \int_{X\times X} \rho(x_1^0, x_2^0) Q^{0,\epsilon}(dx_1^0, dx_2^0). \tag{14.42}$$

Now applying the estimate (14.39) we obtain that

$$\int_{X\times X} \rho(\tilde{x}_1, \tilde{x}_2) \widetilde{Q}^{\epsilon}(d\tilde{x}_1, d\tilde{x}_2) \leq (1+\epsilon)^2 K(P) K(P_1^0, P_2^0). \tag{14.43}$$

Recalling that $\widetilde{Q}^{\epsilon} \in \mathfrak{Q}(T_P P_1^0, T_P P_2^0)$ and the number $\epsilon > 0$ can be arbitrary small, we derive the desired estimate (14.36) from (14.43). $\square$

Proposition 14.4. *Let $P_1 = P_1(x_0, B_1), P_2 = P_2(x_1, B_2), x_0, x_1 \in X, B_1, B_2 \in \mathfrak{B}$, be Markov transition functions and $P_{1,2}$ be their convolution defined as a Markov transition function such that*

$$P_{1,2}(x_0, B_2) = \int_X P_2(x_2, B_2) P_1(x_0, dx_2), \quad x_0 \in X, B_2 \in \mathfrak{B}. \tag{14.44}$$

Then the Kantorovich norm

$$K(P_{1,2}) \leq K(P_1)K(P_2). \tag{14.45}$$

Proof. It is clear from the definition (14.34) that the Markov operator $T_{P_{1,2}} = T_{P_1}T_{P_2}$ and so for any $x_0 \in X$

$$P_{1,2}(x_0, \cdot) = T_{P_2}P_1(x_0, \cdot). \tag{14.46}$$

Now the desired estimate (14.45) follows from the definition (14.34) and Proposition 14.3. □

Now we can derive an ergodic theorem ([Do 70]) for inhomogeneous Markov chains (see §13).

Theorem 14.5. *Let a sequence of transition functions $P_i, i = 1, 2, \ldots$, defining an inhomogeneous Markov chain is such that their Kantorovich norms $K(P_i) < 1, i = 1, 2, \ldots$ and more of it*

$$\sum_{i=1}^{\infty}(1 - K(P_i)) = \infty. \tag{14.47}$$

Let $\overline{P}_1^0, \overline{P}_2^0 \in \mathcal{P}_o(X)$ be a pair of initial disributions and for any $t > 0$

$$\overline{P}_j^t(B) = \int_X \overline{P}_j^0(dx_0)P^{0,a}(x_0, B), \quad B \in \mathfrak{B}, \quad j = 1, 2. \tag{14.48}$$

be the unconditional probabiity in the moment t induced by the initial distribution $\overline{P}_j^0$ (recall (13.4)). Then

$$\lim_{t\to\infty} K(\overline{P}_1^t, \overline{P}_2^t) = 0. \tag{14.49}$$

If for some $\delta > 0$, some $0 \leq t$ and all $s = 1, 2, \ldots, t$ the Kantorovich norm

$$K(P_s) \leq e^{-\delta}, \tag{14.50}$$

then the Kantorovich distance

$$K(\overline{P}_1^t, \overline{P}_2^t)e^{-\delta s}. \tag{14.51}$$

Proof. It follows from the definitions (13.3), (13.4) that that the transition function $P^{0,t}$ is the convolution of the sequence of transitions functions $P_1, P_2, \ldots, P_t$. So it follows from Proposition 14.4 that the Kantorovich norm

$$K(P^{0,t}) \leq \prod_{i=1}^{t} K(P_i). \tag{14.52}$$

So the condition (14.47) implies that

$$\lim_{t \to \infty} K(P^{0,t}) = 0 \tag{14.53}$$

and the statement of Theorem (14.49) follows from Proposition 14.3. The estimate (14.51) is proved in a similiar way. □

Note 14.1. Since, as we have explained, the variation distance is a special case of the Kantorovich distance, the formulated in §13 statement that the condition (13.14) implies the exponential ergodicity with respect to the variation distance follows from the estimate (14.51).

References

[Be] S.N. Bernstein, *Détermination d'une limite inférieure de la dispersion des sommes de grandeurs liées en chaine singuliére*, Math. Sbornik **1** (1936), 29-38.

[Br] D. Brydges, *A short course on cluster expansions*, Critical Phenomena, Random Systems, Gauge Theory (Les Houches, 1984) (K.Ostewalder and R. Stora, eds.), North-Holland, Amsterdam-New York, 1986, pp. 129-183.

[Do 56] R.L. Dobrushin, *Central Limit Theorem for inhomogeneous Markov chains*, Th. Prob. Appl. **1** (1956), 72-88, 365-425.

[Do65] R.L. Dobrushin, *Existence of a phase transition in two and three dimensional Ising models*, Th. Prob. Appl. **10** (1965), 193-213.

[Do70] R.L. Dobrushin, *Description of a system of random variables by the help of conditional distributions*, Th. Prob. Appl. **15** (1970), 469-497.

[Do87] R. Dobrushin, *Induction on volume and no cluster expansions*, Proc. VIII-th Intern. Congress on Math Phys., World scientific, Singapore eds M. Mebkhout and B. Seneor, 1987, pp. 73-91.

[Do87] R. Dobrushin, *Induction on volume and no cluster expansions*, Proc. VIII-th Intern. Congress on Math Phys., World scientific, Singapore eds M. Mebkhout and B. Seneor, 1987, pp. 73-91.

[Do88] R. Dobrushin, *A new approach to the analysis of Gibbs perturbations of Gaussian fields*, Selecta Math. Sov. **7** (1988), 221-277.

[Do94] R.Dobrushin, *Estimates of semiinvariants for the Ising model at low temperatures*, Preprint 125, ESI, Wien, 1994.

[DM] R. Dobrushin and M. Martirosjan, *Non-finite perturbation of Gibbsian fields*, Theor. and Math. Phys. **74** (1988), 10-20.

[DN] R. Dobrushin and B.S. Nahapetian, *Strong convexity of the pressure for lattice systems of classical statistical physics*, Theor. and Math. Phys. **20** (1974), 782-790.

[DN] R. Dobrushin and S. Shlosman, *Consructive criterion for the uniqueness of Gibbs fields*, Stat. Phys. and Dynamical Systems. Rigorous Results, Progr. in Phys., 10 **20** (1974), 782-790.

[DW] R. Dobrushin and V. Warstat, *Completely analytical interactions with infinite values*, Probabl. Th. Rel. Fields **84** (1990), 335-359.

[Du] R.M. Dudley, *Real Analysis and Probability*, Wadsworth & Brooks/Cole, Pacific Grove, California, 1989.

[Ge] H.-O. Georgii, *Gibbs Measures and Phase Transitions*, Walter de Gruyger,, Berlin, 1988..

[GJ] J. Glimm and A. Jaffe, *Quantum Physics. A Functional Integral Point of View*, Springer-Verlag, New York-Berlin, 1987.

[GJS] J. Glimm, A. Jaffe, and T. Spencer, *A convergent expancion about the mean field theory, I*, Ann. Phys **101** (1976), 610–630; *II*, 631–669.

[GK] C. Gruber, H. Kunz, *General properties of polymer systems*, Comm. Math. Phys. **22** (1971), 133–161.

[Gr] Griffiths R.B, *Peierls' proof of spontaneous magnetization in a two-dimensional Ising ferromagnet*, Phys. Rev. **136A** (1964), 437-439.

[IL] I.A. Ibragimov and Y.V. Linnik, *Independent and Stationary Sequences of Random Variables*, Wolters-Noorhoff, Groningen, 1971.

[Ka] L.V. Kantorovich, *On transportation of masses*, Reports Acad. Sci. USSR **37** (1942), 225-226.

[KP] R. Kotecky and D. Preiss, *Cluster expansion for abstract polymer models*, Comm. Math. Phys. **103** (1986), 491-498.

[Ma] V.A. Malyshev, *Cluster expansions in lattice models of statistical physics and quantum theory of fields*, Russian Math. Surveys **35** (1980), no. 2, 1-62.

[MM] V.A. Malyshev and R.A. Minlos, *Gibbs Random Fields. Cluster Expansions*, Kluver Academic Publishers Group, Dordrecht, 1991.

[Pe] R. Peierls, *On Ising's model of ferromagnetism*, Proc. Cambridge Phil. Soc. **32** (1936), 477-481.

[Pre] C.J. Preston, *Gibbs states on Countable Sets*, Cambridge Univ. Press,, London-New York., 1974.

[Pro] J.V. Prokhorov, *Convergence of random processes and limit theorems of the theory of probability*, Th. Prob. Appl. **1** (1956), 177-238.

[Ra] S.T.Rachev, *The Monge-Kantorovich problem on transportation of masses and its applications in stochastics*, Th. Prob. Appl. **29** (1984), 625-653.

[Se] E. Seiler, *Gauge Theories as a Problem of Constructive Quantum Field Theory and Statistical Mechanics. Lectures Notes in Physics, vol 159*, Springer-Verlag, Berlin-New York, 1982.

[Sim] B. Simon, *The Statistical Mechanics of Lattice Gases, Volume 1*, Princeton University Press, Princeton, New Jersey, 1993.

[Sin] Ya. G. Sinai, *The Theory of Phase Transitions: Rigorous results*, Pergamon, London, 1981.

[SS] L. Saulis, V.A. Statulevicius, *Limit Theorems for Large Deviations*, Kluwer Acad. Publ., Dordrecht-Boston-London, 1991.

LECTURES ON INVERSE PROBLEMS

Piet GROENEBOOM

Introduction.

Recently I watched a video recording of a television programme on the BBC about Fermat's last theorem. It reminded me of a conversation at a meeting on statistics and probability theory (not the St-Flour meeting!), where several people were claiming that Andrew Wiles' work on Fermat's last theorem was "unimportant". One of the discussants was even saying, using borrowed authority on the matter, "also pure mathematicians say it is unimportant". Ironically, the video starts with a voice saying that "the most important mathematical problem of the century" has been solved.

Letting aside the question of importance, the thing I personally enjoyed so much in the programme was the directness of the mathematicians that were interviewed, Andrew Wiles still being extremely happy that he finally overcame all difficulties, the atmosphere of joy about all the wonderful ideas and connections that had been generated by the study of the problem. Also striking was the open-mindedness; Shimura saying that Taniyama made the "good mistakes", no doubt meaning by this that he generated a lot of good ideas. All this seemed the exact opposite of people trying to intimidate or bully other people by telling them what they should or shouldn't occupy themselves with. In fact, the atmosphere that came across from the video represented exactly that which made me study mathematics in the first place.

Why would one study Fermat's last theorem? According to me, not because it is believed to be extremely important, or because it is a three centuries old problem or anything like that. No, just because it is a very intriguing and beautiful problem and "there is something there" (actually, "many things are there"), Andrew Wiles denotes it as his "childhood's dream".

In the following, I will mainly talk about one particular inverse problem that has intrigued me for many years. It is very easy to formulate, but very hard to deal with. It quickly leads out of statistics into other fields of mathematics, notably the theory of integral equations. It immediately struck me as a problem that "had something" and could lead to the development of new theories. Although this has not been my motivation for studying the problem, the outcome of the mathematics is actually used now in an AIDS research project and other fields of medical statistics.

At the time of the St-Flour lectures, I still had not solved one of the main problems with which I was struggling, and my laptop computer was running during the night in an investigation of the properties of the relevant integral equations. Recently, however, this problem has actually been solved (not by using the computer, but just by straightforward mathematics, although staring at the pictures on the computer screen has certainly been helpful) and I will discuss this below.

Here is a brief description of the content of my notes. I will discuss inverse problems where one wants to estimate a distribution function or a functional thereof, in situations where random variables, generated by this distribution function, are only indirectly observable. In this situation, the distribution function cannot be estimated at the usual $\sqrt{n}$-rate, in contrast with, for example, the situation for right-censored data. In this sense, the interval censoring problem (see below) is an inverse problem, whereas the right-censoring problem is not. Other examples of inverse problems are discussed in chapter 1.

If one wants to estimate the distribution function by the nonparametric maximum likelihood estimator (NPMLE), one has to use methods from isotonic regression theory and convex optimatization to even compute the estimator in an efficient way. I discuss such methods in chapter 2.

Although the distribution function cannot be estimated at $\sqrt{n}$-rate, certain "smooth functionals" of the model can often be estimated at this rate. The general theory for this is sketched in chapter 3. It is then applied to the interval censoring model, and it is shown that estimators of these functionals, based on the NPMLE, not only attain $\sqrt{n}$-rate, but are also asymptotically efficient.

Finally, it is argued in chapter 4 that progress in developing local asymptotic distribution theory for the NPMLE will depend on progress in developing the smooth functional theory. The "working hypothesis" in GROENEBOOM AND WELLNER (1992) that the local asymptotic distribution of the NPMLE is given by the asymptotic distribution of a toy estimator, obtained by doing one step of the iterative convex minorant algorithm introduced in chapter 2, starting with the underlying distribution function, is verified for one of the versions of the interval censoring model. The result of Theorem 4.4 in section 4.2 has (like the results in section 3.6) defied solution for a long time, and appears for the first time in these notes. The methodology of section 4.2 extends to the other inverse problems mentioned in chapter 1. As an example, the working hypothesis has been verified for a particular case of the deconvolution problem in JONGBLOED (1994).

The discussion of the local asymptotic distribution theory is preceded by a discussion of certain local minimax results, suggesting that the NPMLE also locally attains the best asymptotic variance.

For these inverse problems, the really hard work is the analysis of certain integral equations for which generally no explicit solutions exist. Proofs of Donsker properties have to rely on the success in deriving qualitative properties of the implicitly defined solutions of these integral equations.

I take this opportunity to thank several people for their support in getting this research off the ground. First of all, Richard Gill who started me working on the topic of inverse problems by asking questions about the behavior of the NPMLE for the simplest case of interval censoring ("current status data"). He also got me working on the Wicksell problem and (not surprising in view of his knowledge of ornithology) Hampel's bird migration problem. Next, I want to thank Jon Wellner for his continuous support; Jon has the invaluable gift of "creative listening". I also want to thank my (former) Ph. D. students Bert van Es, Ronald Geskus and Geurt Jongbloed for their interest and enthusiasm in further developing the theory.

Finally, I want to thank Pierre Bernard for the invitation to present these lectures at the St-Flour summer school and Lucien Birgé, Paul Louis Hennequin,

Gérard Kerkyacharian and Dominique Picard for their comments on my lectures (and other support).

Contents

Chapter 1

Some examples of inverse statistical problems

I will not try to define formally what an inverse problem is, but important aspects of what I call an inverse problem are:

1. We only have indirect information about the things we really would like to observe.

2. Estimators of the distribution function of the variables of interest do not (and cannot) have the usual $\sqrt{n}$-rate of convergence. In this sense the interval censoring problem, defined below, is an inverse problem, but in contrast the right-, left- or double censoring problems are *not* inverse problems and indeed have a completely different asymptotic theory. For the latter, see, e.g., ANDERSEN *et. al.* (1993) and GILL (1994).

1.1 The interval censoring problem

I consider the following two cases of interval censoring.

Case 1 (Current status data). Let $(X_1, U_1), \ldots, (X_n, U_n)$ be a sample of random variables in $\mathbb{R}_+^2$, where X_i and U_i are independent (non-negative) random variables with distribution functions F_0 and G, respectively, and where we assume $F_0 << G$ (F_0 is absolutely continuous with respect to G). The only observations which are available are U_i ("observation time") and $\Delta_i = \{X_i \le U_i\}$. Here and (sometimes) in the sequel I will denote the indicator of an event A (such as $\{X_i \le U_i\}$) just by A, instead of 1_A. The (marginal) log likelihood for F_0 is given by the function

$$F \mapsto \sum_{i=1}^{n} \left\{ \Delta_i \log F(U_i) + (1 - \Delta_i) \log\big(1 - F(U_i)\big) \right\}, \qquad (1.1)$$

where F is a right-continuous distribution function. This simplest case of interval censoring is often called the "current status". A *nonparametric maximum likelihood estimator* (NPMLE) $\hat{F}_n$ of F_0 is a (right-continuous) distribution function F, maximizing (1.1).

Case 2. Let $(X_1, U_1, V_1), \ldots, (X_n, U_n, V_n)$ be a sample of random variables in $\mathbb{R}_+^3$, where X_i is a (non-negative) random variable with distribution function F_0, and where U_i and V_i are (non-negative) random variables, independent of X_i, with a joint distribution function H and such that $U_i \le V_i$ with probability one. We assume $F_0 << H_1 + H_2$, where H_1 and H_2 are the marginal distribution functions of U_i and V_i, respectively. The only observations which are available are (U_i, V_i) (the "observation times"), $\Delta_{1i} = \{X_i \le U_i\}$, $\Delta_{2i} = \{X_i \in (U_i, V_i]\}$. In this case the (marginal) log likelihood for F_0 is given by the function

$$F \mapsto \sum_{i=1}^{n} \left\{\Delta_{1i} \log F(U_i) + \Delta_{2i} \log\bigl(F(V_i) - F(U_i)\bigr) + \Delta_{3i} \log\bigl(1 - F(V_i)\bigr)\right\}, \tag{1.2}$$

where $\Delta_{3i} = 1 - \Delta_{1i} - \Delta_{2i}$. In this case, an NPMLE $\hat{F}_n$ of F_0 is a (right-continuous) distribution function F, maximizing (1.2).

One can of course go on, and also define "case k" interval censoring, where instead of a pair (U_i, V_i) of observation times, one has k observation times for each (unobservable) X_i. Case k interval censoring, however, is very much like "case 2", since, for each X_i, only one or two "neighboring" observation times give relevant information. Case k interval censoring is discussed in Geskus and Groeneboom (1996a) and Wellner (1995).

As far as I am aware, the study of interval censoring models started in the fifties with, for example, the papers Ayer *et al.* (1955) and van Eeden (1956). One may wonder why there has been so little progress in the theory of these models in the past forty years, while, at the same time, there has been such a "boom" in the analysis of models with right- or left-censored data.

I think that the reason is to be found in the fact that with right- and left-censoring models, we are much closer to the classical theory. As an example, the Kaplan-Meier estimator (which can be considered to be the nonparametric maximum likelihood estimator for the distribution function in that model) has, at a fixed point inside a region where one has both censored and uncensored observations, a normal limiting distribution and the convergence rate is $\sqrt{n}$. The random process, associated with the Kaplan-Meier estimator is tight and converges (under some mild conditions) in distribution to a Gaussian process.

In contrast with this, as will be shown below, the pointwise convergence rate of the nonparametric maximum likelihood estimator (NPMLE) for the distribution function in the interval censoring problem is slower than $\sqrt{n}$, the asymptotic distribution is non-normal, and the (global) random process, associated with the NPMLE, is not tight. We only will have locally tight random processes after rescaling.

One might think that the fact that the pointwise convergence rate of the NPMLE in the interval censoring problem is slower than $\sqrt{n}$, is a defect of the NPMLE, and that it might be possible to find estimators which converge at a faster rate. This, however, is an illusion. Minimax computations show that the NPMLE indeed attains the optimal rate of convergence, and that only by introducing extra (smoothness) conditions in the model, one can achieve higher rates of convergence. But in that case, one can use these a priori conditions to modify the NPMLE in such a way that it also attains these higher rates of convergence. One could say that the NPMLE actually "catches" the information that is there, without any further conditions, and that if one introduces extra conditions, one can "catch" these too by the NPMLE, by appropriate smoothing.

Perhaps more interesting, though, is the fact that in these models certain smooth functionals, like moments, can be estimated at $\sqrt{n}$-rate by estimators, based on the NPMLE, and that the asymptotic variance of these estimators actually attains the information lower bound.

An up to date account of progress on the present "inverse problem" is HUANG AND WELLNER (1996), where also numerous references to applications in the medical and epidemiological literature can be found.

1.2 Regression models with interval censoring

These models are discussed in detail in HUANG AND WELLNER (1996). As an example, in a regression model under interval censoring, case 1, the observations are (Δ_i, Z_i), $i = 1, \ldots, n$, corresponding to unobservable random variables X_i, where Z_i is a covariate and it is assumed that the distribution function $F_0(x|z)$ of X_i, given the covariate Z_i, satisfies a relation of the type

$$g(F_0(x|z)) = h(x) + \theta' z,$$

where g is a specified function, h an unknown increasing function, and θ a d-dimensional regression parameter. Taking

$$g(s) = -\log\{-\log(1-s)\},\ 0 < s < 1,$$

one obtains the well-known Cox proportional hazards model, proposed in COX (1972), with interval censored data. In that case we have

$$h(x) = -\log \Lambda(x),$$

where Λ is the so-called "baseline cumulative hazard function".

Taking

$$g(s) = \log\{s/(1-s)\},\ 0 < s < 1,$$

one obtains the proportional odds regression model with interval censored data.

There is also an interpretation in terms of the "binary choice model" in econometrics. Here the covariate Z_i denotes socio-economic information on the ith respondent in a questionnaire, and the indicator Δ_i could correspond to an answer "Yes" or "No" of this respondent. See for these applications for example COSSLETT (1983), COSSLETT (1987), MANSKI (1985) and KIM AND POLLARD (1990).

In these models the distribution function F_0 is the nonparametric part of the model and the regression parameter θ the parametric part. Under some conditions, the parameter θ will be a "smooth functional" of the model, and, like the moments in 1.1, can be estimated at $\sqrt{n}$-rate. There is also the general feeling that maximum likelihood estimation will give the estimator with the smallest asymptotic variance, but this has only been established so far for some special cases. As an example, efficiency of the maximum likelihood estimators for the Cox regression model with interval censoring, case 1, is established in HUANG (1996).

1.3 Deconvolution

As an example, consider the deconvolution problem in a model with non-negative random variables and disturbances with a decreasing density. Formally, let $Z_1, \ldots, Z_n$ be a sample from a distribution function H with density

$$h(z) = \int g(z-x)\, dF_0(x),\ z \in I\!R, \tag{1.3}$$

where g is a decreasing density on $[0,\infty)$, and F_0 an unknown distribution function, concentrated on $[0,\infty)$. For example, g could be the exponential density

$$g(x) = e^{-x} 1_{[0,\infty)}(x),\ x \in I\!R,$$

or the Uniform (0,1) density

$$g(x) = 1_{[0,1]}(x),\ x \in I\!R.$$

An NPMLE of F_0 is a distribution function, maximizing

$$\tilde{\phi}(F) = \int \log\Big\{\int g(z-x)\, dF(x)\Big\}\, dH_n(z),$$

as a function of F, where H_n is the empirical distribution function of the sample $Z_1, \ldots, Z_n$.

We have a similar situation as with interval censoring. In this case there is a collection of connected conjectures on the behavior of the NPMLE. These conjectures are that the pointwise rate is of the NPMLE is $n^{-1/3}$, that the limiting distribution is of the same type as that of the NPMLE for interval censoring, and that the NPMLE of smooth functionals (like the mean) will attain the information lower bound. Support for these conjectures can for example be found in JONGBLOED (1994).

1.4 Wicksell's problem

In WICKSELL, (1925) the following stereological problem was studied. Suppose that spheres are embedded in an opaque medium. and suppose that one wants to estimate the distribution function of the sphere radii. Since the medium is opaque, one cannot observe a sample of sphere radii directly. What can be observed is a cross section of the medium, showing circular sections of some spheres.

If we denote the distribution function of the sphere radii by F_S and assume the centres of the spheres to be distributed according to a homogeneous Poisson process, one can show (see, e.g., WATSON (1971)) that the observable circle radii constitute a sample from the density g_C, where

$$g_C(z) = \frac{z}{m_0} \int_z^\infty \frac{dF_S(x)}{\sqrt{x^2 - z^2}}, \tag{1.4}$$

with $0 < m_0 = \int_0^\infty x\,dF_S(x) < \infty$, the expected sphere radius.

For mathematical convenience, it is easier to consider the *squared* radii of both the balls and the circles instead of the radii themselves. In that case the relation between the density g of the observable squared circle radii and the distribution function F of the squared sphere radii follows from (1.4) and is given by

$$g(z) = \frac{1}{2m_0} \int_{(z,\infty)} \frac{dF(x)}{\sqrt{x - z}},$$

where $0 < m_0 = \int \sqrt{x}\,dF(x) < \infty$. This integral equation can be inverted, giving an expression of F in terms of g:

$$F(x) = 1 - \frac{2m_0}{\pi} \int_x^\infty \frac{g(z)}{\sqrt{z - x}}\,dz = 1 - \frac{\int_x^\infty (z - x)^{-1/2} g(z)\,dz}{\int_0^\infty z^{-1/2} g(z)\,dz}.$$

Writing

$$V(x) = \int_x^\infty \frac{g(z)}{\sqrt{z - x}}\,dz,$$

one sees that

$$F(x) = 1 - \frac{V(x)}{V(0)}.$$

Therefore, the problem of estimating F at a fixed point $x_0 > 0$ is equivalent to the problem of estimating V at two fixed points, i.e., x_0 and 0. It is shown in GROENEBOOM AND JONGBLOED (1995) that, under general conditions, the rate for estimating V at a fixed point cannot be faster than $n^{-1/2}\sqrt{\log n}$, and an estimator $\hat{V}_n$ of V is introduced that satisfies the monotonicity constraint. From the monotone estimator of V we get an estimator $\hat{F}_n$ of F, which is actually a distribution function, and is defined by

$$\hat{F}_n(x) = 1 - \hat{V}_n(x)/\hat{V}_n(0).$$

One can also define a (sieved) NPMLE for the distribution function F. This NPMLE is proved to be consistent in JONGBLOED (1995A). At this point the asymptotic distribution of the estimator is not yet known, although it is conjectured to have the same asymptotic distribution as the isotonized naive estimator.

Remarkable features of the Wicksell problem are that smoothness assumptions are not helpful for getting faster rates in pointwise estimation of the distribution function and also, that the $\sqrt{n}$-rate in the estimation of certain 'smooth functionals', like the mean, seems in general to be unattainable. The latter is in contrast with the two preceding examples. In comparison, for the Wicksell problem the pointwise rate is quite good ($n^{-1/2}\sqrt{\log n}$), but there is no improvement for the estimation of moments. Smoothness assumptions on the underlying distribution function do not help here, the difficulties arise from the nature of the inverse problem, in particular from the extra parameter m_0. However, *ratios* of moments can indeed be estimated at the $n^{-1/2}$ rate (in that case we lose the extra parameter in the division).

1.5 Hampel's migrating birds problem (convex densities)

In HAMPEL (1987) the following problem from ornithology is discussed. Suppose that birds regularly visit a certain area. One is interested in the distribution of the amount of time that the birds stay in the area. One obtains information on this by catching birds, marking the birds with a ring and recording the time between two catches of the same bird.

Hampel derives the following relation between the distribution function F of the sojourn time of the bird in the district and the density g of the time between two catches:

$$g(z) = \frac{2}{c}\int_z^\infty (x-z)\,dF(x), \tag{1.5}$$

where it is assumed that $0 < c = \int_0^\infty y^2\,dF(y) < \infty$. It follows from (1.5) that

$$g'(z) = -\frac{2}{c}\left(1 - F(z)\right).$$

This shows that g has to be a bounded decreasing convex density on $[0,\infty)$. Once one has an estimator $\hat{g}_n$ for g, one can then estimate F by

$$\hat{F}_n(x) = 1 - \frac{\hat{g}_n'(x)}{\hat{g}_n'(0)}.$$

So, in this case the "inverse problem" is the estimation of a convex decreasing density. Results in MAMMEN (1991) and JONGBLOED (1995A) show that the optimal pointwise rate of convergence of estimators of a convex density is $O(n^{-2/5})$. The NPMLE of a convex density is well-defined, but at present its (local) asymptotic distribution is still unknown. I should mention, though, that some conjectures on this are known to the insiders. However, for reasons of space and time, I will not be able to discuss these below.

Chapter 2

Algorithms

2.1 The self-consistency equation and the EM algorithm

One of the striking facts in the development (or non-development) of the theory on the problems mentioned in chapter 1 has been the belief that the EM algorithm and the so-called "self-consistency equations" would give the right approach to both the computation of the NPMLE and the distribution theory, see, e.g., TSAI AND CROWLEY (1985) and TSAI AND CROWLEY (1990). This is the more surprising, since, for example, the papers AYER *et al.* (1955) and VAN EEDEN (1956) already pointed in a quite different direction in the case of interval censoring. These early papers, however, have been superseded by the papers TURNBULL (1974) and TURNBULL (1976), and it seemed that the EM algorithm and the self-consistency equation came out as the winners in the approach to these problems. For this reason I first pay attention to these latter methods, before embarking on the approach using Fenchel duality, which I think is actually the right line of attack, and which is connected to the approach in the papers AYER *et al.* (1955) and VAN EEDEN (1956).

I take interval censoring as an example, but the other inverse problems, mentioned in chapter 1, can be treated in a similar way. For interval censoring, case 1, the log likelihood function for F_0 is given by (1.1).

2.2 Self-consistency of the NPMLE for interval censoring, case 1

First note that only the values of F at the observation times occur explicitly in the likelihood. Let $\hat{F}_n$ be a right-continuous distribution function, maximizing (1.1). The order restriction on $\hat{F}_n$ makes it a function that is piecewise constant and uniquely defined on large parts of its domain. Generally the intervals of constancy contain several observation times. The only places where $\hat{F}_n$ is not uniquely defined is between two consecutive ordered observation times for which $\hat{F}_n$ has a different value.

The usual convention is to impose that $\hat{F}_n$ is piecewise constant everywhere, and only has jumps at the observation points. Then it is uniquely determined everywhere between the first and the last point of jump. Let $\tau_1, \ldots, \tau_m$ the ordered points of jump of $\hat{F}_n$. So τ_1 and τ_m are the first and last point of jump of $\hat{F}_n$ respectively. Except for the case that all Δ_i's are one, we always have that $\hat{F}_n(0) = 0$ and that $\hat{F}_n$ is uniquely determined from 0 to τ_1. The function $\hat{F}_n$ may be defective (in which case $\hat{F}_n(\tau_m) < 1$); in that case we do not specify the location of the remaining mass (since we have no information on that).

Now suppose, using a currently fashionable terminology, that an "oracle" would tell us the location of the points of jump τ_i (one could call this an argument "ex oraculo"). Then the maximization problem is simple: the log likelihood function becomes a function of a finite number of parameters θ_i, corresponding to possible values $F(\tau_i)$ of the distribution function at the points τ_i. One can differentiate the log likelihood function with respect to these parameters and set the partial derivatives equal to zero. This gives the equations

$$\theta_i^{-1} \sum_j \Delta_j \{\tau_i \le U_j < \tau_{i+1}\} - (1-\theta_i)^{-1} \sum_j (1-\Delta_j)\{\tau_i \le U_j < \tau_{i+1}\} = 0, \quad (2.1)$$

for $1 \le i < m$. Multiplying (2.1) by $\theta_i(1-\theta_i)$, the equations

$$\theta_i = \frac{\sum_j \Delta_j \{\tau_i \le U_j < \tau_{i+1}\}}{\sum_j \{\tau_i \le U_j < \tau_{i+1}\}}, \quad 1 \le i < m, \quad (2.2)$$

emerge. Hence, if θ_i satisfies (2.2), we can define $\hat{F}_n(\tau_i) = \theta_i$, $1 \le i < m$, and $\hat{F}_n(\tau_i)$ is just given by the number of indicators Δ_i equal to one in the interval $[\tau_i, \tau_{i+1})$, divided by the total number of observations in that interval.

If $\Delta_j = 0$ for some j such that $U_j > \tau_m$, θ_m is also defined as a solution of (2.1), and we then define $\hat{F}_n(\tau_m) = \theta_m$ (the corresponding distribution function is defective in that case). In the other case we define $\hat{F}_n(\tau_m) = 1$. For $t < \tau_1$, we define $\hat{F}_n(t) = 0$. Since $\hat{F}_n$ only has jumps at the points τ_i, $\hat{F}_n(t)$ is now defined for all $t \ge 0$.

Let Q_n be the empirical measure of the pairs (U_j, Δ_j), $j = 1, \ldots, n$. Furthermore, let J_i be the interval $[\tau_i, \tau_{i+1})$, $i = 0, \ldots, m$, where $\tau_0 = 0$ and $\tau_{m+1} = \infty$. Then (2.2) can be conveniently written

$$\int_{J_i} \{\hat{F}_n(u) - \delta\}\, dQ_n(u,\delta) = 0, \; 0 \le i \le m. \quad (2.3)$$

It follows from the definition of $\hat{F}_n$ that (2.3) indeed also holds for $i = 0$ and $i = m$.

In a similar way, (2.1) leads to

$$\int_{J_i} \left\{ \frac{\delta}{\hat{F}_n(u)} - \frac{1-\delta}{1-\hat{F}_n(u)} \right\} dQ_n(u,\delta) = 0, \; 1 \le i < m. \quad (2.4)$$

Since $\hat{F}_n$ is constant on the intervals J_i, we get from (2.4), defining $0/0 = 0$,

$$\int \left\{ \frac{\delta}{\hat{F}_n(u)} - \frac{1-\delta}{1-\hat{F}_n(u)} \right\} \frac{\hat{F}_n(t \wedge u) - \hat{F}_n(t)\hat{F}_n(u)}{\hat{F}_n(t)} \, dQ_n(u,\delta) = 0,$$

for all $t \geq 0$. This can be rewritten as

$$\hat{F}_n(t) = \int \left\{ \frac{\delta \hat{F}_n(t \wedge u)}{\hat{F}_n(u)} + \frac{(1-\delta)\{\hat{F}_n(t) - \hat{F}_n(t \wedge u)\}}{1-\hat{F}_n(u)} \right\} dQ_n(u,\delta), \; t \geq 0. \tag{2.5}$$

Equation (2.5) is in fact the so-called "self-consistency equation" for the NPMLE $\hat{F}_n$, which can also be written:

$$\hat{F}_n(t) = E_{\hat{F}_n}\{F_n(t) \mid (U_i, \Delta_i), \, i = 1, \ldots, n\}, \; t \geq 0, \tag{2.6}$$

where F_n is the ordinary (unobservable) empirical distribution function of the (unobservable) random variables X_i, and where $E_{\hat{F}_n}$ denotes the conditional expectation with respect to the probability measure, induced by $\hat{F}_n$. The crucial point here is that, knowing the points of jump, we have reduced the original problem to a problem where we can look for the solution in the *interior* of a (reduced) parameter space, and then apply the usual techniques of setting partial derivatives equal to zero.

There are several serious difficulties here:

1. We do not know the points τ_i beforehand (we do not have access to an oracle).

2. The self-consistency equation (2.6) does not characterize the maximum likelihood estimator (it is a necessary, but not sufficient condition), and gives no help in finding the points τ_i.

The difficulties arise from the fact that the self-consistency equation is based on differentiation in a reduced parameter space, and that finding this reduced parameter space (or equivalently, the points of jump τ_i of the NPMLE) is the main part of the problem, for which the self-consistency equation provides no help.

The self-consistency equation for the NPMLE, with interval censored, case 2, data, is

$$\hat{F}_n(t) = E_{\hat{F}_n}\{F_n(t) \mid (U_i, V_i, \Delta_{1i}, \Delta_{2i}), \, i = 1, \ldots, n\}, \; t \geq 0, \tag{2.7}$$

where the quadruples $(U_i, V_i, \Delta_{1i}, \Delta_{2i})$, $i = 1, \ldots, n$ consisting of the pairs of observation times (U_i, V_i) and the indicators $(\Delta_{1i}, \Delta_{2i})$ are the available information. The equation can be derived in a similar way as equation (2.6), or one could follow the method on p. 44 of Groeneboom and Wellner (1992).

2.2.1 The EM algorithm for finding the NPMLE

Associated with the self-consistency equation is the EM algorithm. The EM algorithm in this case starts with a distribution function concentrated on the observation points and an extra point to the right of the largest observation point (to allow for a deficient distribution function). If $F^{(m)}$ is the distribution function obtained after m steps of the algorithm, one gets $F^{(m+1)}$ from

$$F^{(m+1)}(t) = E_{F^{(m)}}\{F_n(t) \mid (U_i, \Delta_i),\, i = 1, \ldots, n\},\, t \geq 0. \tag{2.8}$$

A fixed point for the iterations would satisfy the self-consistency equation.

Since the self-consistency equation is only a necessary condition for the non-parametric maximum likelihood estimator, and not a sufficient condition, there are many solutions, satisfying the self-consistency equation, but not maximizing the likelihood. For example, if we start the EM iterations with a distribution, having mass zero at certain observation points, then the fixed point of the iterations will also have zero mass at these points. If these points are points of jump of the NPMLE, then this solution of the self-consistency equation will not maximize the likelihood. For this reason it is necessary to start the EM iterations with a distribution function having positive mass at each observation point.

It turns out that for the present problem there exists a one-step algorithm, giving both the points of jump τ_i and the value of the NPMLE at these points. It is based on isotonic regression theory, and already derived in AYER *et al.* (1955) and VAN EEDEN (1956). We will discuss the algorithm in 2.3.

Generally, however, the convex envelope algorithms, discussed in 2.3, will also need to be applied iteratively. As an example, for case 2 interval censoring, there exists an iterative convex envelope algorithm, introduced in GROENEBOOM (1991) and also discussed in GROENEBOOM AND WELLNER (1992), pages 49 and 69, to be treated in more detail below. This algorithm, which is considerably faster than the EM-algorithm, is based on conditions which are both necessary and sufficient for the NPMLE. In contrast with the EM algorithm and the self-consistency equation, this algorithm is closely connected with the developing theory for all the inverse poblems, mentioned in chapter 1 (the attempts to derive distribution theory for these problems was actually the main motivation for developing the algorithms). For this reason we turn to these methods now. Since I will often have to refer to GROENEBOOM AND WELLNER (1992) in the sequel, I will denote this reference by the abbreviation GW.

2.3 Convex envelope algorithms

2.3.1 Fenchel duality

Let ϕ be a smooth convex function defined on $\mathbb{R}^n$. At the basis of the results in this section lies the lemma below, which gives necessary and sufficient conditions for a vector $\hat{x}$ to be the minimizer of ϕ over a convex cone $\mathcal{K}$ in $\mathbb{R}^n$, where a cone in $\mathbb{R}^n$ is a subset $\mathcal{K}$ of $\mathbb{R}^n$, satisfying

$$x \in \mathcal{K} \Longrightarrow c \cdot x \in \mathcal{K}, \text{ for all } c \geq 0.$$

We give an elementary proof of the lemma which is based on the proof of Lemma 2.1 in JONGBLOED (1995A). It is a special case of Fenchel's duality theorem (see ROCKAFELLAR (1970), Theorem 31.4) and is also used at several places in GW, see, e.g., the proof of Proposition 1.1 on p. 39.

We write $\nabla\phi$ for the vector of partial derivatives of ϕ,

$$\nabla\phi(x) = \left(\frac{\partial}{\partial x_1}\phi(x), \cdots, \frac{\partial}{\partial x_n}\phi(x)\right),$$

and $\langle\cdot,\cdot\rangle$ for the usual inner product in $\mathbb{R}^n$.

Lemma 2.1 *Let $\phi : \mathbb{R}^n \to \mathbb{R} \cup \{\infty\}$ be, in the usual topology of the extended real line $\mathbb{R} \cup \{\infty\}$, a continuous convex function. Let $\mathcal{K} \subset \mathbb{R}^n$ be a convex cone and let $\mathcal{K}_0 = \mathcal{K} \cap \phi^{-1}(\mathbb{R})$. Moreover, suppose that $\mathcal{K}_0$ is non-empty, and that ϕ is differentiable on $\mathcal{K}_0$. Then $\hat{x} \in \mathcal{K}_0$ satisfies*

$$\phi(\hat{x}) = \min_{x \in \mathcal{K}} \phi(x), \tag{2.9}$$

if and only if $\hat{x}$ satisfies

$$\forall x \in \mathcal{K} : \langle x, \nabla\phi(\hat{x})\rangle \geq 0, \tag{2.10}$$

and

$$\langle \hat{x}, \nabla\phi(\hat{x})\rangle = 0. \tag{2.11}$$

Proof: We first prove the if-part. Let $x \in \mathcal{K}$ be arbitrary and let $\hat{x} \in \mathcal{K}_0$ satisfy (2.10) and (2.11). Then we get, using the convexity of ϕ,

$$\phi(x) - \phi(\hat{x}) \geq \langle x - \hat{x}, \nabla\phi(\hat{x})\rangle \geq 0,$$

implying $\phi(\hat{x}) = \min_{x\in\mathcal{K}} \phi(x)$. Note that the inequality is trivially satisfied if $x \in \mathcal{K} \setminus \mathcal{K}_0$.

Conversely, let $\hat{x}$ satisfy (2.9), and first suppose that (2.10) is not satisfied. Then there exists an $x \in \mathcal{K}$ such that $\langle x, \nabla\phi(\hat{x})\rangle < 0$. Since, for each $\epsilon \geq 0$,

$$\hat{x} + \epsilon x = (1+\epsilon)\left\{\frac{1}{1+\epsilon}\hat{x} + \left(1 - \frac{1}{1+\epsilon}\right)x\right\} \in \mathcal{K},$$

we have, for $\epsilon \downarrow 0$, using the continuity of ϕ and the assumption that ϕ is differentiable on $\mathcal{K}_0$,

$$\phi(\hat{x} + \epsilon x) - \phi(\hat{x}) = \epsilon\,\langle x, \nabla\phi(\hat{x})\rangle + o(\epsilon).$$

This shows that for ϵ sufficiently small, $\phi(\hat{x} + \epsilon x) < \phi(\hat{x})$, contradicting the assumption that $\hat{x}$ minimizes ϕ over $\mathcal{K}$.

Now suppose (2.11) is not satisfied. Then $\hat{x} \neq 0$ and, for $|\epsilon| \leq 1$, $(1+\epsilon)\hat{x} \in \mathcal{K}$. Taking the sign of ϵ opposite to that of $\langle \hat{x}, \nabla\phi(\hat{x})\rangle$, we get for $\epsilon \to 0$

$$\phi((1+\epsilon)\hat{x}) - \phi(\hat{x}) = \epsilon \langle \hat{x}, \nabla\phi(\hat{x})\rangle + o(\epsilon),$$

and hence the left-hand side will be negative for $|\epsilon|$ sufficiently small, contradicting again the assumption that $\hat{x}$ minimizes ϕ over $\mathcal{K}$. □

We say that a cone $\mathcal{K}$ is *finitely generated* if there are finitely many vectors $z^{(1)}, \ldots, z^{(k)} \in \mathcal{K}$ such that

$$x \in \mathcal{K} \iff \exists \alpha_1, \alpha_2, \ldots, \alpha_k \geq 0 \text{ such that } x = \sum_{i=1}^{k} \alpha_i z^{(i)}.$$

For finitely generated convex cones we have the following corollary of Lemma 2.1.

Corollary 2.1 *Let ϕ satisfy the conditions of lemma 2.1 and let the convex cone $\mathcal{K}$ be generated by the vectors $z^{(1)}, z^{(2)}, \ldots, z^{(k)}$. Then $\hat{x} \in \mathcal{K}_0 = \mathcal{K} \cap \phi^{-1}(\mathbb{R})$ satisfies*

$$\phi(\hat{x}) = \min_{x \in \mathcal{K}} \phi(x),$$

if and only if

$$\langle z^{(i)}, \nabla\phi(\hat{x})\rangle \geq 0, \text{ for } 1 \leq i \leq k, \tag{2.12}$$

$$\langle z^{(i)}, \nabla\phi(\hat{x})\rangle = 0, \text{ if } \hat{\alpha}_i > 0, \tag{2.13}$$

where the nonnegative numbers $\hat{\alpha}_1, \hat{\alpha}_2, \ldots, \hat{\alpha}_k$ satisfy

$$\hat{x} = \sum_{i=1}^{k} \hat{\alpha}_i z^{(i)}.$$

Proof: If $x \in \mathcal{K}$, then

$$x = \sum_{i=1}^{k} \alpha_i z^{(i)},$$

where the α_i are nonnegative. Hence we can write

$$\langle x, \nabla\phi(\hat{x})\rangle = \sum_{i=1}^{k} \alpha_i \langle z^{(i)}, \nabla\phi(\hat{x})\rangle. \tag{2.14}$$

If (2.12) and (2.13) hold, then (2.10) follows, since all terms in the sum on the right-hand side of (2.14) are nonnegative. If $x = \hat{x}$, (2.11) follows since in that case all terms on the right-hand side of (2.14) are zero.

Suppose (2.10) and (2.11) hold. Then (2.12) follows trivially. Taking $x = \hat{x}$ in (2.14) and observing that all terms in the sum on the right-hand side of (2.14) are nonnegative, it follows that (2.11) can only hold if (2.13) holds. □

2.3.2 A one-step convex envelope algorithm for finding the NPMLE with interval censoring, case 1

We now show how Corollary 2.1 leads to a one-step algorithm for computing the NPMLE for the distribution function F_0 with interval censored data, case 1. It follows from Theorem 1.5.1 in ROBERTSON *et. al.* (1988), applied to the convex function Φ, defined by

$$\Phi(x) = x\log x + (1-x)\log(1-x),\ x \in (0,1),$$

extended to $[0,1]$ by defining $\Phi(0) = \Phi(1) = 0$, that *maximizing* the function

$$x \mapsto \sum_{i=1}^{n}\{\delta_{(i)}\log x_i + (1-\delta_{(i)})\log(1-x_i)\},\ x = (x_1,\ldots,x_n)' \in [0,1]^n,$$

over all vectors $x \in [0,1]^n$ with ordered components $x_1 \le \ldots \le x_n$, is equivalent to *minimizing* the convex function

$$\phi : x \mapsto \sum_{i=1}^{n}\{x_i - \delta_{(i)}\}^2,\ x = (x_1,\ldots,x_n)' \in [0,1]^n, \tag{2.15}$$

over all such vectors. Here $\delta_{(i)}$ is a realization of an indicator Δ_j, corresponding to the ith order statistic $u_{(i)}$ of the (realized) observation times $u_1,\ldots,u_n$, and is equal to zero or one. We can extend ϕ to $\mathbb{R}^n$ by defining

$$\phi(x) = \sum_{i=1}^{n}\{x_i - \delta_{(i)}\}^2,\ x = (x_1,\ldots,x_n)' \in \mathbb{R}^n,$$

and for this extended function the conditions of Lemma 2.1 are satisfied.

Let $\mathcal{K}$ be the convex cone

$$\mathcal{K} = \{x \in \mathbb{R}^n\ :\ x = (x_1,\ldots,x_n)',\ x_1 \le \cdots \le x_n\}. \tag{2.16}$$

Then $\mathcal{K}$ is finitely generated by the vectors $z^{(i)} = \sum_{j=i}^{n} e_j$, $1 \le i \le n$, where the e_j are the unit vectors in $\mathbb{R}_n$, and the vector $z^{(0)} = -z^{(1)}$. This means that any $x \in \mathcal{K}$ can be written in the form

$$x = \sum_{i=1}^{n}\alpha_i z^{(i)},\ \alpha_1 \in \mathbb{R},\ \alpha_i \ge 0,\ i = 2,\ldots,n.$$

By Corollary 2.1, $\hat{x} = \sum_{i=1}^{n}\hat{\alpha}_i z^{(i)}$ minimizes $\phi(x)$ over $\mathcal{K}$, if and only if

$$\langle z^{(i)}, \nabla\phi(\hat{x})\rangle \begin{cases} \ge 0 & \text{for } 1 \le i \le n, \\ = 0 & \text{if } \hat{\alpha}_i > 0 \text{ or } i = 1. \end{cases} \tag{2.17}$$

The condition $\langle z^{(1)}, \nabla\phi(\hat{x})\rangle = 0$ arises from the fact that the inner product of $\nabla\phi(\hat{x})$ with both $z^{(1)}$ and $-z^{(1)}$ has to be non-negative.

Since $z^{(i)} = \sum_{j=i}^{n} e_j$, $i \geq 1$, this gives, by (2.15), that $\hat{x} = \sum_{i=1}^{n} \hat{\alpha}_i z^{(i)}$ has to satisfy

$$\sum_{j=i}^{n} \hat{x}_j \geq \sum_{j=i}^{n} \delta_{(j)},\ i = 1, \ldots, n, \text{ and } \sum_{j=i}^{n} \hat{x}_j = \sum_{j=i}^{n} \delta_{(j)}, \text{ if } \hat{\alpha}_i > 0 \text{ or } i = 1. \tag{2.18}$$

Let $P_0 = (0,0)$ and $P_i = (i, \sum_{j=i}^{n} \delta_{(j)})$, $i = 1, \ldots, n$. Furthermore, let $C : [0, n] \to I\!R$ be the biggest convex function on $[0, n]$, lying below (or touching) the points P_i. The set of points P_i is usually denoted as the *cumulative sum diagram* (or just *cusum diagram*) and the function C as the *(greatest) convex minorant* of this cusum diagram. Then, defining $\hat{x}_i$ as the left derivative of the convex minorant C at i, it is easily verified that the $\hat{x}_i$'s satisfy (2.18).

In fact, the (greatest) convex minorant has to touch the cusum diagram at P_0 and P_n, which means that we will have

$$\sum_{j=1}^{n} \hat{x}_j = \sum_{j=1}^{n} \delta_{(j)}. \tag{2.19}$$

Furthermore, since the convex minorant lies below the points P_i, we must have

$$\sum_{j=1}^{i} \hat{x}_j \leq \sum_{j=1}^{i} \delta_{(j)},\ i = 1, \ldots, n. \tag{2.20}$$

By (2.19) and (2.20) we now get

$$\sum_{j=i}^{n} \hat{x}_j \geq \sum_{j=i}^{n} \delta_{(j)},\ i = 1, \ldots, n,$$

Defining $\hat{x} = (\hat{x}_1, \ldots, \hat{x}_m)'$ and defining the $\hat{\alpha}_i$'s by $\hat{x} = \sum_{i=1}^{n} \hat{\alpha}_i z^{(i)}$, where $\hat{x} = (\hat{x}_1, \ldots, \hat{x}_n)'$, it is seen that $\hat{\alpha}_i > 0$, $i > 1$, means that the slope of C changes at $i - 1$, and this means that C touches the cusum diagram at P_{i-1}. Since $\hat{x}_i$ is the left-continuous slope of C, we get from this

$$\sum_{j=i}^{n} \hat{x}_j = \sum_{j=i}^{n} \delta_{(j)}.$$

It now follows that $\hat{x}$ satisfies (2.18), and since it is also easily seen that $0 \leq \hat{x}_i \leq 1$, for $1 \leq i \leq n$, we get that $\hat{x}$ actually minimizes $\phi(x)$ over all vectors $x \in [0, 1]^n$ with ordered components.

Hence we have the following one-step algorithm for computing the NPMLE: construct the cusum diagram, consisting of the points P_i, and construct its convex minorant. Then the value $\hat{F}_n(u_{(i)})$ of the NPMLE $\hat{F}_n$ at the ith ordered observation time $u_{(i)}$ is given by the left-continuous slope of the convex minorant at i.

2.3.3 The iterative convex minorant algorithm. Necessary and sufficient conditions for the NPMLE

In general, however, we cannot reduce the problem of maximizing the log likelihood to the problem of minimizing a sum of squares as in (2.15). As an example, we consider the problem of finding the NPMLE with interval censoring, case 2. For this case an iterative procedure, generalizing the approach above, was introduced in GROENEBOOM (1991). It is now generally denoted as the *iterative convex minorant* (ICM) algorithm.

Let $(u_i, v_i, \delta_{1i}, \delta_{2i})$, $i = 1, \ldots, n$ be a (realized) sample of observation times and indicators and let $\mathcal{F}_n$ be the set of (possibly defective) distribution functions F on $[0, \infty)$, satisfying

$$\begin{cases} F(u_i) > 0 & \text{if } \delta_{1i} = 1, \\ F(v_i) - F(u_i) > 0 & \text{if } \delta_{2i} = 1, \\ 1 - F(v_i) > 0 & \text{if } \delta_{3i} = 1. \end{cases} \tag{2.21}$$

Note that if F maximizes the log likelihood, scaled by dividing by the sample size:

$$\tilde{\phi}(F) \stackrel{\text{def}}{=} n^{-1} \sum_{i=1}^{n} \left\{ \delta_{1i} \log F(u_i) + \delta_{2i} \log\big(F(v_i) - F(u_i)\big) + \delta_{3i} \log\big(1 - F(v_i)\big) \right\}, \tag{2.22}$$

then F has to satisfy (2.21), since otherwise

$$\tilde{\phi}(F) = -\infty.$$

For distribution functions $F \in \mathcal{F}_n$, the following function $t \mapsto W_F(t)$, $t \in [0, \infty)$, is properly defined:

$$\begin{aligned} W_F(t) &= \int_{u \in [0,t]} \left\{ \frac{\delta_1}{F(u)} - \frac{\delta_2}{F(v) - F(u)} \right\} dQ_n(u, v, \delta_1, \delta_2) \\ &\quad + \int_{v \in [0,t]} \left\{ \frac{\delta_2}{F(v) - F(u)} - \frac{\delta_3}{1 - F(v)} \right\} dQ_n(u, v, \delta_1, \delta_2), \end{aligned} \tag{2.23}$$

where Q_n is the empirical probability measure of the points $(u_i, v_i, \delta_{1i}, \delta_{2i})$, $i = 1, \ldots, n$, and $\delta_3 = 1 - \delta_1 - \delta_2$. As will become clear below, the values of the function $t \mapsto W_F(t)$ contain (partial) sums of partial derivatives of the log likelihood function, if we consider the log likelihood function as a function of a finite number of parameters, representing the values of the distribution function at observation points.

We shall give necessary and sufficient conditions for the NPMLE in terms of the functions $t \to W_F(t)$. We first reduce the set of observation times to a subset of "relevant" observation times. If $\delta_{1i} = 1$, we have the information that the corresponding unobservable x_i is smaller than u_i, and the observation time v_i gives no extra information on its location, since $v_i \geq u_i$. So the observation time v_i is irrelevant in that case and we can remove it from our set of observation times. Likewise, if $\delta_{3i} = 1$, we have the information that the corresponding unobservable x_i is larger than v_i, and u_i can be removed from the set of observation times. As in GW, p. 45, we will denote the "thinned" set of observation times by J_n. Note that only the observation points in J_n give a contribution in the right-hand side of (2.23). We have the following definition.

Definition 2.1 *Let $J_n^{(1)}$ be the set of observation times u_i such that either $\delta_{1i} = 1$ or $\delta_{2i} = 1$, and let $J_n^{(2)}$ be the set of observation times v_i such that either $\delta_{2i} = 1$ or $\delta_{3i} = 1$. Then we define as the set of "relevant observation times" $J_n = J_n^{(1)} \cup J_n^{(2)}$. The jth order statistic of J_n is denoted by $t_{(j)}$.*

In the maximization problem we may now assume that $t_{(1)}$ corresponds to an observation time for which we have the information that the unobservable x_i is to the left of the observation time. The reason for this is that if the unobservable x_i is to the right of this smallest observation time $t_{(1)}$, and if we would put $F(t_{(1)}) > 0$, we could make the likelihood function bigger by putting $F(t_{(1)}) = 0$. We can therefore work with a reduced set of observation times $t_{(i)}$ such that $t_{(1)}$ corresponds to the smallest observation for which we have the information that x_i lies to the left of it, and put the NPMLE $\hat{F}_n$ equal to zero at each $t < t_{(1)}$.

For similar reasons we may assume that the biggest order statistic in J_n, say $t_{(m)}$, corresponds to an observation time for which we have the information that the unobservable x_i is to the right of the observation time. The following proposition (Proposition 1.3, p. 46 of GW, slightly differently formulated) characterizes the NPMLE in this situation.

Proposition 2.1 *Let $t_{(1)}$ correspond to an observation time for which the corresponding unobservable x_i lies to the left, and let the largest order statistic $t_{(m)}$ in J_n correspond to an observation time for which the corresponding unobservable x_i lies to the right. Then $\hat{F}_n$ maximizes (2.22) over all $F \in \mathcal{F}$ if and only if*

$$\int_{[t,\infty)} dW_{\hat{F}_n}(t') \le 0, \qquad \forall t \ge 0, \tag{2.24}$$

$$\int_{[t,\infty)} dW_{\hat{F}_n}(t') = 0, \text{ at each point of jump } t \text{ of } \hat{F}_n, \tag{2.25}$$

and

$$\int_{[0,\infty)} dW_{\hat{F}_n}(t) = 0, \tag{2.26}$$

where $W_{\hat{F}_n}$ is defined by (2.23). Moreover, $\hat{F}_n$ is uniquely determined by (2.24) to (2.26).

Proof: Let the set $S \subset (0,1)^m$ be defined by

$$S = \{x \in (0,1)^m : x = (F(t_{(1)}), \ldots, F(t_{(m)}), \text{ for some } F \in \mathcal{F}\}, \tag{2.27}$$

and let the convex function $\phi : S \to (-\infty, \infty]$ be defined by

$$\phi(x) = -\tilde{\phi}(F), \text{ if } x = (F(t_{(1)}), \ldots, F(t_{(m)})), \text{ for } F \in \mathcal{F}. \tag{2.28}$$

We extend ϕ to $\mathbb{R}^m$, by defining

$$\phi(x) = \infty, \text{ if } x \in \mathbb{R}^m \setminus S.$$

Since $\phi(x^{(n)})$ tends to ∞ if $(x^{(n)})$ is a sequence of points in S, tending to a boundary point of S, and since ϕ is differentiable on the set $\{x \in \mathbb{R}^m : \phi(x) < \infty\}$, ϕ satisfies the conditions of Lemma 2.1.

Let $\mathcal{K}$ be the cone (2.16), with n replaced by m. Then, as above, the cone $\mathcal{K}$ is finitely generated, with generators

$$z^{(0)} = -(1, \ldots, 1)', \; z^{(1)} = (1, \ldots, 1)', \ldots, z^{(m)} = (0, \ldots, 0, 1)'.$$

Hence Corollary 2.1 can be applied, yielding that the vector y minimizes $\phi(x)$ over $\mathcal{K}$, if and only if $y \in S$ and (2.12) and (2.13) are satisfied. This means that $y = \sum_{i=1}^m \alpha_i z^{(i)} \in S$ minimizes $\phi(x)$ over $\mathcal{K}$, if and only if

$$\sum_{j=i}^m \frac{\partial}{\partial y_j} \phi(y) \geq 0, \; i = 1, \ldots, m, \tag{2.29}$$

and

$$\sum_{j=i}^m \frac{\partial}{\partial y_j} \phi(y) = 0, \text{ if } \alpha_i > 0 \text{ or } i = 1. \tag{2.30}$$

But, by definition (2.28) and the identification $\hat{F}_n(t_{(i)}) = y_i$, (2.24) is equivalent to (2.29), and (2.25) and (2.26) are equivalent to (2.30). The latter equivalence is seen by noting that $\alpha_i > 0$ if and only if $t_{(i)}$ is a point of jump of $\hat{F}_n$ and that

$$\int_{[0,\infty)} dW_{\hat{F}_n}(t) = \sum_{j=1}^m \frac{\partial}{\partial y_j} \phi(y).$$

For the proof of the uniqueness we refer to the (end of the) proof of Proposition 1.3 of GW, p. 47. □

Remark. In Proposition 1.3 on p. 46 of GW, the condition

$$\int \hat{F}_n(t) \, dW_{\hat{F}_n}(t) = 0 \tag{2.31}$$

replaces (2.25) and (2.26). The equivalence of these conditions follows in the same way as the equivalence of condition (2.11) in Lemma 2.1 with condition (2.13) in Corollary 2.1 in the case of finitely generated cones. Note that (2.26) is in fact superfluous, since $t_{(1)}$ is a point of jump of $\hat{F}_n$. We nevertheless stated

this condition separately in order to stress the relation with condition (2.25) in Corollary 2.1.
It follows from (2.25) that

$$\int g(t)\, dW_{\hat{F}_n}(t) = 0, \tag{2.32}$$

for any function $g : [0, \infty) \to \mathbb{R}$ that is absolutely continuous w.r.t. $\hat{F}_n$. One can interpret this by saying that "the derivatives of the log likelihood in directions absolutely continuous w.r.t. $\hat{F}_n$ are zero". Furthermore, using (2.26), it is seen that (2.24) is equivalent to

$$\int_{[0,t]} dW_{\hat{F}_n}(t') \geq 0, \qquad \forall t \geq 0, \tag{2.33}$$

For interval censoring, case 1, we could reduce the problem of maximizing the likelihood to the problem of minimizing a simple sum of squares. For interval censoring, case 2, and other examples of inverse problems we will consider, this is not possible. However, the necessary and sufficient conditions (2.24) (or (2.33)), (2.25) and (2.26) suggest that we can try to construct an iterative procedure for which these conditions are satisfied in the limit. This leads to the *iterative convex minorant algorithm.*

First of all, under the conditions of Proposition 2.1, and using the definitions (2.27) and (2.28), where ϕ is extended to $\mathbb{R}^m$ in the same way as in the proof of Proposition 2.1, we reduce the problem of maximizing the likelihood to the problem of minimizing the convex function $\phi : \mathbb{R}^m \to (-\infty, \infty]$. Secondly, we use at each step of the algorithm a quadratic approximation to ϕ. More precisely, we start at an interior point of the set S, where ϕ is finite and differentiable, and use at step k the approximation

$$\phi_k(x) = \phi(x^{(k)}) + \langle x - x^{(k)}, \nabla\phi(x^{(k)})\rangle + \tfrac{1}{2}(x - x^{(k)})' D_k (x - x^{(k)}), \tag{2.34}$$

where D_k is a suitably chosen positive definite diagonal matrix which may change at each iteration, and where we make a step that is sufficiently small to ensure that the next iterate $x^{(k+1)}$ is an interior point of the set S.

Note that the minimization problem at the kth step boils down to the minimization of the quadratic function

$$x \mapsto \left(x - x^{(k)} + D_k^{-1}\nabla\phi(x^{(k)})\right)' D_k \left(x - x^{(k)} + D_k^{-1}\nabla\phi(x^{(k)})\right). \tag{2.35}$$

This minimization problem is similar to the problem of minimizing (2.15) for interval censoring, case 1, and can be solved by constructing a cusum diagram and its convex minorant. The points of the cusum diagram are in this case given by $P_0 = (0, 0)$ and

$$P_i = \left(\sum_{j=1}^{i} d_j^{(k)}, \sum_{j=1}^{i}\left\{d_j^{(k)} x_j^{(k)} - \frac{\partial}{\partial x_j^{(k)}}\phi(x^{(k)})\right\}\right), \; i = 1, \ldots, m.$$

where $d_j^{(k)}$ is the jth diagonal element of the diagonal matrix

$$D_k = \mathrm{diag}\big(d_1^{(k)}, \ldots, d_m^{(k)}\big).$$

The proof that this procedure gives the minimum is similar to the proof we gave for interval censoring, case 1. It runs as follows.

Suppose $x^{(k)} \in S$. Then the partial derivatives of ϕ with respect to $x_j^{(k)}$ are well-defined, and $x^{(k)} = \sum_{i=1}^m \alpha_i z^{(i)}$, where $\alpha_i \geq 0$, $z^{(i)} = \sum_{j=i}^m e_j$. Let $s_i = \sum_{j=1}^i d_j^{(k)}$ and let $C : [0, s_m] \to \mathbb{R}$ be the (greatest) convex minorant $C : [0, s_m] \to \mathbb{R}$ of the cusum diagram. Furthermore, let $\hat{x}_i$ be the left derivative of C at s_i, and let $\hat{x} = (\hat{x}_1, \ldots, \hat{x}_m)'$. Then

$$\sum_{j=1}^i d_j^{(k)} \hat{x}_j \leq \sum_{j=1}^i \Big\{ d_j^{(k)} x_j^{(k)} - \frac{\partial}{\partial x_j^{(k)}} \phi(x^{(k)}) \Big\},$$

since the convex minorant lies below or touches the points P_i. Moreover, defining $\hat{\alpha}_i$ by $\hat{x} = \sum_{i=1}^m \hat{\alpha}_i z^{(i)}$, we get, similarly as before, that $\hat{\alpha}_i > 0$ means that the slope of C changes at s_{i-1}, if $i > 1$, and that

$$\sum_{j=1}^m d_j^{(k)} \hat{x}_j = \sum_{j=1}^m \Big\{ d_j^{(k)} x_j^{(k)} - \frac{\partial}{\partial x_j^{(k)}} \phi(x^{(k)}) \Big\},$$

since the convex minorant touches the cusum diagram at 0 and s_m.

Hence the conditions of Corollary 2.1 are satisfied at $\hat{x}$ for the convex function

$$x \mapsto \sum_{j=1}^m \Big\{ d_j^{(k)} x_j - d_j^{(k)} x_j^{(k)} + \frac{\partial}{\partial x_j^{(k)}} \phi(x^{(k)}) \Big\}^2, \; x = (x_1, \ldots, x_m)' \in \mathbb{R}^m, \quad (2.36)$$

Since (2.36) is another way of writing (2.35), we find that $\hat{x}$ minimizes (2.35).

Example 2.1. As an extremely simple example illustrating the changing cusum diagrams, we consider the sample of size $n = 4$, consisting of the observations

$$(1,3,1,0),\ (2,6,0,1),\ (1,5,0,0),\ (2,9,0,0).$$

In this case there are 5 "relevant" observation times: $1, 2, 5, 6$ and 9, so $m = 5$. The function ϕ is given by

$$\phi(x_1, \ldots, x_5) = \tfrac{1}{4}\{\log x_1 + \log(x_4 - x_2) + \log(1 - x_3) + \log(1 - x_5)\},$$

where the x_i's satisfy $0 < x_1 \leq \ldots \leq x_5 < 1$. This function is maximized by the vector $\hat{x}$, given by

$$\hat{x}_1 = \hat{x}_2 = \hat{x}_3 = \tfrac{1}{4},\ \hat{x}_4 = \hat{x}_5 = \tfrac{5}{8},$$

see Example 1.2 on page 48 of Groeneboom and Wellner (1992). Hence the NPMLE $\hat{F}_n$ has to satisfy

$$\hat{F}_n(1) = \hat{F}_n(2) = \hat{F}_n(5) = \tfrac{1}{4},\ \hat{F}_n(6) = \hat{F}_n(9) = \tfrac{5}{8}.$$

The changing cusum diagrams are shown in Figure 2.1 for the first, second and 16th iteration. The starting values were

$$x^{(i)} = i/6,\ i = 1, \ldots, 5,$$

(allowing remaining mass beyond the last observation point) and the weights $d_j^{(k)}$ are the "second derivatives on the diagonal", see (2.37) below. It is clear that the biggest change is between the first and second iteration. The stopping criterion at the 16th iteration was based on the Fenchel duality relations of Proposition 2.1, with (2.24) replaced by the equivalent (2.33). Using the notation (2.33) and (2.26), the criteria were: stop at step k, if

$$n \int_{[0,t]} dW_{F_n^{(k)}}(t) \geq -10^{-7}, \text{ for all } t \geq 0,$$

(note that this boils down to checking a finite number of inequalities) and

$$n \left| \int_{[0,\infty)} dW_{F_n^{(k)}}(t) \right| \leq 10^{-7},$$

where $F_n^{(k)}$ is the estimate, obtained at the kth iteration step.

So, as an approximation to the NPMLE, one can take $x^{(16)}$, given by the left derivative of the convex minorant of the cusum diagram at the 16th iteration, evaluated at the locations of the cumulative weights

$$\sum_{j \leq i} d_j^{(16)},\ i = 1, \ldots, 5.$$

In this case we got

$$x_1^{(16)} = x_2^{(16)} = x_3^{(16)} = 0.2500000317,\ x_4^{(16)} = x_5^{(16)} = 0.6250000150.$$

Note that the value of $F_n^{(16)}$ at the ith ordered observation point is given by $x_i^{(16)}$.

A slightly less trivial example with a sample of size $n = 10$ is given as Example 1.3 on p. 49 of GW, where the NPMLE has irrational numbers as values (which is the usual situation with case 2 interval censoring).

2.3.4 Jongbloed's modification of the ICM algorithm

There is no guarantee that the vector $\hat{x}$, minimizing (2.36), will belong to S, where S is defined by (2.27). In the original version of the iterative convex minorant algorithm, as discussed in Groeneboom (1991) and GW, the diagonal matrix D_k was defined by

$$d_j^{(k)} = \frac{\partial^2}{\partial x_j^2} \phi(x) \bigg|_{x = x^{(k)}},\ j = 1, \ldots, m, \tag{2.37}$$

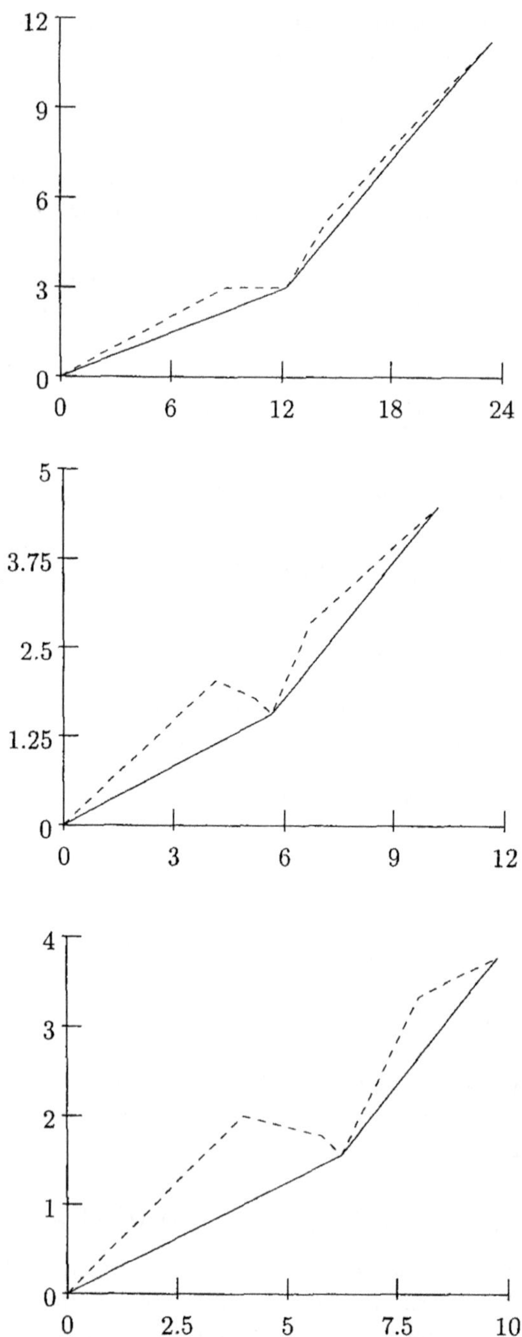

Figure 2.1: From top to bottom: Cusum diagram and convex minorant at resp. the first, second and 16th iteration in Example 2.1. The points of the cusum diagram are connected by the dashed line segments.

and some ad hoc measures were taken to keep the iterates inside S.

Aragón and Eberly (1992) discuss convergence properties of the iterative convex minorant algorithm. As noted in Wellner (1995), although they claim to prove global convergence of their modified algorithm, their proof only shows convergence under the condition that the points of jump do not change anymore (see the "secondary reductions" on p. 132, and note that the dimension r in their Theorem 2, p. 136, is fixed). This assumption is rather unrealistic, since the points of jump may change in very late stages of the algorithm. Recently, Jongbloed (1995b) showed that another modification of the iterative convex minorant algorithm will always (globally) converge from any starting point in S, without making assumptions on the points of jump.

We now discuss Jongbloed's modification. Let, for $x^{(k)} \in S$, the function $x^{(k)} \to A_k(x^{(k)})$ be defined by

$$A_k(x^{(k)}) = \operatorname*{argmin}_{x \in \mathcal{K}} \phi_k(x), \tag{2.38}$$

where ϕ_k is defined by (2.34) (the quadratic approximation to ϕ at the kth iteration step). Since the function (2.34) (or, equivalently: (2.35) or (2.36)) is strictly convex, the argument of the function where the minimum is attained, denoted by $\operatorname{argmin}_{x\in\mathcal{K}} \phi_k(x)$, is uniquely determined. It follows from the preceding that $\operatorname{argmin}_{x\in\mathcal{K}} \phi_k(x)$ can be found by constructing the convex minorant of the corresponding cusum diagram. We have the following lemma (Lemma 1 in Jongbloed (1995b)).

Lemma 2.2 *Let $\phi : \mathbb{R}^n \to \mathbb{R} \cup \{\infty\}$ be a convex function satisfying the conditions of Lemma 2.1, and let let $\mathcal{K}_0 = \mathcal{K} \cap \phi^{-1}(\mathbb{R})$. Assume that ϕ is continuously differentiable on $\mathcal{K}_0$, and that ϕ attains its minimum over the convex cone $\mathcal{K}$ at a unique point $\hat{x} \in \mathcal{K}_0$. Furthermore, let $x \mapsto D(x)$ be a continuous mapping from $\mathcal{K}_0$ into the class of positive definite diagonal matrices, endowed with the usual supremum norm. Finally, let the function $A : \mathcal{K}_0 \to \mathcal{K}$ be defined by*

$$A(x) = \operatorname*{argmin}_{y \in \mathcal{K}} \big(y - x + D(x)^{-1}\nabla\phi(x)\big)' D(x) \big(y - x + D(x)^{-1}\nabla\phi(x)\big). \tag{2.39}$$

Then we have at each $x \in \mathcal{K}_0$ such that $x \neq \hat{x}$:

$$\phi\big(x + \lambda(A(x) - x)\big) < \phi(x), \tag{2.40}$$

for all sufficiently small $\lambda > 0$.

Remark. The content of Lemma 2.2 can be summarized by saying that the mapping A generates a *direction of descent* for ϕ at each $x \in \mathcal{K}_0 \setminus \{\hat{x}\}$.

Proof of Lemma 2.2: Let $x \in \mathcal{K}_0 \setminus \{\hat{x}\}$, and let $\psi : [0,1] \to \mathbb{R} \cup \{\infty\}$ be defined by

$$\psi(\lambda) = \phi\big(x + \lambda(A(x) - x)\big).$$

The derivative of ψ at zero is given by

$$\psi'(0) = \big(A(x) - x\big)'\nabla\phi(x). \tag{2.41}$$

By (2.39) and Lemma 2.1, applied to the function

$$y \mapsto \big(y - x + D(x)^{-1}\nabla\phi(x)\big)'D(x)\big(y - x + D(x)^{-1}\nabla\phi(x)\big), \tag{2.42}$$

we have

$$\langle x, D(x)(A(x) - x) + \nabla\phi(x)\rangle \geq 0, \tag{2.43}$$

and

$$\langle A(x), D(x)(A(x) - x) + \nabla\phi(x)\rangle = 0. \tag{2.44}$$

Subtracting (2.43) from (2.44), we get, using (2.41),

$$\big(A(x) - x)'D(x)(A(x) - x) + \psi'(0) \leq 0. \tag{2.45}$$

If $A(x) = x$, application of Lemma 2.1 on (2.42) yields

$$\langle y, \nabla\phi(x)\rangle \geq 0,\ y \in \mathcal{K},$$

and

$$\langle x, \nabla\phi(x)\rangle = 0,$$

which, again by Lemma 2.1, are just the necessary and sufficient conditions for x to be a minimizer of ϕ, implying $x = \hat{x}$, in contradiction with our assumption. Hence $x \neq \hat{x}$ implies $A(x) \neq x$, and, since $D(x)$ is positive definite, we get from (2.45) that $\psi'(0) < 0$. The result now follows, since, by the continuity of the derivative of ϕ on $\mathcal{K}_0$, ψ' is continuous in a neighborhood of 0. □

We can now describe Jongbloed's modification of the iterative convex minorant algorithm. The idea is to select at step k a point $x^{(k+1)}$ from the segment

$$\text{seg}(x^{(k)}, A(x^{(k)})) = \left\{x^{(k)} + \lambda(A(x^{(k)}) - x^{(k)})\ :\ \lambda \in [0,1]\right\}$$

such that the value of ϕ decreases sufficiently when moving from $x^{(k)}$ to $x^{(k+1)}$. Define the (set-valued) algorithmic map C on $\mathcal{K}_0$

$$C(x) = \left\{\begin{array}{ll} \{A(x)\},\ \text{if } \phi(A(x)) < \phi(x) + (1-\epsilon)\nabla\phi(x)'(A(x) - x), & \\ \{y \in \text{seg}(x, A(x)) : (1-\epsilon)\nabla\phi(x)'(y - x) \leq \phi(y) - \phi(x) & \\ \qquad\qquad \leq \epsilon\nabla\phi(x)'(y - x)\},\ \text{otherwise}, & \end{array}\right. \tag{2.46}$$

where $\epsilon \in (0, 1/2)$ is fixed. See Figure 2.2 for the idea behind the definition of C.

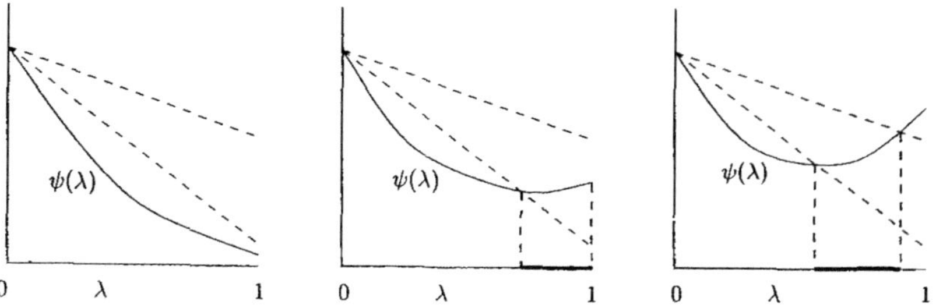

Figure 2.2: The three possible forms of the set returned by the algorithmic map C in the parametrization $\psi(\lambda) = \psi(x + \lambda(A(x) - x))$.

As an initial point we take any $x^{(0)} \in \mathcal{K}_0$. At step k one chooses $x^{(k+1)} = A(x^{(k)})$, whenever it belongs to $C(x^{(k)})$, and otherwise performs a binary search for an element of $C(x^{(k)})$ in the segment $\text{seg}(x^{(k)}, A(x^{(k)}))$. The algorithm is given below in pseudo code which is currently "en vogue" in papers on optimization techniques. It has a suspicious resemblance to the programming language Pascal which, according to C-programmers, causes brain damage. Nevertheless, I will resist the temptation to convert it to pseudo C-code, and give the form below which is also given in JONGBLOED (1995B). A computer program written in C, which one can use to actually analyze data in real-life, is available from the author. The stopping criterion is based on the necessary and sufficient conditions of Corollary 2.1.

Modified iterative convex minorant algorithm

Input:
$\eta > 0$: accuracy parameter;
$\epsilon \in (0, 1/2)$: line search parameter;
$x^{(0)} \in \mathcal{K}_0$: initial point satisfying $\phi(x^{(0)}) < \infty$;

begin
  $x := x^{(0)}$;
  while $|\sum_{i=1}^n x_i \frac{\partial}{\partial x_i}\phi(x)| > \eta$ **or** $|\sum_{i=1}^n \frac{\partial}{\partial x_i}\phi(x)| > \eta$
      or $\min_{1\le j\le n} \sum_{i=j}^n \frac{\partial}{\partial x_i}\phi(x) < -\eta$ **do**
  begin
    $\tilde{y} := \text{argmin}_{y\in\mathcal{K}}(y - x + W(x)^{-1}\nabla\phi(x))'W(x)(y - x + W(x)^{-1}\nabla\phi(x))$;
    if $\phi(\tilde{y}) < \phi(x) + \epsilon\nabla\phi(x)'(\tilde{y} - x)$ **then**
      $x := \tilde{y}$
    else
      begin
        $\lambda := 1$; $s := 1/2$; $z := \tilde{y}$;

```
      while φ(z) < φ(x) + (1 − ε)∇φ(x)'(z − x) (I) or
              φ(z) > φ(x) + ε∇φ(x)'(z − x) I) do
      begin
        if (I) then λ := λ + s;
        if (II) then λ := λ − s;
        z := x + λ(ỹ − x);
        s := s/2;
      end;
      x := z;
    end;
  end;
end.
```

A proof that the algorithm will converge from any starting point inside $\mathcal{K}_0$ is given in JONGBLOED (1995B). The key to the proof is Lemma 2.2, from which it can be deduced that the algorithmic map C is *closed* on $\mathcal{K}_0$, i.e., if (x_k) and (y_k) are sequences in $\mathcal{K}_0$ such that $x_k \to x \in \mathcal{K}_0$, $y_k \to y$ and $y_k \in C(x_k)$, then $y \in C(x)$. The algorithm uses a line search technique which is called "Armijo's rule" in the optimization literature.

As mentioned before, the algorithm can be applied for all inverse problems, mentioned in chapter 1.

Chapter 3

Smooth functionals

A very important aspect of the developing theory on inverse problems in statistics is that a general theory is emerging which will tell us whether certain functionals of the model can be estimated at the usual rate $n^{-1/2}$ and whether the nonparametric maximum likelihood estimator (NPMLE) will attain this rate and be asymptotically efficient, i.e., have the smallest asymptotic variance among all "reasonable" estimators of these functionals. An attractive feature of the NPMLE is that one does not have to perform any preliminary bandwidth choice or choice of some sophisticated "sieve"; one just maximizes the likelihood and in section 2.3 efficient algorithms, based on Fenchel duality theory, for doing that were discussed.

As promised in the introduction, I shall treat the present theory in depth for the interval censoring model and show that smooth functionals of the model can indeed be estimated by the NPMLE at rate $n^{-1/2}$ and that the NPMLE is asymptotically efficient. The results are quite recent and there still remains much to be done. I mainly follow the exposition in Geskus and Groeneboom (1996c).

3.1 Pathwise differentiable functionals; asymptotic efficiency

For more complete and more general information on the relation between pathwise differentiability of functionals and asymptotic efficiency, see part I of Groeneboom and Wellner (1992) or Bickel *et. al.* (1993). Proofs of the theorems in this section or further references can be found in these books as well. Here, I will just introduce the notions of Hellinger differentiable paths and pathwise differentiability of real-valued functionals and state an important theorem relating these notions with asymptotic efficiency. The theory is applied to the interval censoring model. The reference Groeneboom and Wellner (1992) is again denoted by GW in the sequel.

Let the unknown distribution P on $(\mathcal{Y}, \mathcal{B})$ be contained in some class of probability measures $\mathcal{P}$, which is dominated by a σ-finite measure μ. Let P have density p with respect to μ. Since we are interested in estimation of some real-valued function of P, we introduce the functional $\Theta : \mathcal{P} \to \mathbb{R}$.

Let, for some $\delta > 0$, the collection $\{P_t\}$ with $t \in (0, \delta)$ be a one dimensional parametric submodel which is smooth in the following sense:

$$\int \left[t^{-1}(\sqrt{p_t} - \sqrt{p}) - \tfrac{1}{2}a\sqrt{p}\right]^2 d\mu \to 0 \quad \text{as } t \downarrow 0, \text{ for some } a \in L_2(P)$$

Such a submodel is called *Hellinger differentiable.* This property can be seen as an L_2-version of the pointwise (in x) differentiability of $\log p_t(x)$ at $t = 0$ (with $p_0 = p$), with $a(\cdot)$ playing the role of the so-called *score-function* $\frac{d}{dt} \log p_t(\cdot)\big|_{t=0}$ in classical statistics. For we have,

$$\lim_{t\downarrow 0} \frac{\sqrt{p_t} - \sqrt{p_0}}{t} = \frac{1}{2\sqrt{p_0}} \frac{d}{dt} p_t \bigg|_{t=0} = \frac{1}{2} \left(\frac{d}{dt} \log p_t \bigg|_{t=0} \right) \sqrt{p_0}$$

Therefore, a is also called the *score function* or *score.* The following property is shared with the traditional score function $\frac{d}{dt} \log p_t(\cdot)\big|_{t=0}$ and is a well-established fact. I nevertheless present a proof for completeness.

Proposition 3.1 *Each score belonging to some Hellinger differentiable submodel is contained in* $L_2^0(P)$.

Proof: If we can prove that Hellinger differentiability implies $t^{-1}(p_t - p) \to ap$ in the $L_1(\mu)$-sense we are through. For then we have

$$\int a dP = \lim_{t\downarrow 0} t^{-1} \int (p_t - p) d\mu = 0.$$

Let $\|\cdot\|_1$ denote the $L_1(\mu)$-norm, and $\|\cdot\|_2$ denote the $L_2(\mu)$-norm. Then L_1-convergence is proved as follows:

$$\begin{aligned} & \|t^{-1}(p_t - p) - ap\|_1 \\ \leq\ & \|2\sqrt{p}\{t^{-1}(\sqrt{p_t} - \sqrt{p}) - \tfrac{1}{2}a\sqrt{p}\}\|_1 + \|t^{-1}(\sqrt{p_t} - \sqrt{p})^2\|_1 \\ \leq\ & 2\,\|\sqrt{p}\|_2\, \|t^{-1}(\sqrt{p_t} - \sqrt{p}) - \tfrac{1}{2}a\sqrt{p}\|_2 + t\, \left(\|t^{-1}(\sqrt{p_t} - \sqrt{p})\|_2\right)^2 \end{aligned}$$

The first term converges to zero by Hellinger differentiability, the second term is bounded by

$$2t\, \left(\|t^{-1}(\sqrt{p_t} - \sqrt{p}) - \tfrac{1}{2}a\sqrt{p}\|_2\right)^2 + 2t\, \left(\|\tfrac{1}{2}a\sqrt{p}\|_2\right)^2,$$

which also tends to zero as $t \downarrow 0$, again using Hellinger differentiability and $a \in L_2(P)$. □

In order to be able to apply the *convolution theorem,* which provides us with the best possible limit behavior, we need the collection $\mathcal{P}$ to be "rich enough" around P. In our situation the collection of scores a, obtained by considering all possible one-dimensional Hellinger-differentiable parametric submodels, is a linear space. This space is called the *tangent space* at P, denoted by $T(P)$. Notice that $T(P) \subset L_2^0(P)$.

One of the probabilists, attending the St-Flour meeting, asked: "And what will happen if the tangent space consists of one element (i.e., 0)?" The answer to this is very simple: *then there is no statistical problem.* In that case we are back in probability, where we just deal with one underlying probability measure.

Now $\Theta : \mathcal{P} \to \mathbb{R}$ is pathwise differentiable at P if for each Hellinger differentiable path $\{P_t\}$, with corresponding score a, we have

$$\lim_{t \downarrow 0} t^{-1}(\Theta(P_t) - \Theta(P)) = \Theta'_P(a),$$

with $\Theta'_P : T(P) \to \mathbb{R}$ continuous and linear.

Θ'_P can be written in an inner product form. Since $T(P)$ is a subspace of the Hilbert-space $L_2(P)$, the continuous linear functional Θ'_P can be extended to a continuous linear functional $\overline{\Theta}'_P$ on $L_2(P)$. By the Riesz representation theorem, to $\overline{\Theta}'_P$ belongs a unique $\theta_P \in L_2(P)$, called the *gradient*, satisfying

$$\overline{\Theta}'_P(h) = < \theta_P, h >_P \quad \text{for all } h \in L_2(P).$$

Instead of $L_2(P)$, any closed subspace H between $\overline{T(P)}$ and $L_2(P)$ can be chosen as space to which to extend $\overline{\Theta}'_P$ and on which to apply the Riesz representation theorem. Note that θ_P is uniquely determined once the extension of Θ'_P has been made. However, many continuous linear extensions of Θ'_P may be possible, so generally the gradient is not unique. One gradient is playing a special role, which is obtained by extending $T(P)$ to the Hilbert space $\overline{T(P)}$. Then, the extension of Θ'_P *is* unique, yielding the *canonical gradient* or *efficient influence function* $\tilde{\theta}_P \in \overline{T(P)}$. This canonical gradient is also obtained by taking the orthogonal projection of any gradient θ_P, obtained after extension of Θ'_P, into $\overline{T(P)}$. Hence $\tilde{\theta}_P$ is the gradient with minimal norm among all gradients and we have

$$\|\theta_P\|_P^2 = \|\tilde{\theta}_P\|_P^2 + \|\theta_P - \tilde{\theta}_P\|_P^2.$$

In the theorem below these notions are related to estimation of $\Theta(P)$. Let $T_n =$ $t_n(X_1, \ldots, X_n)$ be a real-valued measurable function of the sample. Any Hellinger differentiable path $\{P_t\}$ supplies us with a collection of sequences $\{P_{c/\sqrt{n}}\}$, where $c > 0$. Now we have the following theorem (see, e.g., Theorem 3.11.2, p. 414 of van der Vaart and Wellner(1996)):

Theorem 3.1 *Suppose that:*

(i) Θ is pathwise differentiable at $P \in \mathcal{P}$ along the Hellinger differentiable paths.

(ii) T_n is a regular estimator, meaning that $\sqrt{n}(T_n - \Theta(P_{c/\sqrt{n}}))$ converges, under $P_{c/\sqrt{n}}$, in distribution to a random variable Z, which does not depend on the direction, i.e. the score, of the path $\{P_{c/\sqrt{n}}\}$ to P.

(iii) The set of all directions $\{a\}$ *is a linear space.*

Then there exist random variables Z_0 *and* Δ_0 *such that*

A. Z *has the same distribution as* $Z_0 + \Delta_0$.

B. Z_0 *and* Δ_0 *are independent.*

C. $Z_0 \sim N(0, \|\tilde{\theta}_P\|_P^2)$.

This theorem is a version of the so-called convolution theorem. It says that the limiting distribution of any regular estimator of $\Theta(P)$ is more spread out than the distribution of Z_0. Hence the smallest asymptotic variance we can get for a regular estimator of $\Theta(P)$ is $\|\tilde{\theta}_P\|^2$. An *asymptotically efficient* estimator is a regular estimator for which the limiting distribution equals the distribution of Z_0. Condition (ii) is needed to exclude the occurrence of superefficiency (see e.g. LEHMANN (1983) for a superefficient estimator in a simple model).

3.2 Information bounds under transformations

The interval censoring model is an example of a model with information loss, in which the distribution P is induced by another distribution. In these models the functional to be estimated is also determined by the inducing distribution, but is implicitly defined via P. For the information lower bound theory for such implicitly defined functionals one can consult VAN DER VAART (1991), GW, part 1, and BICKEL *et. al.* (1993). This theory will be applied to case 2 of the interval censoring model. We start with the formulation of the model for case 2. The loss of information is expressed by the fact that, instead of a sample $(X_1, \ldots, X_n)$, we observe $(U_1, V_1, \Delta_1, \Gamma_1), \ldots, (U_n, V_n, \Delta_n, \Gamma_n)$ with $\Delta_i = 1_{\{X_i \le U_i\}}$ and $\Gamma_i = 1_{\{U_i < X_i \le V_i\}}$. We suppose:

(M1) X_i is a non-negative absolutely continuous random variable with distribution function F. Let $S > 0$. F is contained in the class

$$\mathcal{F}_S := \{F \,|\, \text{support}(F) \subset [0, S];\ F \ll \lambda,\ \lambda \text{ being Lebesgue measure}\}.$$

F is the distribution on which we want to obtain information; however, we do not observe X_i directly.

(M2) Instead, we observe the pairs (U_i, V_i), with distribution function H. H is contained in $\mathcal{H}$, the collection of all two-dimensional distributions on $\{(u, v) \,|\, 0 \le u < v\}$, absolutely continuous with respect to two-dimensional Lebesgue measure and such that each H is independent of each F. Let h denote the density of (U_i, V_i), with marginal densities and distribution functions h_1, H_1 and h_2, H_2 for U_i and V_i respectively.

(M3) If both H_1 and H_2 put zero mass on some set A, then F has zero mass on A as well, so $F \ll H_1 + H_2$. This means that F does not have mass on sets in which no observations can occur.

Condition (M3) is needed to ensure consistency. Moreover, without this assumption the functionals we are interested in are not well-defined. So discrete F should be excluded from $\mathcal{F}_S$.
The model formulation for case 1 is completely analogous, with one observation time instead of two observation times per unobservable X_i.
Note that what we *do* observe can be seen as a measurable transformation S of what we *would* observe if there would be no censoring:

$$S(x,u,v) = (u,v,\delta,\gamma),$$

with domain $\{(x,u,v)\,|\,0 \le x,\ 0 \le u < v\}$. This domain will be called the hidden space, and the image space will be called the observation space. In our model P is induced by F and H, and is from now on written as $Q_{F,H}$, having density

$$q_{F,H}(u,v,\delta,\gamma) = h(u,v)F(u)^{\delta}(F(v) - F(u))^{\gamma}(1 - F(v))^{1-\delta-\gamma}$$

with respect to $\lambda_2 \otimes \nu_2$, where ν_2 denotes the counting measure on the set $\{(0,1),(1,0),(0,0)\}$.

We are interested in estimation of some aspect $K(F)$ of F. However, $K(F)$ is only implicitly defined as $\Theta(Q_{F,H})$, with H acting as a nuisance parameter. In particular, we will be concerned with the problem whether the NPMLE $\hat{\Theta}_n$ of $\Theta(Q_{F,H})$ satisfies

$$\sqrt{n}(\hat{\Theta}_n - \Theta(Q_{F,H})) \xrightarrow{\mathcal{D}} N(0, \|\tilde{\theta}_{Q_{F,H}}\|^2).$$

3.3 Implicitly defined pathwise differentiable functionals

We first take a look at the Hellinger differentiable paths. All Hellinger differentiable submodels at $Q_{F,H}$ that can be formed, together with the corresponding score functions, are induced by the Hellinger differentiable paths of densities on the hidden space, according to the following theorem:

Theorem 3.2 *Let $\mathcal{P} \ll \mu$ be a class of probability measures on the hidden space $(\mathcal{Y},\mathcal{B})$. $P \in \mathcal{P}$ is induced by the random vector Y. Suppose that the path $\{P_t\}$ to P satisfies*

$$\int \left[t^{-1}(\sqrt{p_t} - \sqrt{p}) - \tfrac{1}{2}a\sqrt{p}\right]^2 d\mu \to 0 \quad as\ t \downarrow 0$$

for some $a \in L_2^0(P)$.
Let $S : (\mathcal{Y},\mathcal{B}) \to (\mathcal{Z},\mathcal{C})$ be a measurable mapping. Suppose that the induced measures $Q_t = P_t S^{-1}$ and $Q = PS^{-1}$ on $(\mathcal{Z},\mathcal{C})$ are absolutely continuous with respect to μS^{-1}, with densities q_t and q. Then the path $\{Q_t\}$ is also Hellinger differentiable, satisfying

$$\int \left[t^{-1}(\sqrt{q_t} - \sqrt{q}) - \tfrac{1}{2}\overline{a}\sqrt{q}\right]^2 d\mu S^{-1} \to 0 \quad as\ t \downarrow 0$$

with $\overline{a}(z) = E_P(a(Y)|S = z)$.

Proof: See BICKEL *et. al.* (1993).

Note that $\overline{a} \in L_2^0(Q)$. The relation between the scores a in the hidden tangent space $T(P)$ and the induced scores $\overline{a}$ is expressed by the mapping

$$A_P : a(\cdot) \mapsto E_P(a(Y)|S = \cdot).$$

This mapping is called the *score operator*. It is continuous and linear. Its range is the induced tangent space, which is contained in $L_2^0(Q)$.

Since the "hidden" random variables X_i are independent of the observation pairs (U_i, V_i), the one-dimensional submodels in the hidden space can be formed by first looking at the classes $\mathcal{F}_S$ and $\mathcal{H}$ separately. In our assumptions $\mathcal{F}_S$ and $\mathcal{H}$ are large enough to make the tangent spaces $T(F)$ and $T(H)$ as large as possible: $T(F) = L_2^0(F)$ and $T(H) = L_2^0(H)$: for any $a \in L_2^0(F)$ take the path $\{F_t\}$, with densities

$$f_t(x) = \frac{f(x)(1+ta(x))1_{\{x:1+ta(x)\geq 0\}}(x)}{\int f(x)(1+ta(x))1_{\{x:1+ta(x)\geq 0\}}(x)dx}$$

which is Hellinger differentiable at F with score a. In the same way one can find, for any $e \in L_2^0(H)$, a path $\{H_t\} \subset \mathcal{H}$ satisfying

$$t^{-1}(\sqrt{h_t} - \sqrt{h}) \to \tfrac{1}{2}e\sqrt{h} \quad \text{in } L_2(\lambda_2),$$

where λ_2 denotes Lebesgue measure on $\mathbb{R}^2$. Joining these paths gives, using independence and denoting Lebesgue measure on $\mathbb{R}^3$ by λ_3,

$$\int \left[t^{-1}(\sqrt{f_t\, h_t} - \sqrt{f\, h}) - \tfrac{1}{2}(a+e)\sqrt{f\, h}\right]^2 d\lambda_3 \to 0 \quad \text{as } t \downarrow 0,$$

with tangent space $L_2^0(F) \oplus L_2^0(H)$. Now theorem 3.2 yields the tangent space $T(Q_{F,H})$ of the induced Hellinger differentiable paths $\{Q_t\}$ at $Q_{F,H}$ with score operator $A : L_2^0(F) \oplus L_2^0(H) \to T(Q_{F,H})$ given by:

$$[A_{F,H}(a+e)](u,v,\delta,\gamma) = E_{F,H}\{a(X) + e(U,V) \,|\, (U,V,\Delta,\Gamma) = (u,v,\delta,\gamma)\}$$

Having specified the Hellinger differentiable paths in the observation space, one can study differentiability of the functional

$$\Theta(Q_{F,H}) = K(F).$$

along Hellinger differentiable paths. Note that $\Theta(Q_{F,H})$ is defined unambiguously by condition (M3).

If we were able to observe $X_1, \ldots, X_n$ directly, pathwise differentiability of $K(F)$ along the paths $\{F_t\}$ would have to be investigated. Then many functionals are differentiable. Once we have found a gradient $\kappa_F \in L_2(F)$, determining the canonical gradient is an easy task since the tangent space $T(F)$ is equal to $L_2^0(F)$: just subtract $\int \kappa_F dF$ to find the projection into $L_2^0(F)$.

With respect to the functional to be estimated we assume:
(F1)
K is differentiable along Hellinger differentiable paths of distributions from $\mathcal{F}$.

The canonical gradient for this functional is denoted by $\tilde{\kappa}_F$. What functionals satisfy this pathwise differentiability requirement? An important class of functionals are the functionals that are linear in F,

$$K(F) = \int c(x)\, dF(x).$$

All moment functionals $F \mapsto \int x^k\, dF(x)$ belong to this class. Estimation of the distribution function at a fixed point concerns a linear functional as well: for $K(F) = F(t_0)$ we have $c(x) = 1_{[0,t_0]}(x)$. In BICKEL *et. al.* (1993), proposition A.5.2, it is shown that linear functionals on $\mathcal{F}$ with

$$\sup_{F \in \mathcal{F}} E_F\, c(X)^2 < \infty$$

are pathwise differentiable at any $F \in \mathcal{F}$, with canonical gradient

$$\tilde{\kappa}_F(x) = c(x) - \int c(x)\, dF(x).$$

So if we were able to observe the X_i*'s* directly, we would obtain the information lower bound

$$\|c(X) - E_F(c(X))\|_F^2.$$

For nonlinear functionals, there is no general formula for the canonical gradient. An example of a nonlinear functional to which our theory can be applied is

$$K(F) = \int F^2(x)\, w(x)\, dx$$

If the function w is bounded, a slight extension of the proof in BICKEL *et. al.* (1993) shows that this functional has canonical gradient

$$\tilde{\kappa}_F(x) = 2 \int_{s=x}^{M} F(s)\, w(s)\, ds - \int_{x=0}^{M} \int_{s=x}^{M} 2\, F(s)\, w(s)\, ds\, dF(x).$$

In order to show that the NPMLE $K(\hat{F}_n)$ asymptotically attains the lower bound, we have to make the following extra assumption:

(F2) $$K(G) - K(F) = \int \tilde{\kappa}_F(x)\, d(G - F)(x) + \mathcal{O}(\|G - F\|_\lambda^2),$$

for all distribution functions G with support contained in $[0, M]$, with λ denoting Lebesgue measure on $\mathbb{R}$. For linear functionals (F2) holds without the $\mathcal{O}$-term. However, the functional

$$K(F) = \int F^2(x)\, w(x)\, dx$$

also satisfies (F2).

In our censoring model, differentiability of $\Theta(Q_{F,H})$ along the induced Hellinger differentiable paths in the observation space can be proved by looking at the structure of the adjoint $A^*_{F,H}$ of the map $A_{F,H}$ according to theorem 3.3 below, which was first proved in VAN DER VAART (1991) in a more general setting, allowing for Banach space valued functions as estimand. Then the proof is slightly more elaborate.

Recall that the adjoint of a continuous linear mapping $A : G \to H$, with G and H Hilbert-spaces, is the unique continuous linear mapping $A^* : H \to G$ satisfying

$$< Ag, h >_H = < g, A^*h >_G \quad \forall g \in G, h \in H.$$

Of course, any Hilbert space that contains $\overline{\mathcal{R}(A)}$ can be chosen as image space H, creating a different adjoint A^*. However, this does not complicate things: each adjoint A^* has the same behavior on $\overline{\mathcal{R}(A)}$ and its behavior on $\overline{\mathcal{R}(A)}$ determines A^* completely, since A^*h is equal to $A^*(\Pi h)$, with Π the orthogonal projection into $\overline{\mathcal{R}(A)}$.

The score operator from theorem 3.2 is playing the role of A. Its adjoint can be written as a conditional expectation as well. If $Z \sim PS^{-1}$, then:

$$[A_P^* b](y) = E_P(b(Z)|Y = y) \quad \text{a.e.-}[P]$$

Theorem 3.3 *Let $\mathcal{Q} = \mathcal{P}S^{-1}$ be a class of probability measures on the image space of the measurable transformation S. Suppose the functional $\Theta : \mathcal{Q} \to \mathbb{R}$ can be written as $\Theta(Q_P) = K(P)$ with K pathwise differentiable at P in the hidden space, having canonical gradient $\tilde{\kappa}_P$.*
Then Θ is differentiable at $Q_P \in \mathcal{Q}$ along the collection of induced paths in the observation space obtained via theorem 3.2 if and only if

$$\tilde{\kappa}_P \in \mathcal{R}(A_P^*) \tag{3.1}$$

If (3.1) holds, then the canonical gradients $\tilde{\theta}_{Q_P}$ of Θ and $\tilde{\kappa}_P$ of K are related by

$$\tilde{\kappa}_P = A_P^* \tilde{\theta}_{Q_P}$$

Proof: Write A and A^* instead of A_P and A_P^*.
We have

$$\lim_{t\downarrow 0} t^{-1}(\Theta(Q_{P_t}) - \Theta(Q_P)) = \lim_{t\downarrow 0} t^{-1}(K(P_t) - K(P)) = < \tilde{\kappa}_P, g >_P \tag{3.2}$$

Now suppose Θ is pathwise differentiable at Q_P. Then there exists a unique $\tilde{\theta} = \tilde{\theta}_{Q_P} \in \overline{\mathcal{R}(A)}$ satisfying, for any Hellinger differentiable path $\{P_t\}$ with score-function $g \in T(P)$,

$$\begin{aligned}\lim_{t\downarrow 0} t^{-1}(\Theta(Q_{P_t}) - \Theta(Q_P)) &= < Ag, \tilde{\theta} >_{Q_P} \\ &= < g, A^* \tilde{\theta} >_P\end{aligned}$$

Combining this with (3.2) shows that $\tilde{\kappa}_P = A^* \tilde{\theta}_{Q_P}$, hence $\tilde{\kappa}_P \in \mathcal{R}(A^*)$.

Conversely, suppose $\tilde{\kappa}_P = A^*b$ for some b in the domain of A^*. Then we have, for $\{P_t\}$ with score g,

$$\begin{aligned}\lim_{t\downarrow 0} t^{-1}(\Theta(Q_{P_t}) - \Theta(Q_P)) &= <\tilde{\kappa}_P, g>_P \\ &= <A^*b, g>_P \\ &= <b, Ag>_{Q_P}\end{aligned}$$

Hence Θ is pathwise differentiable with gradient b. □

This is applied to the interval censoring model. First, the score operator $A(a+e)$ is split into two parts:

$$A(a+e) = L_1 a + L_2 e.$$

Since the functional Θ does not depend on H and since $(U_i, V_i) \sim H$ and $X_i \sim F$ are independent, it can be shown that pathwise differentiability of Θ at $Q_{F,H}$ is equivalent to

$$\tilde{\kappa}_F \in \mathcal{R}(L_1^*)$$

and if this holds, then the canonical gradient is the unique element $\tilde{\theta}$ in $\overline{\mathcal{R}(L_1)}$ satisfying

$$L_1^* \tilde{\theta} = \tilde{\kappa}_F. \tag{3.3}$$

The operators L_1 and L_2 have the following form:

$$\begin{aligned}[L_1 a](u, v, \delta_1, \delta_2) &= \frac{\delta_1 \int_0^u a\,dF}{F(u)} + \frac{\delta_2 \int_u^v a\,dF}{F(v)-F(u)} + \frac{(1-\delta_1-\delta_2)\int_v^M a\,dF}{1-F(v)} \quad \text{a.e.} - [Q_{F,H}] \\ [L_2 e](u, v, \delta_1, \delta_2) &= e(u,v) \quad \text{a.e.} - [Q_{F,H}]\end{aligned} \tag{3.4}$$

The adjoint of L_1 can be written as $[L_1^* b](x) = E_P(b(U, V, \Delta, \Gamma) | X = x)$ and we get

$$\begin{aligned}[L_1^* b](x) = &\int_{u=x}^M \int_{v=u}^M b(u, v, 1, 0)\, h(u,v)\, dv\, du + \\ &\int_{u=0}^x \int_{v=x}^M b(u, v, 0, 1)\, h(u,v)\, dv\, du + \\ &\int_{u=0}^x \int_{v=u}^x b(u, v, 0, 0)\, h(u,v)\, dv\, du \quad \text{a.e.-}[F].\end{aligned} \tag{3.5}$$

Note that the structure of L_1^* does not depend on F.

Many functionals that are pathwise differentiable in the model without censoring, lose this property in the interval censoring model. Any functional K with a canonical gradient that is not a.e. equal to a continuous function cannot be obtained under L_1^*. So not all linear functionals remain pathwise differentiable. For example, $\kappa(F) = F(t_0)$, with canonical gradient $1_{[0,t_0]}(\cdot) - F(t_0)$, loses this property. This is in correspondence with $F(t_0)$ not being estimable at $\sqrt{n}$-rate. However, functionals of the form $K(F) = \int c(x)\,dF(x)$, with c sufficiently smooth, will be shown to remain differentiable under censoring. Hence for these functionals the above information lower bound theory holds.

In the interval censoring model, both case 1 and case 2, the function

$$\phi(x) := \int_x^M a(t)\,dF(t) \text{ with } a \in L_2^0(F).$$

appears explicitly in the score operator L_1. Therefore it will play an important role. It will be called the *integrated score function.* From its definition we know that ϕ satisfies $\phi(0) = \phi(M) = 0$ and that ϕ is continuous for $F \in \mathcal{F}_S$.

Before continuing with case 2, I will first have a short look at case 1.

3.4 Interval censoring, case 1 (current status data)

For case 1, an explicit formula for the information lower bound for smooth functionals can be obtained under conditions comparable to conditions (M1) to (M3) in section 3.2. Suppose the unobservable X_i's to have distribution function F, with support contained in $[0, M]$ and density f, and consider the same class $\mathcal{F}_S$ as in (M1). The observation times T_i have an absolutely continuous distribution function G with density g. X_i and T_i are assumed to be independent and F is dominated by G. Then the score operator L_1 has the form

$$[L_1 a](t,\delta) = E\{a(X)|T = t, \Delta = \delta\} = \frac{\delta \int_0^t a\,dF}{F(t)} + \frac{(1-\delta)\int_t^M a\,dF}{1 - F(t)} \quad \text{a.e.}-[Q_{F,G}]$$

with adjoint

$$[L_1^* b](x) = E\{b(T,\Delta)|X = x\} = \int_{t=x}^M b(t,1)\,g(t)\,dt + \int_{t=0}^x b(t,0)\,g(t)\,dt \quad \text{a.e.-}[F].$$

We consider differentiability of functionals at the point F.

First consider the case $\tilde{\theta} \in \mathcal{R}(L_1)$. Then the score equation $L_1^* L_1 a = \tilde{\kappa}_F$ has to be solved in $a \in L_2^0(F)$. It can be written as an equation in ϕ:

$$\int_{t=0}^x \frac{\phi(t)}{1 - F(t)}\,g(t)\,dt - \int_{t=x}^M \frac{\phi(t)}{F(t)}\,g(t)\,dt = \tilde{\kappa}_F(x) \qquad \text{a.e.-}[F].$$

Suppose $\tilde{\kappa}_F(x)$ to be continuously differentiable. Then we get, by taking derivatives,

$$\frac{\phi(x)}{1 - F(x)} g(x) + \frac{\phi(x)}{F(x)} g(x) = \tilde{\kappa}_F'(x)$$

Hence, if $g > 0$,

$$\phi(x) = \tilde{\kappa}_F'(x) \frac{F(x)\{1 - F(x)\}}{g(x)}, \tag{3.6}$$

yielding the canonical gradient

$$\tilde{\theta}_F(t,\delta) = \begin{cases} -\tilde{\kappa}'_F(t)\frac{1-F(t)}{g(t)} & \text{if } \delta = 1 \\ \tilde{\kappa}'_F(t)\frac{F(t)}{g(t)} & \text{if } \delta = 0 \end{cases}$$

and information lower bound

$$\|\tilde{\theta}\|^2_{Q_{F,G}} = \int_0^M [\tilde{\kappa}'_F(x)]^2 \; \frac{F(x)[1-F(x)]}{g(x)}\, dx.$$

This is subject to the condition that $\phi(x)$ can be obtained as the integral $\int_x^M a(t)\,dF(t)$ over some $L_2^0(F)$-function a. $\phi(0) = 0$ implies $\int a\,dF = 0$. The derivative $a = d\phi/dF$ is equal to

$$\frac{\tilde{\kappa}'}{g}(1-2F) + F(1-F)\frac{d}{dF}\left[\frac{\tilde{\kappa}'}{g}\right].$$

If $(\tilde{\kappa}'/g) \circ F^{-1}$ is Lipschitz on $[0,1]$, then $d\phi/dF$ is square integrable.

The same Lipschitz condition is used by Huang and Wellner (1995a) to prove asymptotic optimality of the NPMLE for linear functionals, such as the mean. Since an explicit formula for ϕ does not exist in case 2, we have to rely on qualitative properties of the solution which complicates the analysis considerably.

In Chapter 5 of part 2 of GW a different method of proof is followed. It is proved there that the NPMLE of the mean asymptotically attains the efficiency bound under the stronger conditions that both f and g are bounded away from zero on $[0, M]$ with g having bounded derivative g'. This can be compared with assuming existence of $d\tilde{\kappa}'/dF$ and dg/dF, for then we get

$$\frac{d\phi}{dF} = \frac{\tilde{\kappa}'}{g}(1-2F) + \frac{F(1-F)}{g}\left[\frac{d\tilde{\kappa}'}{dF} - \frac{\tilde{\kappa}'}{g}\frac{dg}{dF}\right].$$

which is an $L_2(F)$-function at least under these conditions on f and g, and if $\frac{d}{dx}\tilde{\kappa}'$ is bounded. (For the mean this holds trivially: $\frac{d}{dx}\tilde{\kappa}' \equiv 0$.)

3.5 Interval censoring, case 2; lower bound

I will consider here the case $\tilde{\theta} \in \mathcal{R}(L_1)$, and study solvability of

$$\tilde{\kappa}_F(x) = [L_1^* L_1 a](x) \quad \text{a.e.} - F \tag{3.7}$$

in the variable $a \in L_2^0(F)$ (in section 4.2 I will deal with a situation where $\tilde{\theta} \in \overline{\mathcal{R}(L_1)} \setminus \mathcal{R}(L_1)$).

The support of F may consist of a finite number of disjoint intervals. However, (3.7) is not defined on intervals where F does not put mass, and these intervals do not play any further role. So without loss of generality we may assume the support of F to consist of one interval $[0, M]$. By the structure of the score operator L_1 this can be reformulated as an equation in ϕ. If one supposes equation (3.7) to hold for all $x \in [0, M]$, taking derivatives on both sides yields the following integral equation:

$$\phi(x) + d_F(x) \int_{t=0}^{M} \frac{\phi(x) - \phi(t)}{|F(x) - F(t)|} h^*(t,x)\, dt = k(x) d_F(x), \tag{3.8}$$

with $h^*(t,x) = h(t,x) + h(x,t)$ and $d_F(x)$ defined by

$$d_F(x) = \frac{F(x)[1 - F(x)]}{h_1(x)[1 - F(x)] + h_2(x)F(x)},$$

writing $k(x)$ instead of $\tilde{\kappa}'_F(x)$. Although k may depend on F, we do not explicitly express this dependence. The reason for this is that, in proving asymptotic efficiency of the NPMLE, we have to consider convex combinations $F = (1 - \alpha)F_0 + \alpha \hat{F}_n$, where F_0 (the unknown distribution) is continuous and $\hat{F}_n$ (the NPMLE of F_0) is purely discrete. Solvability and structure of the solution to (3.8) will also be investigated for such combinations, with k still determined by the underlying distribution F_0 (so $k = \tilde{\kappa}'_{F_0}$).

Since I will need the notation h^*, used in (3.8) a lot in the future, I define it separately below for easy reference.

Definition 3.1 *Let H be the distribution function of an observation pair (U, V), with density h. Then H^* is defined by*

$$H^*(u,v) = \begin{cases} H(u,v) & \text{if } u \le v \\ H(v,u) & \text{if } u > v \end{cases}, \tag{3.9}$$

and h^* is defined by

$$h^*(u,v) = \begin{cases} h(u,v) & \text{if } u \le v \\ h(v,u) & \text{if } u > v \end{cases}. \tag{3.10}$$

Note that $h^*(u,v) = h(u,v) + h(v,u)$, a.e. with respect to Lebesgue measure.

Apart from the model conditions $(M1)$ to $(M3)$, some extra conditions will have to be introduced in order to make the proofs in this chapter possible. For the distributions we assume

(D1) $h_1(x) + h_2(x) > 0$ for all $x \in [0, M]$.

(D2) $h(u,v)$ is continuous. The partial derivatives $\Delta_x^1(t) = \frac{\partial}{\partial x} h(t,x)$ and $\Delta_x^2(t) = \frac{\partial}{\partial x} h(x,t)$ exist, except for at most a finite number of points x, where left and right derivatives exist. The derivatives are bounded, uniformly in t and x.

(D3) F is a nondegenerate distribution function with at most finitely many points of jump $x_i \in (0, M)$. Let $D = \{x_0 = 0, x_1, \ldots, x_m, x_{m+1} = M\}$ denote the ordered set of jump points of F, augmented with the endpoints of the interval $[0, M]$. We assume that F is differentiable between jumps, except for at most a finite number of points, where left and right derivatives exist. Everywhere outside D, the derivative is bounded and $\geq c$ for some $c > 0$. The set of points of jump may be empty. Note that if F has jumps, we assume that F has derivative $\geq c$ also on the (non-empty) intervals $(0, x_1)$ and (x_m, M) (where we allow $x_1 = x_m$, though).

For the functional we should have

(F3) k is differentiable, except for at most a finite number of points x, where left and right derivatives exist. The derivative is bounded, uniformly in x.

Note that the letter "D" in conditions (D1) to (D3) stands for "distribution" and the letter "F" in (F1) to (F3) for "functional".

Of course, (D2) implies continuity of h_1 and h_2. (D1) is the equivalent of $g > 0$ in case 1 and is needed. It implies that d_F is bounded. In case 1, the function ϕ has an explicit representation of the form

$$\phi = k \, \frac{F[1-F]}{g},$$

whereas in case 2, ϕ can only be expressed implicitly as solution to (3.8). If (3.8) is solvable, its solution ϕ can be shown to contain a factor $F(1-F)$, just like in case 1. The structure of d_F already suggests this factor to be present. Validity of the factorization is shown by inserting

$$\phi = F(1-F)\,\xi$$

in (3.8). Some reordering yields an integral equation in ξ, which will be shown to be solvable. This ξ-equation has the following form:

$$\xi(x) + c_F(x) \int_{t=0}^{M} \frac{\xi(x) - \xi(t)}{|F(x) - F(t)|} \, h^*(t,x) F(t)(1 - F(t)) \, dt = k(x) c_F(x), \quad (3.11)$$

with $c_F(x)$ given by

$$\begin{aligned} c_F^{-1}(x) &= \int_{t=0}^{x} [1 - F(t)] \, h(t,x) \, dt + \int_{t=x}^{M} F(t) \, h(x,t) \, dt \\ &= h_2(x) \, E\{1 - F(U) | V = x\} + h_1(x) \, E\{F(V) | U = x\}. \quad (3.12) \end{aligned}$$

This equation is similar in structure to the ϕ-equation. So the lemmas and theorems in the remainder of this chapter apply to both the ϕ-equation (3.8) and the ξ-equation (3.11). Most of the proofs will only be given for the ϕ-equation.

Unlike the situation treated in GESKUS AND GROENEBOOM (1996A), I will allow the observation density to have mass along the diagonal. This has the consequence that the integral equation may no longer be a Fredholm integral equation. However, I first will consider a "desingularized" integral equation, to which the theory on Fredholm integral equations of the second kind can be applied.

If F has jumps, the solution of the integral equation will in general also have jumps. However, the key observation in analyzing the integral equation and in proving the efficiency of the NPMLE is that, even when F has discontinuities, we can make a change of scale in such a way that the solution of the integral equation can be extended to a Lipschitz function in the transformed scale.

First some notation. Let $G(t) = F^{-1}(t)$, $t \in [0,1]$, with a derivative g which exists except for at most a finite number of points, where, however, G has left and right derivatives. Furthermore, let $\bar{k}(t) = k(G(t))$, $\bar{H}(t,u) = H(G(t),G(u))$ and likewise $\bar{h}(t,u) = h(G(t),G(u))$, and let $\bar{d}_F$ be defined by

$$\bar{d}_F(t) = \frac{t(1-t)}{(1-t)\bar{h}_1(t) + t\,\bar{h}_2(t)}, \tag{3.13}$$

where $\bar{h}_i = h_i \circ G$, $i = 1,2$. Note that, if F has jumps, $\bar{d}_F \neq d_F \circ G$. Also note that $\bar{k}$, $\bar{d}$ and $\bar{h}$ are continuous. In a similar way, we define $\bar{c}_F = c_F \circ G$, where c_F is defined by (3.12). Finally, $\bar{H}^*$ and $\bar{h}^*$ are defined by

$$\bar{H}^*(t,u) = H^*(G(t),G(u)) \text{ and } \bar{h}^*(t,u) = h^*(G(t),G(u)). \tag{3.14}$$

We now have the following lemma.

Lemma 3.1 *(i) The integral equation*

$$\bar{\phi}_\epsilon(t) = \bar{d}_F(t)\left\{\bar{k}(t) - \int_0^M \frac{\bar{\phi}_\epsilon(t) - \bar{\phi}_\epsilon(t')}{|t-t'| \vee \epsilon}\,\bar{h}^*(t',t)\,dG(t')\right\} \tag{3.15}$$

has a unique continuous solution $\bar{\phi}_\epsilon$, satisfying

$$\inf_{x\in[0,M]} d_F(x)k(x) \le \bar{\phi}_\epsilon(t) \le \sup_{x\in[0,M]} d_F(x)k(x), \tag{3.16}$$

for all $t \in [0,1]$ and $\epsilon > 0$.
For points t in the range of F, say $t = F(x)$, we have $\bar{\phi}_\epsilon(t) = \phi_\epsilon(x)$

(ii) The integral equation

$$\bar{\xi}_\epsilon(t) = \bar{c}_F(t)\left\{\bar{k}(t) - \int_0^1 \frac{\bar{\xi}_\epsilon(t) - \bar{\xi}_\epsilon(t')}{|t-t'| \vee \epsilon}\,\bar{h}^*(t',t)\,t'(1-t')\,dG(t')\right\} \tag{3.17}$$

has a unique continuous solution $\bar{\xi}_\epsilon$, satisfying

$$\inf_{x\in[0,M]} c_F(x)k(x) \le \bar{\xi}_\epsilon(t) \le \sup_{x\in[0,M]} c_F(x)k(x), \tag{3.18}$$

for all $t \in [0,1]$ and $\epsilon > 0$.
For points t in the range of F, say $t = F(x)$, we have $\bar{\xi}_\epsilon(t) = \xi_\epsilon(x)$

Proof:
ad (i) By the Fredholm theory, as used e.g. in Theorem 6, p. 87, of GESKUS AND GROENEBOOM (1996A), the $\bar{\phi}_\epsilon$-equation (3.15) can be shown to have a unique

continuous solution, for each $\epsilon > 0$. Note that the integration in (3.15) is only with respect to $dG(t')$ and therefore only involves values belonging to the range of F. So for points t in the range of F we have

$$\bar{\phi}_\epsilon(t) = \phi_\epsilon(G(t)).$$

Let $m = \arg\min[\bar{\phi}_\epsilon]$ and $s = \arg\max[\bar{\phi}_\epsilon]$. We have

$$\bar{\phi}_\epsilon(s) \le \bar{d}_F(s)\bar{k}(s) \le \sup_{x\in[0,M]} d_F(x)k(x),$$

since $\bar{\phi}_\epsilon(s) - \bar{\phi}_\epsilon(t) \ge 0$, $t \in [0,1]$, and similarly

$$\bar{\phi}_\epsilon(m) \ge \bar{d}_F(m)\bar{k}(m) \ge \inf_{x\in[0,M]} d_F(x)k(x),$$

since $\bar{\phi}_\epsilon(m) - \bar{\phi}_\epsilon(t) \le 0$, $t \in [0,1]$. Hence we have (3.16).

ad (ii) The argument is completely similar to the argument given for (i). □

The following lemma is the crux of the proof of the existence of the solution to the original integral equation. Since this lemma is the key lemma for the proof of the existence of the solution of the integral equation, also when the integral equation has a singular kernel, I give the full proof (also given in GESKUS AND GROENEBOOM (1996C)) below.

Lemma 3.2 *The functions $\bar{\phi}_\epsilon$ are Lipschitz on $[0,1]$, uniformly in $\epsilon > 0$.*

Proof: We will use similar notation as in Lemma 3.1. Let $x_1, \ldots, x_m$ be the points of jump of F and let $x_0 = 0$, $x_{m+1} = M$. Furthermore, let $\tau_i = F(x_i)$, $i = 0, \ldots, m+1$. For $i = 0, \ldots, m$, the interval $[\tau_i, \tau_{i+1}]$ can be divided into two parts:

(1) the interval $[\tau_i, \tau_i')$, where $\tau_i' = F(x_{i+1}-)$. The interval $[\tau_i, \tau_i')$ corresponds to the interval $[x_i, x_{i+1})$ in the original scale. The function G is strictly increasing and differentiable on the interval (τ_i, τ_i'), and is right and left differentiable at τ_i and τ_i' respectively.

(2) the interval $[\tau_i', \tau_{i+1}]$. This interval corresponds to the jump of F at x_{i+1}. Here the function G is constant, again having right and left derivatives at the respective endpoints.

If $i = m$, the second interval only consists of the point 1. Let

$$\begin{aligned} D' &= \{\tau_0, \ldots, \tau_{m+1}\} \cup \{\tau_0', \ldots, \tau_m'\} \\ &\quad \cup \{\text{discontinuity points of } \bar{k}'(t),\ \bar{d}_F'(t) \text{ and } \sup_{t'\in[0,1]} \left|\tfrac{\partial}{\partial t}\bar{h}^*(t',t)\right|\}. \end{aligned}$$

Then $\bar{\phi}_\epsilon(t)$ is differentiable for $t \notin D'$, and has left and right derivatives for $t \in D'$. Using

$$\begin{aligned}\bar{k}(t) - \int_0^1 \frac{\bar{\phi}_\epsilon(t) - \bar{\phi}_\epsilon(t')}{|t-t'| \vee \epsilon} \bar{h}^*(t',t)\, dG(t') &= \bar{\phi}_\epsilon(t)/\bar{d}_F(t) \\ &= \xi_\epsilon(t)[(1-t)\bar{h}_1(t) + t\,\bar{h}_2(t)],\end{aligned}$$

and using left or right derivatives when $t \in D'$, we have:

$$\begin{aligned}\bar{\phi}'_\epsilon(t) &= \bar{d}'_F(t)\,\xi_\epsilon(t)\,[(1-t)\bar{h}_1(t) + t\,\bar{h}_2(t)] \\ &\quad + \bar{d}_F(t)\Big\{\bar{k}'(t) - \int_0^1 \frac{\bar{\phi}_\epsilon(t) - \bar{\phi}_\epsilon(t')}{|t-t'| \vee \epsilon} \frac{\partial}{\partial t}\bar{h}^*(t',t)\, dG(t') \\ &\quad - \bar{d}_F(t)\Big\{\int_{t':|t-t'|>\epsilon} \Big\{\frac{\bar{\phi}'_\epsilon(t)}{|t-t'|} - \frac{\bar{\phi}_\epsilon(t) - \bar{\phi}_\epsilon(t')}{(t-t')^2}\Big\}\, d\bar{H}^*(t',t) \\ &\quad - \bar{d}_F(t)\,\bar{\phi}'_\epsilon(t)\,\epsilon^{-1} \int_{t-\epsilon}^{t+\epsilon} \bar{h}^*(t',t) g(t')\, dt'. \end{aligned} \tag{3.19}$$

Note that $\frac{\partial}{\partial t}\bar{H}^*(t,u) = \bar{h}^*(t,u)g(t)$. Moving the terms containing $\bar{\phi}'_\epsilon$ to the left-hand side of (3.19), shows that $\bar{\phi}'_\epsilon(t)$ has a finite upper bound, using Lemma 3.1. Moreover, $\bar{\phi}'_\epsilon$ is piecewise continuous on the closed intervals from one point in D' to the subsequent one. So $\bar{\phi}'_\epsilon$ attains a maximum value, which may be a right or left derivative.

Let $M_\epsilon \stackrel{\text{def}}{=} \sup_{t\in[0,1]} \bar{\phi}'_\epsilon(t)$ and suppose that $\bar{\phi}'_\epsilon$ attains its supremum at a point s. Note that $M_\epsilon \geq 0$, since $\bar{\phi}_\epsilon(0) = \bar{\phi}_\epsilon(1) = 0$ and $\bar{\phi}_\epsilon$ is continuous. Then, if $|t-s| > \epsilon$,

$$\frac{\bar{\phi}'_\epsilon(s)}{|s-t|} - \frac{\bar{\phi}_\epsilon(s) - \bar{\phi}_\epsilon(t)}{(s-t)^2} \geq \frac{\int_t^s \{\bar{\phi}'_\epsilon(s) - \bar{\phi}'_\epsilon(u)\}\, du}{(s-t)^2} \geq 0.$$

Now let $K_\epsilon(t)$ be defined by

$$\begin{aligned}K_\epsilon(t) &\stackrel{\text{def}}{=} \bar{d}_F(t)\bar{k}'(t) + \bar{d}'_F(t)\,\xi_\epsilon(t)\,[(1-t)\bar{h}_1(t) + t\,\bar{h}_2(t)] \\ &\quad - \bar{d}_F(t) \int_0^1 \frac{\bar{\phi}_\epsilon(t) - \bar{\phi}_\epsilon(t')}{|t-t'| \vee \epsilon} \frac{\partial}{\partial t}\bar{h}^*(t',t)\, dG(t')\end{aligned}$$

and let $C_\epsilon(t)$ be defined by

$$C_\epsilon(t) \stackrel{\text{def}}{=} 1 + \bar{d}_F(t)\,\epsilon^{-1} \int_{t-\epsilon}^{t+\epsilon} \bar{h}^*(t',t) g(t')\, dt',\ t \in [0,1]. \tag{3.20}$$

Then we have

$$\bar{\phi}'_\epsilon(s)\, C_\epsilon(s) \leq K_\epsilon(s), \tag{3.21}$$

implying

$$M_\epsilon \leq \sup_{t\in[0,1]} K_\epsilon(t)/C_\epsilon(t). \tag{3.22}$$

In a similar way, if $m_\epsilon \stackrel{\text{def}}{=} \inf_{t\in[0,1]} \phi'_\epsilon(t)$, we get

$$m_\epsilon \geq \inf_{t\in[0,1]} K_\epsilon(t)/C_\epsilon(t). \tag{3.23}$$

Let the function A_δ be defined by

$$A_\delta(t) \stackrel{\text{def}}{=} \bar{d}_F(t) \int_{t-\delta}^{t+\delta} \left|\frac{\partial}{\partial t}\bar{h}^*(t',t)\right| dG(t'),\ t \in [0,1]$$

Fix $\delta > 0$ such that, for all $t \in [0,1]$,

$$A_\delta(t)/C_\epsilon(t) \leq \tfrac{1}{2}. \tag{3.24}$$

Note that $\delta > 0$ can be chosen independently of $\epsilon > 0$, since

$$\lim_{\epsilon\downarrow 0} C_\epsilon(t) = 1 + 2\bar{d}_F(t)\bar{h}(t,t)g(t),\ t \in (0,1).$$

Then we get from (3.24), for each $t \in [0,1]$, by applying the mean value theorem on the ratios $\{\bar{\phi}_\epsilon(t) - \bar{\phi}_\epsilon(t')\}/|t-t'| \vee \epsilon$,

$$\begin{aligned} &\bar{d}_F(t) \int_{t-\delta}^{t+\delta} \left|\frac{\bar{\phi}_\epsilon(t) - \bar{\phi}_\epsilon(t')}{|t-t'| \vee \epsilon}\, \frac{\partial}{\partial t}\bar{h}(t',t)\right| dG(t') \Big/ C_\epsilon(t) \\ \leq\ & A_\delta(t) \max\{M_\epsilon, |m_\epsilon|\}/C_\epsilon(t) \leq \tfrac{1}{2}\max\{M_\epsilon, |m_\epsilon|\}. \end{aligned}$$

Defining $B_\delta(t)$ by

$$\begin{aligned} B_\delta(t) \stackrel{\text{def}}{=}\ & \bar{d}_F(t)|\bar{k}'(t)| + |\bar{d}'_F(t)|\,[(1-t)\bar{h}_1(t) + t\,\bar{h}_2(t)] \sup_{t'\in[0,1]}\{\bar{c}_F(t')|\bar{k}(t')|\} \\ & + \frac{2\bar{d}_F(t)}{\delta} \sup_{t'\in[0,1]}\{\bar{d}_F(t')|\bar{k}(t')|\} \sup_{t'\in[0,1]}\left|\frac{\partial}{\partial t}\bar{h}^*(t',t)\right|, \end{aligned}$$

we get, for $t \in [0,1]$,

$$\begin{aligned} & d_F(t)|\bar{k}'(t)| + |\bar{d}'_F(t)\,\bar{\xi}_\epsilon(t)|\,[(1-t)\bar{h}_1(t) + t\,\bar{h}_2(t)] \\ & + \bar{d}_F(t) \int_{|t'-t|>\delta} \frac{|\bar{\phi}_\epsilon(t)| + |\bar{\phi}_\epsilon(t')|}{|t-t'|} \left|\frac{\partial}{\partial t}\bar{h}^*(t',t)\right| dG(t') \\ \leq\ & \bar{d}_F(t)|\bar{k}'(t)| + |\bar{d}'_F(t)\,\bar{\xi}_\epsilon(t)|\,[(1-t)\bar{h}_1(t) + t\,\bar{h}_2(t)] \\ & + \frac{2\bar{d}_F(t)}{\delta} \sup_{t'\in[0,1]} |\bar{\phi}_\epsilon(t')| \int_{t':|t'-t|>\delta} \left|\frac{\partial}{\partial t}\bar{h}^*(t',t)\right| dG(t') \\ \leq\ & B_\delta(t) \leq c, \end{aligned} \tag{3.25}$$

for some constant c, independent of ϵ and t. Hence, for each $t \in [0,1]$,

$$|\bar{\phi}'_\epsilon(t)| \leq A_\delta(t)/C_\epsilon(t) + B_\delta(t)/C_\epsilon(t) \leq \tfrac{1}{2}\max\{M_\epsilon, |m_\epsilon|\} + B_\delta(t)/C_\epsilon(t),$$

implying

$$\tfrac{1}{2}\max\{M_\epsilon, |m_\epsilon|\} \le \sup_{t\in[0,1]} B_\delta(t)/C_\epsilon(t) \le \sup_{t\in[0,1]} c/C_\epsilon(t) \le c', \tag{3.26}$$

for some constant c' independent of ϵ.
Hence $\bar\phi'_\epsilon(t)$ is bounded on $[0,1]$, uniformly in ϵ and t, implying that $\bar\phi_\epsilon$ is Lipschitz, uniformly in $\epsilon > 0$. □

We now have the following theorem.

Theorem 3.4 *Let the following conditions on F_0, H and $\tilde\kappa_{F_0}$ be satisfied: (M1) to (M3); (D1) to (D3); (F1) to (F3). Let $G(t) = F^{-1}(t)$, $\bar k(t) = k(G(t))$, $\bar H^*(t,u) = H^*(G(t), G(u))$, $\bar h^*(t,u) = h^*(G(t), G(u))$, and let $\bar d_F$ be defined by*

$$\bar d_F(t) \stackrel{\text{def}}{=} \frac{t(1-t)}{(1-t)\bar h_1(t) + t\bar h_2(t)}, \tag{3.27}$$

where $\bar h_i = h_i \circ G$, $i = 1, 2$. Then

(i) The integral equation

$$\bar\phi(t) = \bar d_F(t)\Big\{\bar k(t) - \int_0^1 \frac{\bar\phi(t) - \bar\phi(t')}{|t-t'|}\, d\bar H^*(t',t)\Big\},\ t\in[0,1], \tag{3.28}$$

has a unique solution which is Lipschitz on $[0,1]$.

(ii) The Lipschitz norm for the functions $\bar\phi$ in (i) has the following upper bound. Let $C(t)$ be defined by

$$C(t) \stackrel{\text{def}}{=} 1 + 2\bar d_F(t) g(t) \bar h(t,t). \tag{3.29}$$

Moreover, let $A_\delta(t)$ and $B_\delta(t)$ be defined by

$$A_\delta(t) \stackrel{\text{def}}{=} \bar d_F(t) \int_{t-\delta}^{t+\delta} \Big|\frac{\partial}{\partial t}\bar h^*(t',t)\Big|\, dG(t'), \tag{3.30}$$

and

$$\begin{aligned} B_\delta(t) \stackrel{\text{def}}{=}\ & \bar d_F(t)|\bar k'(t)| \\ & + |\bar d_F'(t)|\,[(1-t)\bar h_1(t) + t\,\bar h_2(t)] \sup_{t'\in[0,1]} \{\bar c_F(t')|\bar k(t')|\} \\ & + \frac{2\bar d_F(t)}{\delta} \sup_{t'\in[0,1]} \{\bar d_F(t')|\bar k(t')|\} \sup_{t'\in[0,1]} \Big|\frac{\partial}{\partial t}\bar h^*(t',t)\Big| \end{aligned} \tag{3.31}$$

At the points in

$$\begin{aligned} D' =\ & \{\text{discontinuity points of } g(t), \text{ augmented with 0 and 1}\} \\ & \cup \Big\{\text{discontinuity points of } \bar k'(t),\ \bar d_F'(t) \text{ and } \sup_{t'\in[0,1]} \Big|\frac{\partial}{\partial t}\bar h(t',t)\Big|\Big\}, \end{aligned}$$

A_δ and B_δ have two versions, one corresponding to taking left derivatives and one corresponding to taking right derivatives.
Then there exists a $\delta > 0$ such that

$$\sup_{t\in[0,1]} A_\delta(t)/C(t) \le 1/2,$$

and

$$|\bar\phi(u) - \bar\phi(t)| \le c(u-t),\ 0 \le t < u \le 1, \tag{3.32}$$

where c is given by

$$c = 2 \sup_{t\in[0,1]} B_\delta(t)/C(t). \tag{3.33}$$

(iii) The integral equation (3.8) has a unique solution ϕ.

Proof:
ad (i) By the preceding two lemma's, the set $\{\bar\phi_\epsilon : \epsilon \le \epsilon_0\}$ (for some $\epsilon_0 > 0$) is bounded and equicontinuous. Hence, by the Arzelà-Ascoli theorem, each sequence $\bar\phi_{\epsilon_n}$, $\epsilon_n \downarrow 0$, has a subsequence $(\bar\phi_{\epsilon_m})$, converging in the supremum metric to a continuous function $\bar\phi$ on $[0,1]$. By Lebesgue's dominated convergence theorem we get, for such a subsequence $(\bar\phi_{\epsilon_m})$,

$$\begin{aligned}\bar\phi(x) &= \lim_{m\to\infty} \bar\phi_{\epsilon_m}(x)\\ &= \bar d_F(x)\Big\{\bar k(x) - \int_0^1 \frac{\bar\phi(x)-\bar\phi(t)}{|x-t|}\,\bar h^*(t,x)\,dG(t)\Big\}.\end{aligned} \tag{3.34}$$

Uniqueness of the solution follows in the same way as in lemma 3.1.
ad (ii) It was shown in (3.26) in the proof of Lemma 3.2 that

$$\sup_{t\in[0,1]} |\phi'_\epsilon(t)| \le 2 \sup_{t\in[0,1]} B_\delta(t)/C_\epsilon(t),$$

where C_ϵ is defined by (3.20). But since

$$\lim_{\epsilon\downarrow 0} C_\epsilon(t) = 1 + 2\bar d_F(t)\bar h(t,t)g(t),$$

for $t \in [0,1]$, (3.32) now follows.
ad (iii) We define ϕ by $\phi(x) = \bar\phi(F(x))$. If $t = F(x)$, we get, by a change of variables,

$$\begin{aligned}\phi(x) &= \bar\phi(t)\\ &= \bar d_F(t)\Big\{\bar k(t) - \int_0^t \frac{\bar\phi(t)-\bar\phi(t')}{|t-t'|}\,d\bar H^*(t',t)\Big\}\\ &= d_F(x)\Big\{k(x) - \int_0^M \frac{\phi(x)-\phi(x')}{|F(x)-F(x')|}\,dH^*(x',x)\Big\},\end{aligned}$$

and hence ϕ satisfies the original integral equation. Uniqueness of ϕ follows from uniqueness of $\bar\phi$ (since a solution ϕ conversely defines a solution $\bar\phi$ on the inverse scale).

Solvability of $\tilde{\kappa}_F = L_1^* L_1 a$ can now immediately be seen.

Corollary 3.1 *The equation $\tilde{\kappa}_F = L_1^* L_1 a$ is solvable.*

Proof: By the Lipschitz property of $\bar{\phi}$ we have, for any $0 \le x < y \le M$,

$$\frac{|\phi(y) - \phi(x)|}{F(y) - F(x)} = \frac{|\bar{\phi}(F(y)) - \bar{\phi}(F(x))|}{F(y) - F(x)} \le K,$$

for some constant K. Thus the Radon-Nikodym derivative $d\phi/dF$ is a.e.-$[F]$ bounded by K. □

For the canonical gradient we get, if $u < v$,

$$\begin{aligned}\tilde{\theta}_F(u, v, \delta_1, \delta_2) &= [L_1 a](u, v, \delta_1, \delta_2) \\ &= -\delta_1 \frac{\phi(u)}{F(u)} - \delta_2 \frac{\phi(v) - \phi(u)}{F(v) - F(u)} + (1 - \delta_1 - \delta_2) \frac{\phi(v)}{1 - F(v)}. \end{aligned} \tag{3.35}$$

3.6 Asymptotic efficiency of the NPMLE

In this section I will denote the unknown distribution function of the unobservable random variables X_i by F_0 and assume that F_0 is continuous. Let $\hat{F}_n$ be the NPMLE of F_0, based on the sample of observations $(U_1, V_1, \Delta_{1,1}, \Delta_{2,1}), \ldots,$ $(U_n, V_n, \Delta_{1,n}, \Delta_{2,n})$. It is obtained by maximizing the likelihood

$$\prod_{i=1}^{n} F(U_i)^{\Delta_{1i}} (F(V_i) - F(U_i))^{\Delta_{2i}} (1 - F(V_i))^{\Delta_{3i}}\, h(U_i, V_i) \tag{3.36}$$

over the class of piecewise constant right-continuous (sub-)distribution functions on $[0, M]$, having jumps only at a subset of the points U_i and V_i, $i = 1, \ldots, n$.

The NPMLE does not depend on $\prod h(U_i, V_i)$, so we do not have to perform any preliminary density estimation or bandwidth choice. As noted earlier, the fact that we do not have to solve a bandwidth problem is one of the great advantages of the nonparametric maximum likelihood approach in the present problem.

By the restriction that $\hat{F}_n$ only has mass at the observation times, we get a piecewise constant function. Let $x_0 = 0$, $x_{m+1} = M$ and let $x_1 < \ldots < x_m$ be the points of jump of F in the interval $(0, M)$. Then $\hat{F}_n$ satisfies the following two properties:

Proposition 3.2 *Any function σ that is constant on the same intervals $J_i = [x_{i-1}, x_i)$ as $\hat{F}_n$ satisfies*

$$\begin{aligned}&\int_{u \in J_i} \sigma(t) \left\{ \frac{\delta_1}{\hat{F}_n(u)} - \frac{\delta_2}{\hat{F}_n(v) - \hat{F}_n(u)} \right\} dQ_n(u, v, \delta_1, \delta_2) \\ &+ \int_{v \in J_i} \sigma(u) \left\{ \frac{\delta_2}{\hat{F}_n(v) - \hat{F}_n(u)} - \frac{1 - \delta_1 - \delta_2}{1 - \hat{F}_n(v)} \right\} dQ_n(u, v, \delta_1, \delta_2) = 0\end{aligned}$$

for $i = 2, \ldots, m$.

Proof: This is the same as (2.32) (which was based on Proposition 2.1). □

Proposition 3.3

$$\|\hat{F}_n - F_0\|_{H_i} = \mathcal{O}_p\big(n^{-1/3}(\log n)^{1/6}\big) \text{ as } n \to \infty, \text{ for } i = 1, 2$$

Proof: See VAN DE GEER (1996), example 3.2. □

The following result is needed in the proof of Lemma 3.3.

Proposition 3.4

$$\lim_{n\to\infty} Pr\{\hat{F}_n \text{ is defective}\} = 0.$$

Proof: See GESKUS AND GROENEBOOM (1996B), Proposition 3.

The behavior of the NPMLE will crucially depend on the nature of the joint density of the observation pairs (U_i, V_i). I will discuss the "hard case", where it is assumed that the density h is "positive on the diagonal". This means that the kernel of the integral equation will have a real singularity. As an example, if $h(u,v) = 2,\ 0 \le u \le v \le 1$, we have exactly this situation. The "hard case" was looked at first, but solved last. The efficiency of the NPMLE in the "strict separation case", where there is a strip with no mass along the diagonal, is proved in GESKUS AND GROENEBOOM (1996B).

So, in addition to the smoothness conditions (D1) to (D3), given in section 3.5, I assume

$$\text{(D4)} \qquad h(t,t) = \lim_{u \downarrow t} h(t,u) \ge c > 0,$$

for all $t \in (0, M)$ and some $c > 0$.

The canonical gradient $\tilde{\theta}$ will be extended to piecewise constant distribution functions F with finitely many discontinuities. However, since $F(v) - F(u)$ can be zero on regions where H puts mass, we have to use methods which are different from those in section 3.5. In order to stress dependence on F, I will write ϕ_F instead of ϕ.

Instead of *one* function ϕ_F we now need a *pair* of functions (ϕ_F, ψ_F), satisfying

$$\phi_F(x) = d_F(x)\left\{k(x) - \int_0^M r_F(t,x)\, h^*(t,x)\, dt\right\}, \tag{3.37}$$

where $r_F(u,v)$ is defined by

$$r_F(u,v) = \begin{cases} \frac{\phi_F(u)-\phi_F(t)}{F(u)-F(t)} &, \text{ if } F(t) < F(u), \\ \frac{\psi_F(u)-\psi_F(t)}{F_0(u)-F_0(t)} &, \text{ if } F(t) = F(u),\ t < u, \end{cases} \tag{3.38}$$

and where ϕ_F is constant on the same intervals as F.

The ψ_F-part is needed because the left-hand side of (3.37) is constant between jumps, whereas d_F and k are not. The key to the proof of the existence of a solution pair (ϕ_F, ψ_F) and also to the other proofs in this section are a representation of the equation for ϕ_F on an inverse scale and the construction of a continuous extension of the equation for ϕ_F on this inverse scale (similar techniques were used in section 3.5). Using a similar notation as in section 3.5, we denote by G the inverse of F, where, for purely discrete distribution functions F, we take the right-continuous version of the inverse, defined by

$$G(t) = \inf\{x \in [0, M] : F(x) > t\},\, t \geq 0.$$

Furthermore, we define

$$\bar{k}_F = k \circ G,\, \bar{h}_{1,F} = h_1 \circ G,\, \bar{h}_{2,F} = h_2 \circ G,$$

$$\bar{d}_F(t) = \frac{t(1-t)}{(1-t)\bar{h}_{1,F}(t) + t\,\bar{h}_{2,F}(t)},$$

and likewise $\bar{H}^*(t, u) = H(G(t), G(u))$, $0 \leq t \leq u \leq 1$.

For part (iii) of theorem 3.5 we will also need the following notation

$$\tilde{d}_i \stackrel{\text{def}}{=} \frac{z_i(1-z_i)}{\Delta_i(h_1)\,(1-z_i) + \Delta_i(h_2)\,z_i}, \tag{3.39}$$

$$\Delta_i(g) \stackrel{\text{def}}{=} \int_{x_i}^{x_{i+1}} g(t)\,dt, \tag{3.40}$$

$$\Delta_{ij}(h) \stackrel{\text{def}}{=} \int_{u=x_i}^{x_{i+1}} \int_{v=x_j}^{x_{j+1}} h(u,v)\,dv\,du. \tag{3.41}$$

The following theorem shows the existence of the solution pair. Moreover it gives a uniform Lipschitz condition for the functions $\bar{\phi}_F$ and ψ_F, which will be a crucial tool in showing the Donsker property for $\tilde{\theta}_F$.

Theorem 3.5 *Let the following conditions on F_0, H and $\bar{\kappa}_{F_0}$ be satisfied: (M1) to (M3); (D1) to (D4); (F1) to (F3). Furthermore, let $\mathcal{F}_{[0,M]}$ be the set of discrete non-defective distribution functions on $[0, M]$ with finitely many points of jump, contained in $(0, M)$. Then there exists an $\epsilon > 0$ such that, for $F \in \mathcal{F}$, where $\mathcal{F}$ is defined by $\{F \in \mathcal{F}_{[0,M]} : \sup_{x\in[0,M]} |F(x) - F_0(x)| \leq \epsilon\}$,*

(i) There exists a unique Lipschitz function $\bar{\phi}_F : [0,1] \to \mathbb{R}$ such that, for $t \in [0,1] \setminus D$,

$$\bar{\phi}_F(t) = \bar{d}_F(t)\Big\{\bar{k}_F(t) - \int_{t'\in[0,t)} \frac{\bar{\phi}_F(t) - \bar{\phi}_F(t')}{|t-t'|}\, d\bar{H}^*(t',t)\Big\}, \tag{3.42}$$

where D is the (finite) set of discontinuities of the right-continuous inverse $G = F^{-1}$ in $(0,1)$, augmented with 0 and 1. The function $\bar{\phi}_F$ is Lipschitz, uniformly for $F \in \mathcal{F}$.

(ii) *There exists a pair* (ϕ_F, ψ_F), *solving the integral equation (3.37), where* ϕ_F *is absolutely continuous with respect to* F *and the function* ψ_F *is Lipschitz on each interval between jumps of* F, *uniformly for* $F \in \mathcal{F}$, *with a Lipschitz norm not depending on the interval.*

(iii) *Let* $z_i = F(x_i)$ *and* $y_i = \phi_F(x_i)$, $i = 1, \ldots, m$. *Then, using the definitions (3.39) to (3.41), we have that the vector* $y = (y_1, \ldots, y_m)'$ *is the unique solution of the set of linear equations*

$$y_i \left\{ \tilde{d}_i^{-1} + \sum_{j<i} \frac{\Delta_{ji}(h)}{z_i - z_j} + \sum_{j>i} \frac{\Delta_{ij}(h)}{z_j - z_i} \right\}$$
$$= \Delta_i(k) + \sum_{j<i} \frac{\Delta_{ji}(h)}{z_i - z_j} y_j + \sum_{j>i} \frac{\Delta_{ij}(h)}{z_j - z_i} y_j, \quad i = 1, \ldots, m. \tag{3.43}$$

Theorem 3.5 is proved by approximating the purely discrete distribution function F by the function $F_\alpha = (1-\alpha)F_0 + \alpha F$ and by studying the behavior of the corresponding function ϕ_{F_α}, as $\alpha \uparrow 1$, using the upper bound for the Lipschitz norms of the functions ϕ_{F_α}, $\alpha < 1$, given Theorem 3.4. The rather technical proof is given in GESKUS AND GROENEBOOM (1996C). By Theorem 3.5, the definition of the function $\tilde{\theta}_F$ can be extended to piecewise constant distribution functions $F \in \mathcal{F}$ by defining

$$\tilde{\theta}_F(u, v, \delta_1, \delta_2) = -\frac{\delta_1 \phi_F(u)}{F(u)} - \delta_2 \, r_F(u, v) + \frac{(1 - \delta_1 - \delta_2)\phi_F(v)}{1 - F(v)}, \tag{3.44}$$

where ϕ_F and ψ_F solve equation (3.37), and where $\frac{\phi_F(u)}{F(u)}$ and $\frac{\phi_F(v)}{1-F(v)}$ are defined to be zero if $F(u) = 0$ or if $1 - F(v) = 0$, respectively. Note that $\tilde{\theta}_F$ no longer has an interpretation as canonical gradient.

In the sequel we will write Q_F instead of $Q_{F,H}$. The following theorem contains the efficiency result for the NPMLE.

Theorem 3.6 *Let the conditions of Theorem 3.5 be satisfied. Then*

$$\sqrt{n}(K(\hat{F}_n) - K(F_0)) \xrightarrow{\mathcal{D}} N(0, \|\tilde{\theta}_{F_0}\|^2_{Q_{F_0}}) \quad as \quad n \to \infty \tag{3.45}$$

Skeleton of the proof: It is sufficient to show the following:

$$\sqrt{n}(K(\hat{F}_n) - K(F_0)) = \sqrt{n} \int \tilde{\theta}_{F_0} \, d(Q_n - Q_{F_0}) + o_p(1). \tag{3.46}$$

Using the uniform consistency of $\hat{F}_n$, we may assume that $\hat{F}_n \in \mathcal{F}$, for all large n, where $\mathcal{F}$ is defined as in Theorem 3.5. The proof consists of the following steps.

I. By conditions (D1) and (F2), and Proposition 3.3 we have

$$\sqrt{n}(K(\hat{F}_n) - K(F_0)) = \sqrt{n} \int \tilde{\kappa}_{F_0} \, d(\hat{F}_n - F_0) + o_p(1).$$

II. In lemma 3.3 the following will be shown:

$$\int \tilde{\kappa}_{F_0}\, d(F - F_0) = -\int \tilde{\theta}_F\, dQ_{F_0},$$

if $F \in \mathcal{F}$.

III. By Theorem 3.5, part (ii), $\phi_{\hat{F}_n}$ is constant on the same intervals as $\hat{F}_n$. Since $\delta_2 = 0$, if $\hat{F}_n(u) = \hat{F}_n(t)$, Proposition 3.2 can be used to obtain

$$\int \tilde{\theta}_{\hat{F}_n}\, dQ_n = 0,$$

yielding

$$-\sqrt{n}\int \tilde{\theta}_{\hat{F}_n}\, dQ_{F_0} = \sqrt{n}\int \tilde{\theta}_{\hat{F}_n}\, d(Q_n - Q_{F_0}).$$

IV. This is further split into

$$\begin{aligned}\sqrt{n}\int \tilde{\theta}_{\hat{F}_n} d(Q_n - Q_{F_0}) &= \sqrt{n}\int \tilde{\theta}_{F_0} d(Q_n - Q_{F_0})\\ &\quad + \sqrt{n}\int (\tilde{\theta}_{\hat{F}_n} - \tilde{\theta}_{F_0}) d(Q_n - Q_{F_0}).\end{aligned}$$

The last term will be shown to be $o_p(1)$ (Theorem 3.7).

□

Lemma 3.3 *Let $\mathcal{F}$ be defined as in Theorem 3.5. Then we have under the conditions of Theorem 3.5, for all $F \in \mathcal{F}$,*

$$\int \tilde{\kappa}_{F_0}\, d(F - F_0) = -\int \tilde{\theta}_F\, dQ_{F_0}.$$

Proof: Let, for $F \in \mathcal{F}$, $L_F : L_2(F) \to L_2(Q_F)$ be defined by (3.4) with transpose L^*, defined by (3.5). As noted after (3.5), the structure of L^* does not depend on F. Furthermore, let $1 \in L_2(F)$ denote the constant function $1(x) \equiv 1$, $x \in \mathbb{R}$. Under L_F this transforms into the constant function $1(t, u, \delta, \gamma) \equiv 1$ on $L_2(Q_F)$. We have

$$\begin{aligned}\int \tilde{\theta}_F\, dQ_{F_0} &= <\tilde{\theta}_F, 1>_{Q_{F_0}}\\ &= <\tilde{\theta}_F, L_{F_0}(1)>_{Q_{F_0}}\\ &= <L^*(\tilde{\theta}_F), 1>_{F_0}\\ &= \int L^*(\tilde{\theta}_F)\, dF_0.\end{aligned}$$

If we can prove

$$L^*(\tilde{\theta}_F) = \tilde{\kappa}_{F_0} - \int \tilde{\kappa}_{F_0}\, dF \qquad \text{a.e.-}F_0,$$

we are done. This is shown as follows:
Recall that the integral equation was obtained by taking derivatives in the equation $\tilde{\kappa}_{F_0}(x) = [L^*\tilde{\theta}_F](x)$ for all $x \in [0, M]$, with $\tilde{\theta}_F = \tilde{\theta}_{F_0}$. Hence we get by integrating:

$$[L^*\tilde{\theta}_F](x) = [\tilde{\kappa}_{F_0}](x) + C, \qquad \text{for all } x \in [0, M].$$

For the constant C we have

$$\begin{aligned} C &= \int C\,dF \\ &= \int L^*(\tilde{\theta}_F)\,dF - \int \tilde{\kappa}_{F_0}\,dF =< L^*(\tilde{\theta}_F), 1 >_F - \int \tilde{\kappa}_{F_0}\,dF. \end{aligned}$$

It is easily seen that $\tilde{\theta}_F$ is contained in $L_2^0(Q_F)$. We also have

$$< L^*(\tilde{\theta}_F), 1 >_F =< \tilde{\theta}_F, L_F(1) >_{Q_F} =< \tilde{\theta}_F, 1 >_{Q_F} = 0.$$

The result now follows. □

The hard part of the proof of Theorem 3.6 is to show that

$$\sqrt{n} \int (\tilde{\theta}_{\hat{F}_n} - \tilde{\theta}_{F_0}) d(Q_n - Q_{F_0}) = o_p(1). \tag{3.47}$$

I will sketch the general idea of the proof. Full details are given in GESKUS AND GROENEBOOM (1996c).

We start by defining a neighborhood, shrinking with n, such that the probability that $\hat{F}_n$ belongs to $\mathcal{F}_n$ tends to 1, as $n \to \infty$. Let, for $F \in \mathcal{F}$,

$$q_F(u, v, \delta_1, \delta_2) = \delta_1 F(u) + \delta_2\{F(v) - F(u)\} + (1 - \delta_1 - \delta_2)\{1 - F(v)\} \tag{3.48}$$

and

$$q_{F_0}(u, v, \delta_1, \delta_2) = \delta_1 F_0(u) + \delta_2\{F_0(v) - F_0(u)\} + (1 - \delta_1 - \delta_2)\{1 - F_0(v)\}. \tag{3.49}$$

It is proved in VAN DE GEER (1996), that

$$h^2(q_{\hat{F}_n}, q_{F_0}) = \mathcal{O}_p(n^{-2/3}(\log n)^{1/3}), \tag{3.50}$$

where $h(q_F, q_{F_0})$ is the Hellinger distance between the densities q_F and q_{F_0} w.r.t. the product of the measure, induced by H, and counting measure on $\{0,1\}^2 \setminus \{(1,1)\}$. The result (3.50) has already been used above, since Proposition (3.3) is based on it. Now, if $\mathcal{F}_n$ is the set of distribution functions $F \in \mathcal{F}$, satisfying

$$h^2(q_F, q_{F_0}) \le n^{-2/3} \log n, \tag{3.51}$$

we have:

$$Pr\{\hat{F}_n \in \mathcal{F}_n\} \to 1, \text{ as } n \to \infty.$$

In fact, the upper bound $n^{-2/3}\log n$, defining the class $\mathcal{F}_n$, could be replaced by

$$c_n n^{-2/3}(\log n)^{1/3},$$

where we only need $c_n \to \infty$, as $n \to \infty$, but I follow the method of GESKUS AND GROENEBOOM (1996C) of being a little bit wasteful with powers of $\log n$ in an attempt to avoid an accumulation of constants in the upper bounds.

We need to study properties of the empirical integrals

$$\sqrt{n}(Q_n - Q_{F_0})(\tilde{\theta}_F - \tilde{\theta}_{F_0}),$$

for $F \in \mathcal{F}_n$. The denominators in $\tilde{\theta}_{F_0}$ and $\tilde{\theta}_F$ can be arbitrarily close to zero. If $F \in \mathcal{F}$, then $F(v) - F(u)$ will be zero on a region of positive Lebesgue measure, in which case we get for the "middle part" $\delta_2 \cdot r_F$ of $\tilde{\theta}_F$:

$$\delta_2 \cdot r_F(t,u) = \delta_2 \cdot \{\psi_F(u) - \psi_F(t)\}/(F_0(u) - F_0(t)\}.$$

These difficulties are faced in GESKUS AND GROENEBOOM (1996C) by considering three regions of integration:

$$C_{n,\eta}(F) = \{w : q_F(w) > \eta\, q_{F_0}(w),\, q_{F_0}(w) > n^{-1/3}\}, \tag{3.52}$$

$$D_\eta(F) = \{w : q_F(w) \le \eta q_{F_0}(w)\}, \tag{3.53}$$

and

$$C_n(F_0) = \{w : q_{F_0}(w) \le n^{-1/3}\}, \tag{3.54}$$

for some $\eta \in (0,1)$, where the elements w of the sets, defined above, are of the form $w = (u, v, \delta_1, \delta_2)$. On the region $C_{n,\eta}(F)$, $\tilde{\theta}_F$ has a behavior which is comparable to the behavior of $\tilde{\theta}_{F_0}$; on the other region the uniform boundedness of $\tilde{\theta}_F$ is used, together with the fact that the integrals over these regions become sufficiently small.

The following two lemma's, proved in GESKUS AND GROENEBOOM (1996C), take care of the regions (3.53) and (3.54), repectively.

Lemma 3.4 *Let, for $\eta \in (0,1)$, the set $D_\eta(F)$ be defined by (3.53). Then*

$$\sup_{F\in\mathcal{F}_n} Q_n D_\eta(F) = \mathcal{O}_p(n^{-2/3}\log n). \tag{3.55}$$

Lemma 3.5 *Let the function b_n be defined by*

$$b_n = 1_{\{q_{F_0} \le n^{-1/3}\}}.$$

Then

$$Q_n b_n = \mathcal{O}_p(n^{-2/3}) \tag{3.56}$$

Using the uniform boundedness of the families of functions $\{\tilde{\theta}_F : F \in \mathcal{F}_n\}$, it follows from lemma's 3.4 and 3.5 that the integrals

$$\sqrt{n}\int_{D_\eta(F)} \{\tilde{\theta}_F - \tilde{\theta}_{F_0}\}\, d(Q_n - Q_{F_0})$$

and

$$\sqrt{n}\int_{C_n(F_0)} \{\tilde{\theta}_F - \tilde{\theta}_{F_0}\}\, d(Q_n - Q_{F_0})$$

are both $o_p(1)$, uniformly for $F \in \mathcal{F}_n$.

For the entropy calculations on the remaining region $C_{n,eta}$ an approximating class of ratios $r_{F_k,G_k,\bar{\phi}_k}$ of the form

$$\frac{\bar{\phi}_k(G_k(v)) - \bar{\phi}_k(F_k(u))}{G_k(v) - F_k(u)},$$

are used, where F_k and G_k are distribution functions such that $F_k \le F \le G_k$ ((F_k, G_k) is a "bracket" for F) and where $\bar{\phi}_k$ is a Lipschitz function approximating $\bar{\phi}_F$. In this way the good behavior of the ratios r_F on the region $C_{n,\eta}(F)$ is preserved on the same region by the approximating ratio $r_{F_k,G_k,\bar{\phi}_k}$. Next the chaining lemma is applied, using a chain with a kind of "funnel" structure, preserving this good behavior on the region $C_{n,\eta}(F)$. Note that the approximating ratios are outside the original class of ratios r_F.

We will now look in particular at the behavior of the "middle part" of $\tilde{\theta}_F$. Consider triples $(F_k, G_k, \bar{\phi}_k)$, where F_k and G_k are distribution functions belonging to $\mathcal{F}$ and $\bar{\phi}_k$ belongs to a uniform class of Lipschitz functions on $[0, 1]$, with the same uniform Lipschitz norm c_{Lip} and upper bound as the functions $\bar{\phi}_F$, $F \in \mathcal{F}$, of Theorem 3.5. For these triples we define

$$r_{F_k,G_k,\bar{\phi}_k}(t,u) = \frac{\bar{\phi}_k(G_k(u)) - \bar{\phi}_k(F_k(t))}{G_k(u) - F_k(t)}, \text{ if } G_k(u) > F_k(t), \tag{3.57}$$

and, for pairs (t, u) such that $F_k(t) = G_k(u)$, we define $r_{F_k,G_k,\bar{\phi}_k}(t,u) = 0$. Moreover, we define the semi-metric

$$\begin{aligned}
&d_n\big((F_k, G_k, \bar{\phi}_k), (F_l, G_l, \bar{\phi}_l)\big)\\
&= 2\max_{t\in[0,1]} |\bar{\phi}_k(t) - \bar{\phi}_l(t)| \left\{\int_{F_0(u)-F_0(t)>n^{-1/3}} \frac{1}{\{F_0(u) - F_0(t)\}^2}\,\gamma\, dQ_n\right\}^{1/2}\\
&+ \left\{\int_{F_0(u)-F_0(t)>n^{-1/3}} \left\{\frac{F_k(t) - F_l(t)}{F_0(u) - F_0(t)}\right\}^2 \gamma\, dQ_n\right\}^{1/2}\\
&+ \left\{\int_{F_0(u)-F_0(t)>n^{-1/3}} \left\{\frac{G_k(u) - G_l(u)}{F_0(u) - F_0(t)}\right\}^2 \gamma\, dQ_n\right\}^{1/2}
\end{aligned} \tag{3.58}$$

For this semi-metric we have the following result.

Lemma 3.6 *Let, for distribution functions F and G such that $F \le G$, the set $C_{n,\eta}(F,G)$ be defined by*

$$C_{n,\eta}(F,G) = \big\{(t,u) : F_0(u) - F_0(t) > n^{-1/3},\ G(u) - F(t) \ge \eta\{F_0(u) - F_0(t)\}\big\}. \tag{3.59}$$

Then we have for all pairs of distribution functions (F_k, G_k) and (F_l, G_l) such that $F_k \le G_k$ and $F_l \le G_l$,

$$\begin{aligned}&Q_n(r_{F_k,G_k,\bar\phi_k} - r_{F_l,G_l,\bar\phi_l})^2\gamma 1_{C_{n,\eta}(F_k,G_k)\cap C_{n,\eta}(F_l,G_l)}\\ &\le C^2 d_n\big((F_k,G_k,\bar\phi_k),(F_l,G_l,\bar\phi_l)\big)^2,\end{aligned}$$

where $r_{F,G,\bar\phi}$ is defined by (3.57), and where $C > 0$ is a constant, only depending on $\eta \in (0,1)$ and the Lipschitz norm c_{Lip}, corresponding to the uniform Lipschitz class of functions $\bar\phi_F$, $F \in \mathcal{F}$.

Proof: If $G_k(u) - F_k(t) > 0$ and $G_l(u) - F_l(t) > 0$, we have the decomposition

$$\begin{aligned}&\{r_{F_k,G_k,\bar\phi_k}(t,u) - r_{F_l,G_l,\bar\phi_l}(t,u)\}\gamma\\ =\ &\left\{\frac{\bar\phi_k(G_k(u)) - \bar\phi_k(F_k(t))}{G_k(u) - F_k(t)} - \frac{\bar\phi_l(G_l(u)) - \bar\phi_l(G_l(t))}{G_l(u) - F_l(t)}\right\}\gamma\\ =\ &\frac{\bar\phi_k(G_k(u)) - \bar\phi_k(F_k(t))}{G_k(u) - F_k(t)}\{G_l(u) - F_l(t) - \{G_k(u) - (F_k(t)\}\}\frac{\gamma}{G_l(u) - F_l(t)}\\ &+\{\bar\phi_k(G_k(u)) - \bar\phi_k(F_k(t)) - \{\bar\phi_l(G_l(u)) - \bar\phi_l(F_l(t))\}\}\frac{\gamma}{G_l(u) - F_l(t)}.\end{aligned}$$

This implies (for full details, see GESKUS AND GROENEBOOM (1996C)):

$$\begin{aligned}&Q_n(r_{F_k,G_k,\bar\phi_k} - r_{F_l,G_l,\bar\phi_l})^2\gamma 1_{C_{n,\eta}(F_k,G_k)\cap C_{n,\eta}(F_l,G_l)}\\ &\le C^2 d_n\big((F_k,G_k,\bar\phi_k),(F_l,G_l,\bar\phi_l)\big)^2,\end{aligned} \tag{3.60}$$

where $C > 0$ is a constant, only depending on η and the Lipschitz norm c_{Lip}, corresponding to the uniform Lipschitz class of functions $\bar\phi_F$, $F \in \mathcal{F}$. □

For the set of functions $\mathcal{F}_n$ we then get the following theorem which finishes the proof of Theorem 3.6.

Theorem 3.7 *Let $\mathcal{F}_n$ be the set of distribution functions $F \in \mathcal{F}$, defined by (3.51). Then we have, under the conditions of Theorem 3.6, for each $\epsilon > 0$,*

$$Pr\Big\{\sup_{F\in\mathcal{F}_n}\big|\sqrt{n}(Q_n - Q_{F_0})(\tilde\theta_F - \tilde\theta_{F_0})\big| > \epsilon\Big\} \to 0,\ \text{as } n\to\infty. \tag{3.61}$$

Proof: I will (using the notation of POLLARD (1984), page 150) denote the empirical process $\sqrt{n}(Q_n - Q_{F_0})$ by E_n and the symmetrized empirical process E_n by E_n^0. Fix an (arbitrary) $\epsilon > 0$. By the symmetrization lemma we have

$$\begin{aligned}&Pr\Big\{\big|E_n(r_F - r_{F_0})\gamma\big| > \epsilon \text{ for some } F \in \mathcal{F}_n\Big\}\\ &\le 4Pr\Big\{\big|E_n^0(r_F - r_{F_0})\gamma\big| > \frac{1}{4}\epsilon \text{ for some } F \in \mathcal{F}_n\Big\}.\end{aligned}$$

Let $\epsilon > 0$ and $\eta \in (0,1)$ be fixed in the following. We are going to show that

$$Pr\Big\{|E_n^0(r_F - r_{F_0})\gamma| > \frac{1}{4}\epsilon \text{ for some } F \in \mathcal{F}_n \mid \xi_n\Big\} \to 0, \text{ as } n \to \infty, \qquad (3.62)$$

for all

$$\xi_n = \big((T_1, U_1, \delta_1, \gamma_1), \ldots, (T_n, U_n, \delta_n, \gamma_n)\big),$$

such that

$$\int_{q_{F_0} \le n^{-1/3}} dQ_n \le n^{-2/3} \log n, \qquad (3.63)$$

$$\int_{q_{F_0} > n^{-1/3}} q_{F_0}^{-2}\, dQ_n \le (\log n)^2, \qquad (3.64)$$

and

$$\sup_{F \in \mathcal{F}_n} Q_n D_\eta(F) = \sup_{F \in \mathcal{F}_n} \int_{q_F \le \eta q_{F_0}} dQ_n \le n^{-2/3} (\log n)^2, \qquad (3.65)$$

are satisfied for the empirical measure Q_n, corresponding to ξ_n. By the preceding lemmas, the probability that these conditions are *not* satisfied for the sample ξ_n tends to zero, as $n \to \infty$. In (3.63) to (3.65) we again use out method of absorbing constants into extra powers of $\log n$.

Let $\epsilon_n = \frac{1}{16}\epsilon/\sqrt{n}$ and let, for each $\delta > 0$, S_δ be a (minimal) net of triples $(F_k, G_k, \bar\phi_k)$ such that for any $F \in \mathcal{F}_n$ there exists a triple $(F_k, G_k, \bar\phi_k) \in S_\delta$ satisfying $F_k \le F \le G_k$, $F_k, G_k \in \mathcal{F}$ and

$$d_n\big((F_k, G_k, \bar\phi_k), (F, F, \bar\phi_F)\big) < \delta/C.$$

where the constant $C > 0$ is as in Lemma 3.6, and where the Lipschitz norm of $\bar\phi_k$ is bounded above by the Lipschitz norm of the class $\{\bar\phi_F : F \in \mathcal{F}\}$. Then (3.62) will hold if we can show that, for some $\epsilon_n' \le \epsilon_n$,

$$Pr\Big\{|E_n^0(r_{F_k, G_k, \bar\phi_k} - r_{F_0})\gamma| > \frac{1}{8}\epsilon \text{ for some } (F_k, G_k, \bar\phi_k) \in S_{\epsilon_n'} \mid \xi_n\Big\} \to 0, \; (3.66)$$

as $n \to \infty$, since, conditionally on ξ_n,

$$\begin{aligned}
&|E_n^0(r_F - r_{F_0})\gamma| \le \\
&\le |E_n^0(r_F - r_{F_k, G_k, \bar\phi_k})\gamma| + |E_n^0(r_{F_k, G_k, \bar\phi_k} - r_{F_0})\gamma| \\
&\le n^{1/2} C\, d_n\big((F_k, G_k, \bar\phi_k), (F, F, \bar\phi_F)\big) + |E_n^0(r_{F_k, G_k, \bar\phi_k} - r_{F_0})\gamma| \\
&\qquad\qquad + \mathcal{O}\big(n^{-1/6} (\log n)^2\big) \\
&\le |E_n^0(r_{F_k, G_k, \bar\phi_k} - r_{F_0})\gamma| + n^{1/2}\epsilon_n + o(1) \\
&\le |E_n^0(r_{F_k, G_k, \bar\phi_k} - r_{F_0})\gamma| + \frac{1}{16}\epsilon + \frac{1}{16}\epsilon = |E_n^0(r_{F_k, G_k, \bar\phi_k} - r_{F_0})\gamma| + \frac{1}{8}\epsilon,
\end{aligned}$$

for all large n.

We now construct a chain in the following way. Let

$$\delta_i = 3^{-i} n^{-1/12}, \; i = 0, 1, \ldots,$$

and let k be the smallest integer such that $3^{-k}n^{-1/12} \le \epsilon_n$. Define $\mathcal{T}_{\delta_k} = \mathcal{S}_{\delta_k}$ and let, recursively, $\mathcal{T}_{\delta_{i-1}}$ be a minimal δ_{i-1}-net for the semi-distance d_n, such that for each triple $(F_{\delta_i}, G_{\delta_i}, \bar{\phi}_{\delta_i}) \in \mathcal{T}_{\delta_i}$ there exists a triple $(F_{\delta_{i-1}}, G_{\delta_{i-1}}, \bar{\phi}_{\delta_{i-1}}) \in \mathcal{T}_{\delta_{i-1}}$ satisfying $F_{\delta_{i-1}}, G_{\delta_{i-1}} \in \mathcal{F}$, $F_{\delta_{i-1}} \le F_{\delta_i} \le G_{\delta_i} \le G_{\delta_{i-1}}$, and

$$d_n\big((F_{\delta_{i-1}}, G_{\delta_{i-1}}, \bar{\phi}_{\delta_{i-1}}), (F_{\delta_i}, G_{\delta_i}, \bar{\phi}_{\delta_i})\big) < \delta_{i-1}/C,$$

where the constant $C > 0$ is as in Lemma 3.6. The cardinality $N_n(\delta_i)$ of $\mathcal{T}_{\delta_i}$ satisfies

$$\log N_n(\delta_i) \le c\,\delta_i^{-1} \log n,\ i = 0, \dots, k,$$

for some constant $c > 0$, using (3.64) and the entropy results in Birman and Solomjak (1967). Hence,

$$\sum_{i=0}^{k-1} \delta_i \sqrt{\log N_n(\delta_{i+1})} \le c(\log n)^{1/2} n^{-1/24}, \tag{3.67}$$

for some constant $c > 0$.

Furthermore, let recursively, starting with an element in $\mathcal{S}_{\delta_k} = \mathcal{T}_{\delta_k}$, $(F_{\delta_{i-1}}, G_{\delta_{i-1}}, \bar{\phi}_{\delta_{i-1}})$ be the closest point to $(F_{\delta_i}, G_{\delta_i}, \bar{\phi}_{\delta_i})$ in $\mathcal{T}_{\delta_{i-1}}$ for the semi-distance d_n, such that $F_{\delta_{i-1}} \le F_{\delta_i} \le G_{\delta_i} \le G_{\delta_{i-1}}$. Defining

$$H(\delta_i) = \{2\log(N_n(\delta_i)^2/\delta_i)\}^{1/2} \text{ and } \eta_i = \delta_i H_n(\delta_{i+1}),$$

we get, for all large n, using the fact that, by (3.67), $\sum_{i=0}^{k-1} \eta_i = o(1)$,

$$\begin{aligned} &Pr\left\{\max_{(F_{\delta_k}, G_{\delta_k}, \bar{\phi}_{\delta_k}) \in \mathcal{T}_{\delta_k}} \left|E_n^0(r_{F_{\delta_0}, G_{\delta_0}, \bar{\phi}_{\delta_0}} - r_{F_{\delta_k}, G_{\delta_k}, \bar{\phi}_{\delta_k}})\gamma\right| > \epsilon/32 \mid \xi_n\right\} \\ &\le 2\sum_{i=0}^{k} \delta_i = \mathcal{O}(n^{-1/12}). \end{aligned}$$

We also have, if A_0 is the set of triples that can occur at the (coarse) end of a chain as constructed above, and if $(F_{\delta_0}, G_{\delta_0}, \bar{\phi}_{\delta_0}) \in A_0$,

$$\begin{aligned} \left|E_n^0(r_{F_{\delta_0}, G_{\delta_0}, \bar{\phi}_{\delta_0}} - r_{F_0})\gamma\right| \;\le\; & \left|E_n^0(r_{F_{\delta_0}, G_{\delta_0}, \bar{\phi}_{\delta_0}} - r_{F_{\delta_k}, G_{\delta_k}, \bar{\phi}_{\delta_k}})\gamma\right| \\ & + \left|E_n^0(r_{F_{\delta_k}, G_{\delta_k}, \bar{\phi}_{\delta_k}} - r_{F_0})\gamma\right|, \end{aligned}$$

for some $(F_{\delta_k}, G_{\delta_k}, \bar{\phi}_{\delta_k})$ at the beginning (fine end) of the chain. Moreover,

$$\begin{aligned} &\left|E_n^0(r_{F_{\delta_k}, G_{\delta_k}, \bar{\phi}_{\delta_k}} - r_{F_0})\gamma\right| \\ &\le \left|E_n^0(r_{F_{\delta_k}, G_{\delta_k}, \bar{\phi}_{\delta_k}} - r_F)\gamma\right| + \left|E_n^0(r_F - r_{F_0})\gamma\right| \\ &\le n^{1/2}\epsilon_n + \left|E_n^0(r_F - r_{F_0})\gamma\right| = \epsilon/32 + \left|E_n^0(r_F - r_{F_0})\gamma\right|, \end{aligned}$$

for some $F \in \mathcal{F}_n$, by the properties of $\mathcal{T}_{\delta_k}$ and the construction of the chain.

But for an $F \in \mathcal{F}_n$ we have, by Hoeffding's inequality, for all large n,

$$\begin{aligned} Pr\{|E_n^0(r_F - r_{F_0})\gamma| > \epsilon/32 \mid \xi_n\} &\le 2\exp\{-\tfrac{1}{2}(\epsilon/32)^2/Q_n(r_F - r_{F_0})^2\} \\ &\le 2\exp\{-cn^{1/12}\epsilon^2/(\log n)^2\}, \end{aligned}$$

for a constant $c > 0$, since it can be shown, using Lemma 3.6, (3.63) to (3.65), and Lemmas 5.1 and 5.2 in the appendix of GESKUS AND GROENEBOOM (1996C), that

$$\sup_{F \in \mathcal{F}_n} Q_n\big(r_F - r_{F_0}\big)^2 \gamma \le kn^{-1/6}(\log n)^2,$$

for some $k > 0$ and all large n. Hence, using the correspondence between $(F_{\delta_0}, G_{\delta_0}, \bar\phi_{\delta_0})$ and $(F_{\delta_k}, G_{\delta_k}, \bar\phi_{\delta_k})$ along the chain, but this time doing the counting at the "coarse end" of the chain, we get

$$\begin{aligned} &Pr\Big\{\sup_{(F_{\delta_0}, G_{\delta_0}, \bar\phi_{\delta_0}) \in A_0} \big|E_n^0(r_{F_{\delta_0}, G_{\delta_0}, \bar\phi_{\delta_0}} - r_{F_0})\gamma\big| > \frac{3}{32}\epsilon \mid \xi_n\Big\} \\ &\le Pr\Big\{\sup_{(F_{\delta_0}, G_{\delta_0}, \bar\phi_{\delta_0}) \in A_0} \big|E_n^0(r_{F_{\delta_0}, G_{\delta_0}, \bar\phi_{\delta_0}} - r_{F_{\delta_k}, G_{\delta_k}, \bar\phi_{\delta_k}})\gamma\big| > \epsilon/32 \mid \xi_n\Big\} \\ &\quad + Pr\Big\{\sup_{(F_{\delta_0}, G_{\delta_0}, \bar\phi_{\delta_0}) \in A_0} \big|E_n^0(r_{F_{\delta_k}, G_{\delta_k}, \bar\phi_{\delta_k}} - r_{F_0})\gamma\big| > \epsilon/16 \mid \xi_n\Big\} \\ &\le \exp\{c_1 n^{1/12}\log n - c_2\epsilon^2 n^{1/6}/(\log n)^2\} + o(1) \to 0, \text{ as } n \to \infty, \end{aligned}$$

for constants $c_1, c_2 > 0$, since the number of triples in the δ_0-net for the semi-distance d_n is $\exp\{\mathcal{O}(n^{1/12}\log n)\}$. Thus we get

$$\begin{aligned} &Pr\Big\{\max_{(F_{\delta_k}, G_{\delta_k}, \bar\phi_{\delta_k}) \in \mathcal{T}_{\delta_k}} \big|E_n^0(r_{F_{\delta_k}, G_{\delta_k}, \bar\phi_{\delta_k}} - r_{F_0})\big|\gamma > \epsilon/8 \mid \xi_n\Big\} \\ &\le Pr\Big\{\max_{(F_{\delta_k}, G_{\delta_k}, \bar\phi_{\delta_k}) \in \mathcal{T}_{\delta_k}} \big|E_n^0(r_{F_{\delta_k}, G_{\delta_k}, \bar\phi_{\delta_k}} - r_{F_{\delta_0}, G_{\delta_0}, \bar\phi_{\delta_0}})\gamma\big| > \epsilon/32 \mid \xi_n\Big\} \\ &\qquad + Pr\Big\{\sup_{(F_{\delta_0}, G_{\delta_0}, \bar\phi_{\delta_0}) \in A_0} \big|E_n^0(r_{F_{\delta_0}, G_{\delta_0}, \bar\phi_{\delta_0}} - r_{F_0})\gamma\big| > \frac{3}{32}\epsilon \mid \xi_n\Big\} \\ &\to 0,\ n \to \infty. \end{aligned}$$

This proves (3.66).

In a very similar way, it is shown that

$$Pr\Big\{\Big|E_n\Big\{\frac{\phi_{\hat F_n}(t)}{\hat F_n(t)} - \frac{\phi_{F_0}(t)}{F_0(t)}\Big\}\delta\Big| > \epsilon\Big\} \to 0, \tag{3.68}$$

and

$$Pr\Big\{\Big|E_n\Big\{\frac{\phi_{\hat F_n}(u)}{1-\hat F_n(u)} - \frac{\phi_{F_0}(u)}{1-F_0(t)}\Big\}(1-\gamma-\delta)\Big| > \epsilon\Big\} \to 0, \tag{3.69}$$

as $n \to \infty$. For example, to prove (3.68), we condition on a sample

$$\xi_n = \big((T_1, U_1, \delta_1, \gamma_1), \ldots, (T_n, U_n, \delta_n, \gamma_n)\big),$$

such that conditions (3.63) to (3.65) are satisfied for the empirical measure Q_n, corresponding to ξ_n. We then approximate pairs $(F, \bar\phi_F)$ by pairs $(F_k, \bar\phi_k)$, where $F_k \geq F$. The remaining part of the argument is the same (and in fact easier). The reason for treating the ratios r_F separately was mainly notational.

The result now follows from (3.62), (3.68) and (3.69). □

3.7 Simulations

3.7.1 Computation of $\phi_{\hat F_n}$ and $\bar\phi_{\hat F_n}$

If F is a purely discrete distribution function, we know from theorem 3.5 that ϕ_F, as given by equation 3.37, is a piecewise constant function as well. In this equation, we do not need the ψ_F-part in order to obtain the ϕ-solution. We know from part (iii) of Theorem 3.5 that the values of ϕ_F can be found from a finite set of linear equations $Ay = b$, where (as noted in the proof of part (iii) of Theorem 3.5) the matrix A is a symmetric, strictly diagonally dominant positive definite M-matrix (also called a Stieltjes matrix).

The solution of the integral equation in the transformed scale is easily obtained from this, since the integral parts are with respect to a measure that has mass restricted to the values $\tau_i = \hat F_n(x_i)$. In Figure 3.1 a picture of the NPMLE is shown and in figures 3.2 to 3.4 the solutions $\bar\phi_{\hat F_n}$, $\bar\xi_{\hat F_n}$ and $\bar\phi'_{\hat F_n}$ are exhibited, based on a random sample of size $n = 300$ from a uniform distribution on $[0, 1]$, censored by two uniformly distributed observation times (so H is the uniform distribution on the upper triangle of the unit square), where $k \equiv 1$, which is the derivative of the canonical gradient for the mean functional.

These solutions are compared with the solution $\bar\phi_{F_0}$, with $F_0 = U(0,1)$, implying $\bar\phi_{F_0} \equiv \phi_{F_0}$. This solution is obtained in Geskus (1992), based on a power series expansion using Legendre polynomials. The function $\bar\xi_{\hat F_n}$ is defined by $\bar\xi_{\hat F_n}(t) = \bar\phi_{\hat F_n}(t)/(t(1-t))$, $t \in (0,1)$, and likewise $\bar\xi_{F_0}(t) = \bar\phi_{F_0}(t)/(t(1-t))$, $t \in (0,1)$.

The number of jumps of the NPMLE was 15 and the locations of the jumps are indicated by small vertical bars (slightly smaller than the tickmarks at 0.25, etc.) on the x-axis in Figure 3.1. On the other hand, in figures 3.2 to 3.4 the small vertical bars on the x-axis denote the values of $\hat F_n$ at these points of jump. The derivative $\bar\phi'_{\hat F_n}$ is actually continuous in this case (this will generally not be the case), and has cusps at the points $\hat F_n(x_i)$. It can be shown that the cusps of the derivative $\bar\phi'_{\hat F_n}$ are located on the curve $t \mapsto \frac{1}{2}(1-2t)\bar\xi_{\hat F_n}(t)$, $t \in (0,1)$.

3.7.2 A simulation of $K(\hat F_n)$

For the same uniform case as above, we did a computer experiment of 10.000 samples of magnitude 1000, and estimated the mean $\mu(F_0)$ by the NPMLE $\mu(\hat F_{1000})$. Estimating the variance of $\sqrt{1000}(\mu(\hat F_{1000}) - \mu(F_0))$ by the unbiased

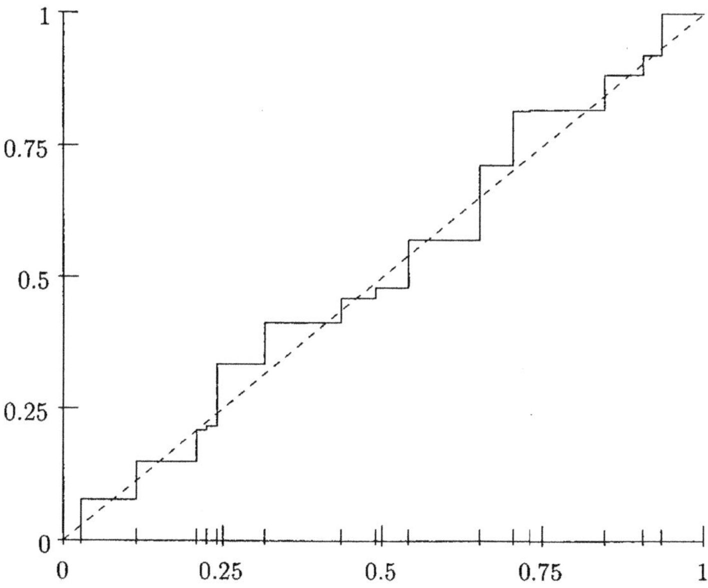

Figure 3.1: $\hat{F}_n$, based on sample size 300, and F_0 (dashed)

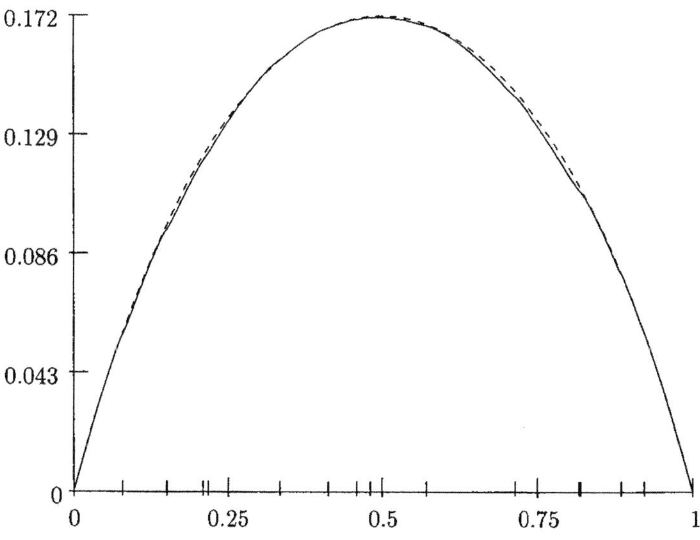

Figure 3.2: $\bar{\phi}_{\hat{F}_n}$ and $\bar{\phi}_{F_0}$(dashed)

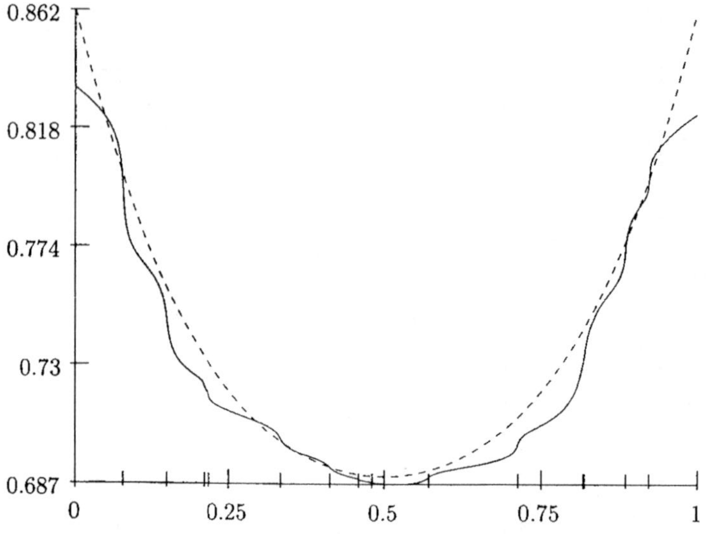

Figure 3.3: $\bar{\xi}_{\hat{F}_n}$ and $\bar{\xi}_{F_0}$(dashed)

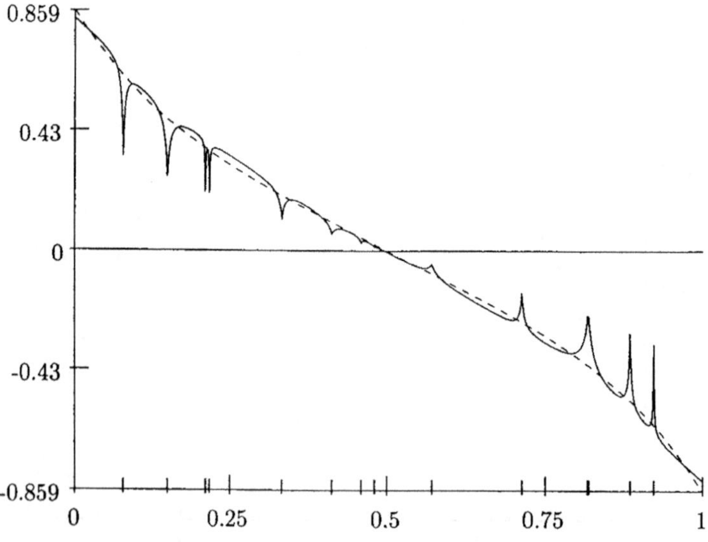

Figure 3.4: $\bar{\phi}'_{\hat{F}_n}$ and $\bar{\phi}'_{F_0}$(dashed)

estimator $S^2_{10.000}$ yielded the number 0.11917, while analytic computations as in GESKUS (1992) yield 0.1198987 for the information lower bound. So the estimate is very close to the information lower bound.

Chapter 4

Local behavior

As in the preceding chapters, I will denote the reference GROENEBOOM AND WELLNER (1992) by GW.

4.1 Minimax results

In this section I briefly discuss some local minimax results. The main virtue of these results is that they give the best possible local convergence rate and even the best constant, as far as this constant depends on the fixed distribution for which one considers a sequence of perturbations. In fact, the computation of such a lower bound is almost a routine matter now, very much in contrast with the attempts to prove that certain estimators (in particular the NPMLE) are locally optimal!

Suppose that we want to estimate a real-valued parameter θ which can be written Tq, where q is a probability density with respect to a σ-finite measure μ on a measurable space $(\Omega, \mathcal{A})$. Let (T_n), $n \geq 1$, be a sequence of estimators, based on samples of size n, i.e. T_n has a representation $T_n = t_n(X_1, \ldots, X_n)$, where $X_1, \ldots, X_n$ is a sample, generated by q, and $t_n : \Omega^n \to \mathbb{R}$ is a Borel measurable function. Furthermore, let $l : [0, \infty) \to \mathbb{R}$ be an increasing convex loss function, with $l(0) = 0$. The *risk* of the estimator T_n in estimating Tq, using the loss function l, is then defined by

$$E_{n,q} l(|T_n - Tq|),$$

where $E_{n,q}$ denotes the expectation with respect to the product measure $q^{\otimes n}$, corresponding to the sample $X_1, \ldots, X_n$.

Lemma 4.1 below can then be used to derive an asymptotic lower bound for the minimax risk.

Lemma 4.1 *Let $\mathcal{G}$ be a set of probability densities on a measurable space $(\Omega, \mathcal{A})$ with respect to a σ-finite dominating measure μ, and let T be a real-valued functional on $\mathcal{G}$. Moreover, let $l : [0, \infty) \to \mathbb{R}$ be an increasing convex loss function, with $l(0) = 0$. Then, for any $q_1, q_2 \in \mathcal{G}$ such that the Hellinger distance*

$H(q_1, q_2) < 1$:

$$\inf_{T_n} \max\{E_{n,q_1} l(|T_n - Tq_1|), E_{n,q_2} l(|T_n - Tq_2|)\}$$
$$\geq l\Big(\tfrac{1}{4}|Tq_1 - Tq_2|\{1 - H^2(q_1, q_2)\}^{2n}\Big). \tag{4.1}$$

The proof uses the following inequality, which is sometimes referred to as Le Cam's inequality.

Lemma 4.2 *Let q_1 and q_2 be two probability densities on a measurable space $(\Omega, \mathcal{A})$ with respect to a σ-finite dominating measure μ. Let $H(q_1, q_2)$ denote the Hellinger distance between q_1 and q_2. Then*

$$\{1 - H^2(q_1, q_2)\}^2 \leq 1 - \left\{1 - \int q_1 \wedge q_2 \, d\mu\right\}^2 \leq 2 \int q_1 \wedge q_2 \, d\mu,$$

where $(q_1 \wedge q_2)(x) = q_1(x) \wedge q_2(x)$.

Proof: Trivial. □

Proof of Lemma 4.1: By Jensen's inequality, we have

$$\liminf_{n} \inf_{T_n} \max\{E_{n,q_1} l(|T_n - Tq_1|), E_{n,q_2} l(|T_n - Tq_2|)\}$$
$$\geq l\Big\{\inf_{T_n} \max\{E_{n,q_1}|T_n - Tq_1|, E_{n,q_2}|T_n - Tq_2|\}\Big\}.$$

Hence, if we can prove the theorem for $l(x) = x$, we get, by the monotonicity of l,

$$\inf_{T_n} \max\{E_{n,q_1} l(|T_n - Tq_1|), E_{n,q_2} l(|T_n - Tq_2|)\}$$
$$\geq l\Big(\tfrac{1}{4}|Tq_1 - Tq_2|\{1 - H^2(q_1, q_2)\}^{2n}\Big),$$

so it suffices to prove the result for $l(x) = x$. We have, using the triangle inequality and Lemma 4.2,

$$\begin{aligned}
&\max\{E_{n,q_1}|T_n - Tq_1|, E_{n,q_2}|T_n - Tq_2|\} \\
\geq\ & \tfrac{1}{2}\{E_{n,q_1}|T_n - Tq_1| + E_{n,q_2}|T_n - Tq_2|\} \\
\geq\ & \tfrac{1}{2}\int \{|t_n(x_1, \ldots, x_n) - Tq_1| + |t_n(x_1, \ldots, x_n) - Tq_2|\} \\
& \qquad \cdot \prod_{i=1}^n q_1(x_i) \wedge \prod_{i=1}^n q_2(x_i) \, d\mu(x_1) \ldots d\mu(x_n) \\
\geq\ & \tfrac{1}{2}|Tq_1 - Tq_1| \int \prod_{i=1}^n q_1(x_i) \wedge \prod_{i=1}^n q_2(x_i) \, d\mu(x_1) \ldots d\mu(x_n) \\
\geq\ & \tfrac{1}{4}|Tq_1 - Tq_2|\{1 - H^2(q_1^{\otimes n}, q_2^{\otimes n})\}^2 \geq \tfrac{1}{4}|Tq_1 - Tq_2|\{1 - H^2(q_1, q_2)\}^{2n}.
\end{aligned}$$

This proves the result. $\square$

With the help of Lemma 4.1, it is generally relatively straightforward to derive minimax lower bounds. It was shown in GROENEBOOM (1987), essentially by using Lemma 4.1, that for interval censoring, case 1, one has a minimax bound of the form

$$\begin{aligned}&\liminf_{n\to\infty} n^{1/3}\max\{E_{n,q_0}|T_n - F_0(t_0)|, E_{n,q_n}|T_n - F_n(t_0)|\}\\ \ge\ & k\{F_0(t_0)(1-F_0(t_0))f_0(t_0)/g(t_0)\}^{1/3},\end{aligned} \tag{4.2}$$

for some $k > 0$, where F_0 is a distribution function with a derivative $f_0(t_0) > 0$ at t_0. Here $g(t_0) > 0$ is the density of the observation times at t_0, the densities q_0 and q_n are defined by

$$q_0(u,\delta) = g(u)F_0(u)^\delta(1-F_0(u))^{1-\delta},\ q_n(u,\delta) = g(u)F_n(u)^\delta(1-F_n(u))^{1-\delta},$$

with respect to the measure $\mu = \lambda_1 \otimes \lambda_2$ on $\Omega = I\!R_+ \times \{0,1\}$, where λ_1 is Lebesgue measure and λ_2 is counting measure, and F_n is a distribution function defined by

$$F_n(t) = F_0(t) - c_1 f_0(t_0)\int_0^t \{1_{[t_0 - c_2 n^{-1/3}, t_0)}(u) - 1_{[t_0, t_0 + c_2 n^{-1/3}]}(u)\}\,du \tag{4.3}$$

for suitably chosen $c_1, c_2 > 0$. The functional T is in this case given by

$$Tq = F_0(t_0) = \frac{d}{du}\int_{x:g(x)>0} 1_{[0,u]}(x)\frac{\delta}{g(x)}\,q(x,\delta)\,d\lambda_1(x)d\lambda_2(\delta)\,\Big|_{u=t_0}. \tag{4.4}$$

That T is a non-smooth functional on the set of probability measures defined on the observation space is again suggested by its representation as a derivative in (4.4). The constant in the lower bound (4.2), as far as it depends on F_0, f_0 and g, corresponds to the asymptotic variance of the NPMLE in Theorem 4.2 below.

In GROENEBOOM (1987) also another approach to the local minimimax lower bounds is given, based on techniques used in IBRAGIMOV AND HAS'MINSKII (1981), where instead of two distributions a one-dimensional family of distribution functions is considered. The result is essentially the same. This approach is detailed in GW, exercises 2.4.1 to 2.4.4, pages 20 and 21.

Still another (and possibly easiest) approach is to consider the perturbation

$$F_n(x) = \begin{cases} F_0(x) & \text{if } x < t_0 - cn^{-1/3}\\ F_0(t_0 - cn^{-1/3}) & \text{if } x \in [t_0 - cn^{-1/3}, t_0)\\ F_0(t_0 + cn^{-1/3}) & \text{if } x \in [t_0, t_0 + cn^{-1/3})\\ F_0(x) & \text{if } x \ge t_0 + cn^{-1/3},\end{cases} \tag{4.5}$$

for a suitably chosen $c > 0$. Using the perturbation (4.5), we get

$$H^2(q_n, q_0) = \tfrac{1}{2}\int\{\sqrt{q_n} - \sqrt{q_0}\}^2\,d\mu \sim \tfrac{1}{12}n^{-1}\frac{f_0(t_0)^2 g(t_0)c^3}{F_0(t_0)(1-F_0(t_0))},\ n\to\infty.$$

So one could say that the Hellinger distance of order $n^{-1/2}$ between q_n and q_0 corresponds to a distance of order $n^{-1/3}$ between Tq_n and Tq_0. Lemma 4.1 now yields:

$$\begin{aligned} & n^{1/3} \inf_{T_n} \max\{E_{n,q_0}|T_n - F_0(t_0)|, E_{n,q_n}|T_n - F_n(t_0)|\} \\ \geq\ & \tfrac{1}{4} n^{1/3} |F_n(t_0) - F_0(t_0)| \{1 - H^2(q_n, q_0)\}^{2n} \\ \rightarrow\ & \tfrac{1}{2} c f_0(t_0) \exp\left\{-\tfrac{1}{6} \frac{f_0(t_0)^2 g(t_0) c^3}{F_0(t_0)(1 - F_0(t_0))}\right\}, \ n \rightarrow \infty. \end{aligned} \tag{4.6}$$

Minimizing the last expression in (4.6) over c yields the minimax lower bound (4.2), where k does not depend on f_0, F_0 or g.

Using the same type of perturbation (4.5), but with $n^{-1/3}$ replaced by $(n \log n)^{-1/3}$, it is proved in BAKKER (1988) that

$$\begin{aligned} & \liminf_{n \rightarrow \infty} (n \log n)^{1/3} \max\{E_{n,q_0}|T_n - F_0(t_0)|, E_{n,q_n}|T_n - F_n(t_0)|\} \\ \geq\ & k\{f_0(t_0)^2 / h(t_0, t_0)\}^{1/3}, \end{aligned} \tag{4.7}$$

where it is assumed that F_0 is continuously differentiable in a neigborhood of t_0 with a strictly positive derivative $f_0(t_0)$ at t_0, and moreover that $t \mapsto h(t,t)$ is continuous at t_0, with $h(t_0, t_0) > 0$. The same rate is derived in GILL AND LEVIT (1995), using the van Trees inequality; in their result the constant, reflecting the dependence on the underlying "perturbed" fixed distribution, does not seem be optimal, though.

Under the hypotheses of (4.7), Theorem 5.3 on p. 100 of GW gives the local limit behavior of a one-step "toy estimator" $F_n^{(1)}$.

Theorem 4.1 *(Theorem 5.3, p. 100, GW) Let* $0 < F_0(t_0), H(t_0, t_0) < 1$, *and let* $F_n^{(1)}$ *be the toy estimator, obtained at the first step of the iterative convex minorant algorithm, starting the iterations with* F_0. *Then*

$$(n \log n)^{1/3} \{F_n^{(1)}(t_0) - F_0(t_0)\} / \{\tfrac{3}{4} f_0(t_0)^2 / h(t_0, t_0)\}^{1/3} \xrightarrow{\mathcal{D}} 2Z,$$

where Z *is the last time where standard Brownian motion minus the parabola* $y(t) = t^2$ *reaches its maximum.*

It is conjectured in GW (the "working hypothesis", see section 4.2) that the NPMLE has the same local limit behavior. This conjecture, however, is at present still unproved. Lucien Birgé (personal communication) constructed a kernel type estimator which achieves the right rate, but has a non-optimal asymptotic variance (if the conjecture is true), nor is this estimator optimal in estimating smooth functionals (the latter does not depend on the conjecture, but foolows from the results in chapter 3.

So, although the local minimax result has been known for 8 years now, the proof that the asymptotic variance of the NPMLE or any other (data-driven) estimator achieves the right rate and right constant is still not available. One of the reasons to believe that the NPMLE achieves the optimal rate and constant, is the fact that it is spatially adaptive, and that in all cases of interval censoring where the local asymptotic distribution could be derived, it actually achieves this "distribution-dependent" optimal constant. The question whether it also achieves the best asymptotic "universal constant" (apart from achieving the constant depending on the underlying distribution) is still completely unsolved, as is the related (simpler) question for the Grenander estimator of a monotone density. In fact, in these cases the optimal universal constant in the minimax risk is unknown.

We leave it as an exercise to the reader, again using the perturbation (4.5), that, if H has no mass in a strip along the diagonal ("strict separation"), we get a minimax lower bound of the form

$$\begin{aligned} &\liminf_{n\to\infty} n^{1/3} \max\{E_{n,q_0}|T_n - F_0(t_0)|, E_{n,q_n}|T_n - F_n(t_0)|\} \\ \geq\ & k\{a(t_0)/f_0(t_0)\}^{1/3}, \end{aligned} \tag{4.8}$$

where $a(t_0)$ is defined as in Theorem 4.3 below. This corresponds to the variance of the same type of toy estimator as in Theorem 4.1, but also, as I will show below, to the asymptotic variance of the NPMLE (which in this case can be proved to have the same local asymptotic behavior as the toy estimator).

For the deconvolution problem in a model with non-negative random variables and disturbances with a decreasing density, discussed in section 1.3, and the estimation of a convex density (Hampel's migrating birds problem), discussed in section 1.5, there are similar minimax lower bounds, see VAN ES (1991) and JONGBLOED (1995A).

Finding the right perturbations in Wicksell's problem is somewhat trickier; in GROENEBOOM AND JONGBLOED (1995) this is done by defining the perturbations in the Fourier domain and transforming back. Recent results in GOLUBEV AND LEVITT (1996) show that in this case the isotonized naive estimator of GROENEBOOM AND JONGBLOED (1995) has the optimal asymptotic variance, even up to the universal constant (if the Hölder smoothness parameter γ, defined in their manuscript, is equal to 1) . Since the simulation experiments, reported in JONGBLOED (1995A), strongly indicate that a (sieved) NPMLE has the same asymptotic behavior as the isotonized naive estimator, it is again suggested that the NPMLE has the optimal local asymptotic variance.

It is clear that there is a lot of movement in the local minimax theory at the moment, and I will not try to catch all this movement here. Instead I will proceed to the rather subtle problems, having to do with the local asymptotic distribution theory of the NPMLE.

4.2 Local asymptotic distribution theory

The local asymptotic distribution of the NPMLE with "case 1" interval censoring (current status data) was established in GROENEBOOM (1987). The result is given below.

Theorem 4.2 *Let G be the distribution function of U and let $g = G'$ be the corresponding density function (at points where it is well-defined). Let t_0 be such that $0 < F_0(t_0), G(t_0) < 1$, and let F_0 and G be differentiable at t_0 with strictly positive derivatives $f_0(t_0)$ and $g(t_0)$, respectively. Furthermore, let $\hat{F}_n$ be the NPMLE of F_0. Then we have, as $n \to \infty$,*

$$n^{1/3}\Big\{2g(t_0)/\{f_0(t_0)F_0(t_0)(1 - F_0(t_0))\}\Big\}^{1/3}\{\hat{F}_n(t_0) - F_0(t_0)\} \xrightarrow{\mathcal{D}} 2Z,$$

where Z is the last time where standard Brownian motion minus the parabola $y(t) = t^2$ reaches its maximum.

A proof different from the proof in Groeneboom (1987) is given in Groeneboom and Wellner (1992). It was meant as a "skeleton" for more general proofs of the same type, in particular for interval censoring, case 2, and deconvolution problems. The idea of the proof in GW is to introduce a "toy estimator", obtained by taking one step in the iterative convex minorant algorithm, discussed in chapter 2, starting with the real underlying distribution function F_0. The word "toy" refers to the fact that of course in practice one cannot start the iterations with the underlying distribution (nor would there be any need to start the iterations, if the underlying distribution were known!)

It was conjectured in GW that this toy estimator has the same asymptotic distribution as the NPMLE. This conjecture is called there "the working hypothesis". The motivation for studying the toy estimator is the fact that it is much easier to study the behavior of an estimator arising from doing one step of the iterative convex minorant algorithm than to study the asymptotic behavior of the NPMLE directly. Moreover, there are general reasons to believe that the asymptotic equivalence of the toy estimator and the NPMLE will hold, as will become clear from the example I will study below. The example below also shows that progress in the development of the local asymptotic distribution theory will depend on progress in the smooth functional theory. The main result of this section is Theorem 4.4, which gives the asymptotic behavior of the NPMLE for interval censoring, case 2, under the hypothesis of strict separation of the observation times.

Trusting the working hypothesis, Wellner (1995) derived the asymptotic distribution of the toy estimator for interval censoring, case 2, under the hypothesis that the joint density $h(u, v)$ of an observation pair (U_i, V_i) converges to zero as $u < v$ approaches the diagonal $u = v$. This hypothesis is in particular satisfied if the so-called "strict separation hypothesis" holds, i.e., H has zero mass on some strip along the diagonal $u = v$.

He introduces the following conditions.

(C1) The support of F_0 is an interval $[0, M]$, where $M < \infty$.

(C2) F_0 and H have densities f_0 and h w.r.t. Lebesgue measure on $\mathbb{R}$ and $\mathbb{R}^2$, respectively.

(C3) Let the functions $k_{1,\epsilon}$ and $k_{2,\epsilon}$ be defined by

$$k_{1,\epsilon}(u) = \int_u^M \frac{h(u,v)}{F_0(v) - F_0(u)} \{F_0(v) - F_0(u) < \epsilon^{-1}\}\, dv,$$

and

$$k_{2,\epsilon}(v) = \int_0^v \frac{h(u,v)}{F_0(v) - F_0(u)} \{F_0(v) - F_0(u) < \epsilon^{-1}\}\, du.$$

Then, for $i = 1, 2$ and each $\epsilon > 0$,

$$\lim_{\alpha\to\infty} \alpha \int_{(t_0, t_0+t/\alpha]} k_i(u, \epsilon\alpha)\, du = 0.$$

(C4) $0 < F_0(t_0) < 1$ and $0 < H(t_0, t_0) < 1$.

Under these conditions, he proves the following result.

Theorem 4.3 *(Wellner, 1995). Suppose that assumptions (C1) to (C4) hold. Let k_i, $i = 1, 2$, be defined by*

$$k_1(u) = \int_u^M \frac{h(u,v)}{F_0(v) - F_0(u)}\, dv, \text{ and } k_2(v) = \int_0^v \frac{h(u,v)}{F_0(v) - F_0(u)}\, du,$$

and suppose that f_0, h_1, h_2, k_1 and k_2 are continuous at t_0, where h_1 and h_2 are the first and second marginal densities of h, respectively. Moreover, assume $f_0(t_0) > 0$. Then, if $F_n^{(1)}$ is the estimator of the distribution function F_0, obtained after one step of the iterative convex minorant algorithm, starting the iterations with F_0, we have

$$n^{1/3}\{2a(t_0)/f_0(t_0)\}^{1/3}\{F_n^{(1)}(t_0) - F_0(t_0)\} \xrightarrow{\mathcal{D}} 2Z,$$

where Z is the last time where standard two-sided Brownian motion minus the parabola $y(t) = t^2$ reaches its maximum, and where

$$a(t_0) = \frac{h_1(t_0)}{F_0(t_0)} + k_1(t_0) + k_2(t_0) + \frac{h_2(t_0)}{1 - F_0(t_0)}.$$

I shall prove, under stronger conditions, that the NPMLE $\hat{F}_n$ has the same limiting distribution as the toy estimator $F_n^{(1)}$ in Wellner's theorem. Before embarking on the proof, I will sketch the general idea. Let the process W_F be defined as in chapter 2 (see (2.23)), that is,

$$\begin{aligned} W_F(t) &= \int_{u\le t} \left\{\frac{\delta_1}{F(u)} - \frac{\delta_2}{F(v) - F(u)}\right\} dQ_n \\ &\quad + \int_{v\le t} \left\{\frac{\delta_2}{F(v) - F(u)} - \frac{\delta_3}{1 - F(v)}\right\} dQ_n. \end{aligned} \tag{4.9}$$

Furthermore, let the process G_F be defined by

$$\begin{aligned} G_F(t) &= \int_{u\le t}\left\{\frac{\delta_1}{F(u)^2}+\frac{\delta_2}{(F(v)-F(u))^2}\right\}dQ_n \\ &\quad+\int_{v\le t}\left\{\frac{\delta_2}{(F(v)-F(u))^2}+\frac{\delta_3}{(1-F(v))^2}\right\}dQ_n, \end{aligned} \tag{4.10}$$

and let the process V_F be defined by

$$V_F(t)=W_F(t)+\int_{[0,t]}F(s)\,dG_F(s). \tag{4.11}$$

Finally, let J_n be the set of "relevant observation times", as defined by Definition 2.1 in chapter 2. Then we have the following proposition, proved in GROENEBOOM (1991) (the proof is reproduced in GW, Proposition 1.4, p. 49):

Proposition 4.1 *Let $T_{(1)}$ correspond to an observation time U_i such that $\Delta_{1i}=1$, and let the largest observation time $T_{(m)}$ correspond to an observation V_i such that $\Delta_{3i}=1$. Then $\hat F_n$ is the NPMLE of F_0 if and only if $\hat F_n$ is the left derivative of the convex minorant of the self-induced cumulative sum diagram consisting of the points*

$$P_j=\left(G_{\hat F_n}(T_{(j)}),V_{\hat F_n}(T_{(j)})\right),\ j=1,\dots,m,$$

and $P_0=(0,0)$.

The following trivial consequence of Proposition 4.1 will be important later, in the derivation of the asymptotic distribution.

Corollary 4.1 *Let the conditions of Proposition 4.1 be satisfied and let $t_0\in(T_{(i)},T_{(i+1)}]\subset(0,M)$, for some $i\ge 0$. Then $\hat F_n(t_0)-F_0(t_0)$ is the left derivative on the interval $(G_{\hat F_n}(T_{(i)}),G_{\hat F_n}(T_{(i+1)})]$ of the convex minorant of the self-induced cumulative sum diagram consisting of the points*

$$P_j=\left(G_{\hat F_n}(T_{(j)}),V^{(0)}_{\hat F_n}(T_{(j)})\right),\ j=1,\dots,m,$$

and $P_0=(0,0)$, where, for distribution functions F, the process $V_F^{(0)}$ is defined by

$$V_F^{(0)}(t)=W_F(t)+\int_{[0,t]}\{F(s)-F_0(t_0)\}\,dG_F(s). \tag{4.12}$$

The derivation of the local asymptotic behavior of the NPMLE is based on the idea of replacing $\hat F_n$ by F_0 in the processes $V^{(0)}_{\hat F_n}$ and $G_{\hat F_n}$ in Corollary 4.1, and showing that the left derivatives of the corresponding cumulative sum diagrams have the same local asymptotic behavior. This is in fact the motivation for studying the "toy estimator".

In showing that the left derivatives of the cumulative sum diagrams, corresponding to the points $P_j=\left(G_{\hat F_n}(T_{(j)}),V^{(0)}_{\hat F_n}(T_{(j)})\right)$ and the points $P_j^{(0)}=$

$\big(G_{F_0}(T_{(j)}), V_{F_0}^{(0)}(T_{(j)})\big)$ have the same local asymptotic behavior, the interplay between the local and global theory enters, since the remainder terms are behaving as "smooth functionals", and can therefore be shown to give a negligible contribution to the local asymptotic behavior of $\hat{F}_n$.

We now introduce conditions which are stronger than Wellner's conditions. They are:

(S1) h_1 and h_2 are continuous, with $h_1(x) + h_2(x) > 0$ for all $x \in [0, M]$.

(S2) $(u, v) \mapsto h(u, v)$ is continuous, with uniformly bounded partial derivatives, except at a finite number of points, where left and right (partial) derivatives exist.

(S3) $\text{Prob}\{V - U < \epsilon_0\} = 0$ for some ϵ_0 with $0 < \epsilon_0 \leq 1/2\,M$, so h does not have mass close to the diagonal.

(S4) F is a distribution function with support $[0, M]$ and a finite number of jumps, all in $(0, M)$; at points $x \in (0, M)$ where F does not have a jump, F has a derivative f which is continuous at x and satisfies

$$f(x) \geq c,$$

for a constant $c > 0$ independent of x.

Remark. Note that Wellner's condition (C3) is replaced by the stronger ("strict separation") hypothesis (S3).

For proving the local result for the NPMLE, I now introduce the set $\mathcal{F}^d_{[0,M]}$ of discrete non-defective distribution functions on $[0, M]$ with finitely many points of jump, contained in $(0, M)$, and the set $\mathcal{F}_\delta$, defined by

$$\mathcal{F}_\delta = \{F \in \mathcal{F}^d_{[0,M]} : \sup_{x \in [0,M]} |F(x) - F_0(x)| \leq \delta\}, \tag{4.13}$$

for a fixed $\delta > 0$ to be determined below. I assume that the underlying distribution function F_0 has a bounded derivative f_0 satisfying $f_0(x) \geq c_0 > 0$, for all $x \in (0, M)$ and some $c_0 > 0$. So it is assumed that F_0 satisfies (S4), but an additional assumption is that F_0 is continuous.

I start by proving the following lemma.

Lemma 4.3 *Let the conditions (S1) to (S4) be satisfied. Let $k : [0, M] \to \mathbb{R}$ be a bounded right-continuous function with at most a finite set of discontinuities $D = \{t_1, \ldots, t_m\} \subset (0, M)$. Furthermore, assume that k has a bounded derivative at the points $x \in [0, M] \setminus (D \cup \{0\} \cup \{M\})$ and let this derivative k' have bounded left and right limits at D, a bounded right limit at 0 and a bounded left limit at M. Then we have, defining h^* as in Definition 3.1 of Chapter 3:*

(*i*) *The equation*

$$\phi(x) = d_F(x)\left\{k(x) - \int_0^M \frac{\phi(x)-\phi(x')}{|F(x)-F(x')|}\,h^*(x',x)\,dx'\right\}, \tag{4.14}$$

has a unique right-continuous solution $\phi = \phi_F$ *with at most a finite number of jump discontinuities. The set of jump discontinuities is contained in the union of the sets of jump discontinuitues of* F *and* k.

(*ii*) *For points* $x < y$ *in an interval not containing jumps of* F *or* k, ϕ_F *satisfies*

$$|\phi_F(y) - \phi_F(x)| \le c_1(y-x), \tag{4.15}$$

where $c_1 > 0$ *is a constant, only depending on the constant* c *in (S4), and not (otherwise) on* F.

(*iii*) *At points of jump* x *of* F *and/or* k, ϕ_F *satisfies*

$$|\phi_F(x) - \phi_F(x-)| \le c_2\{F(x) - F(x-) + k(x) - k(x-)\}, \tag{4.16}$$

where $c_2 > 0$ *is a constant, only depending on the constant* c *in (S4), and not (otherwise) on* F *or* k.

Remark. Note that, if k has discontinuities not coinciding with discontinuities of F, the canonical gradient $\tilde\theta_F$, defined by

$$\tilde\theta_F(u,v,\delta_1,\delta_2) = -\delta_1\frac{\phi_F(u)}{F(u)} - \delta_2\frac{\phi_F(v)-\phi_F(u)}{F(v)-F(u)} + (1-\delta_1-\delta_2)\frac{\phi_F(v)}{1-F(v)},$$

does not have the representation $[L_1a](u,v,\delta_1,\delta_2)$, as in (3.35) in chapter 3. The function $\tilde\theta_F$ belongs in this case to the *closure* of the range of L_1, but not to the range itself! For this reason we have to extend the theory of chapter 3.

Proof of Lemma 4.3: As in chapter 3, we look at the equation (4.14) in a transformed scale. Let G be the (continuous) inverse of F defined by

$$G(t) = \inf\{x : F(x) > t\},$$

and let $\bar d_F(t)$ be defined by

$$\bar d_F(t) = \frac{t(1-t)}{\bar h_1(t)(1-t) + \bar h_2(t)t},$$

where $\bar h_i(t) = h_i(G(t))$, $i = 1,2$. Furthermore, let $\bar H(t,u) = H(G(t),G(u))$, $\bar h(t,u) = h(G(t),G(u))$, and let $\bar k_F(t) = k(G(t))$. We then consider the equation

$$\bar\phi_F(t) = \bar d_F(t)\left\{\bar k_F(t) - \int_{t'\in(0,t)} \frac{\bar\phi_F(t) - \bar\phi_F(t')}{|t-t'|}\,d\bar H^*(t',t)\right\}. \tag{4.17}$$

By hypothesis (S3) and the results in GESKUS AND GROENEBOOM (1996A), this integral equation has a unique solution $\bar{\phi}_F$, satisfying

$$\inf_{u\in(0,1)} \bar{d}_F(u)\bar{k}_F(u) \le \bar{\phi}_F(t) \le \sup_{u\in(0,1)} \bar{d}_F(u)\bar{k}_F(u), \tag{4.18}$$

for each $t \in [0,1]$. Again using hypothesis (S3), we can write this (4.17) as

$$\bar{\phi}_F(t)\Big\{1 + \bar{d}_F(t)\int_{t'\in(0,1)} \frac{1}{|t-t'|}\, d\bar{H}^*(t',t)\Big\}$$
$$= \bar{d}_F(t)\Big\{\bar{k}_F(t) + \int_{t'\in(0,1)} \frac{\bar{\phi}_F(t')}{|t-t'|}\, d\bar{H}^*(t',t)\Big\}. \tag{4.19}$$

A solution of the original equation (4.14) is obtained from this by defining

$$\phi_F(x) = \bar{\phi}_F(F(x)),\ x \in [0,M].$$

Note that the uniqueness of ϕ_F follows from the uniqueness of $\bar{\phi}_F$. Part (i) now follows.
ad (ii) Let $t \in (0,1)$ be a point of where both G and $\bar{k}_F$ are differentiable. Then we have:

$$\bar{\phi}_F'(t)\Big\{1 + \bar{d}_F(t)\int_{t'\in(0,1)} \frac{1}{|t-t'|}\, d\bar{H}^*(t',t)\, dG(t')\Big\}$$
$$= \bar{d}_F'(t)\Big\{\bar{k}_F(t) - \int_{t'\in(0,1)} \frac{\bar{\phi}_F(t)-\bar{\phi}_F(t')}{|t-t'|}\, d\bar{H}^*(t',t)\Big\}$$
$$+\bar{d}_F(t)\Big\{\bar{k}_F'(t) + \int_{t'\in(0,1)} \frac{\bar{\phi}_F(t)-\bar{\phi}_F(t')}{(t-t')^2}\, d\bar{H}^*(t',t)\Big\}$$
$$-\bar{d}_F(t)\Big\{\int_{t'\in(0,1)} \frac{\bar{\phi}_F(t)-\bar{\phi}_F(t')}{|t-t'|}\frac{\partial}{\partial t}\bar{h}^*(t',t)\, dG(t') \tag{4.20}$$

Writing

$$\bar{\phi}_F'(t) = \phi_F'(x)/f(x),$$

for points $t = F(x) \in (0,1)$ in the range of F, and x such that both F and k are continuous at x, we get from (4.20) that $\phi_F'(x)$ is uniformly bounded at such points, implying (ii), since the right-hand side of (4.20) yields a bound which is uniform for all distribution functions F satisfying (S4), noting that the denominators in the integrands stay uniformly bounded away from zero, and the numerators are uniformly bounded, using (4.18).
ad (iii) For points t not in the closure of the range of F, uniform boundedness of $\bar{\phi}_F'(t)$ again follows from (4.20). Note that G is locally constant at such points, meaning that the equation for the derivative simplifies to

$$\bar{\phi}_F'(t)\Big\{1 + \bar{d}_F(t)\int_{t'\in(0,1)} \frac{1}{|t-t'|}\, d\bar{H}^*(t',t)\Big\}$$

$$
\begin{aligned}
&= \bar{d}'_F(t)\Big\{\bar{k}_F(t) - \int_{t'\in(0,1)} \frac{\bar{\phi}_F(t)-\bar{\phi}_F(t')}{|t-t'|}\,d\bar{H}^*(t',t)\Big\}\\
&\quad +\bar{d}_F(t)\int_{t'\in(0,1)} \frac{\bar{\phi}_F(t)-\bar{\phi}_F(t')}{(t-t')^2}\,d\bar{H}^*(t',t)\Big\}\\
&= \Big\{1-2t+\bar{d}_F(t)\big(\bar{h}_1(t)-\bar{h}_2(t)\big)\Big\}\bar{\phi}_F(t)/\{t(1-t)\}\\
&\quad +\bar{d}_F(t)\int_{t'\in(0,1)} \frac{\bar{\phi}_F(t)-\bar{\phi}_F(t')}{(t-t')^2}\,d\bar{H}^*(t',t). \qquad (4.21)
\end{aligned}
$$

The function $\bar{\xi}_F$, defined by

$$
\bar{\xi}_F(t) = \bar{\phi}_F(t)/\{t(1-t)\},\ t\in(0,1),
$$

satisfies the equation

$$
\bar{\xi}_F(t) = \bar{c}_F(t)\Big\{\bar{k}_F(t) - \int_{t'\in(0,1)} \frac{\bar{\phi}_F(t)-\bar{\phi}_F(t')}{|t-t'|}\,t'(1-t')\,d\bar{H}^*(t',t)\Big\}, \qquad (4.22)
$$

where

$$
\bar{c}_F(t) = 1/\Big\{\int_{t'\in(0,t)}(1-t')\,d\bar{H}(t',t) + \int_{u\in(t,u)} u\,d\bar{H}(t,u)\Big\}. \qquad (4.23)
$$

Since $\bar{c}_F(t)$ can be written

$$
\begin{aligned}
\bar{c}_F(t) = 1/\Big\{\int_{x':F(x')<t} &(1-F(x'))h(x',G(t))\,dx'\\
&+\int_{y:F(y)>t} F(y)h(G(t),y)\,dy\Big\}, \qquad (4.24)
\end{aligned}
$$

we get from condition (S1) that $|\bar{c}_F(t)|$ is uniformly bounded by a constant only depending on H. Hence we get for $\bar{\xi}_F$ (similarly as (4.18) for $\bar{\phi}_F$):

$$
\inf_{u\in(0,1)} \bar{c}_F(u)\bar{k}_F(u) \le \bar{\xi}_F(t) \le \sup_{u\in(0,1)} \bar{c}_F(u)\bar{k}_F(u), \qquad (4.25)
$$

for all $t\in(0,1)$.

Hence we can write (4.21) as

$$
\begin{aligned}
&\bar{\phi}'_F(t)\Big\{1+\bar{d}_F(t)\int_{t'\in(0,1)} \frac{1}{|t-t'|}\,d\bar{H}^*(t',t)\Big\}\\
&= \Big\{1-2t+\bar{d}_F(t)\big(\bar{h}_1(t)-\bar{h}_2(t)\big)\Big\}\bar{\xi}_F(t)\\
&\qquad +\bar{d}_F(t)\int_{t'\in(0,1)} \frac{\bar{\phi}_F(t)-\bar{\phi}_F(t')}{(t-t')^2}\,d\bar{H}^*(t',t), \qquad (4.26)
\end{aligned}
$$

where $|\bar{\xi}_F(t)|$ is bounded by a constant only depending on H.

This implies for points of jump x of F such k is continuous at x:

$$|\phi_F(x) - \phi_F(x-)| = |\bar{\phi}_F(F(x)) - \bar{\phi}_F(F(x-))| \le c'(F(x) - F(x-)), \qquad (4.27)$$

where c' is a constant only depending on H and k. Note that

$$\begin{aligned} &\left| \int_{t' \in (0,1)} \frac{\bar{\phi}_F(t) - \bar{\phi}_F(t')}{(t-t')^2}\, d\bar{H}^*(t',t) \right| \\ &\le 2 \sup_{t \in (0,1)} \int_0^M \frac{|\bar{\phi}_F(t)|}{(t - F(x'))^2} h^*(x', G(t))\, dx' \\ &\le 2 \sup_{t \in (0,1)} |\bar{d}_F(t) \bar{k}_F(t)| \sup_{x \in (0,M)} \frac{h_1(x) + h_2(x)}{c^2 \epsilon_0^2}, \end{aligned}$$

by conditions (S3) and (S4), where $\sup_{t \in (0,1)} |\bar{d}_F(t)\bar{k}_F(t)|$ is bounded by a constant, only depending on H and k.

If k also has a jump at x, the jump of ϕ_F at x is the sum of $\bar{\phi}_F(u) - \bar{\phi}_F(t)$ and the jump of $\bar{\phi}_F$ at t, where $u = F(x)$ and $t = F(x-)$. The part $\bar{\phi}_F(u) - \bar{\phi}_F(t)$ is again bounded by $c'(F(x) - F(x-))$. For the other part, we note that, by (4.19), the jump of $\bar{\phi}_F$ at t is given by

$$\{\bar{k}_F(t) - \bar{k}_F(t-)\}\bar{d}_F(t) \Big/ \left\{1 + \bar{d}_F(t) \int_0^1 \frac{1}{|t-t'|} \bar{h}^*(t',t)\, dt' \right\}. \qquad (4.28)$$

The same relation holds if F is continuous at x and k has a jump at x. Since the right-hand side of (4.28) is the product of the jump of k at x and a factor which is uniformly bounded for the distribution functions F satisfying (S4), (iii) follows. □

For the set of distribution functions $\mathcal{F}_\delta$ we have the following Corollary.

Corollary 4.2 *Let the functions k and H satisfy the same conditions as in Lemma 4.3. Then there exists an $\delta > 0$ such that, for all $F \in \mathcal{F}_\delta$:*

(i) The equation

$$\phi(x) = d_F(x) \left\{ k(x) - \int_0^M \frac{\phi(x) - \phi(x')}{|F(x) - F(x')|} h(x', x)\, dx' \right\}, \qquad (4.29)$$

has a unique right-continuous solution $\phi = \phi_F$ with at most a finite number of jump discontinuities. The set of jump discontinuities is contained in the union of the sets of jump discontinuitues of F and k.

(ii) For points $x < y$ in an interval not containing jumps of F or k, ϕ_F satisfies

$$|\phi_F(y) - \phi_F(x)| \le c_1(y - x), \qquad (4.30)$$

where $c_1 > 0$ is a constant independent of k and $F \in \mathcal{F}_\delta$.

(iii) At points of jump x of F and/or k, ϕ_F satisfies

$$|\phi_F(x) - \phi_F(x-)| \le c_2\{F(x) - F(x-) + k(x) - k(x-)\}, \tag{4.31}$$

where $c_2 > 0$ is a constant independent of k and $F \in \mathcal{F}_\delta$.

Proof: ad (i) By choosing $\delta > 0$ sufficiently small we can make sure that

$$F(y) - F(x) > c > 0, \text{ if } y - x > \epsilon_0,$$

for some $c > 0$ and all $F \in \mathcal{F}_\delta$, where ϵ_0 is as in condition (S3). Here we use the condition $f_0 > c_0 > 0$. The lemma now follows from the corresponding properties in Lemma 4.3 by considering, as in chapter 3, solutions ϕ_{F_α} where $F_\alpha = \alpha F + (1-\alpha)F_0$ and where we let $\alpha \to 1$. The distribution functions ϕ_{F_α} satisfy the conditions of Lemma 4.3, if $\alpha \in [0,1)$. Applying the Arzelà-Ascoli theorem to the functions ϕ_{F_α}, $\alpha < 1$, restricted to intervals not containing jumps of F or k, it is seen that there exist a pointwise limit ϕ_F, defined by

$$\phi_F(x) = \lim_{\alpha \uparrow 1} \phi_{F_\alpha}(x),$$

at each continuity point x of F and k. At discontinuity points of F or k the function ϕ_F can be defined by taking limits from the right. Since k and F are continuous from the right, the integral equation is then also satisfied at such points. Uniqueness follows in the usual way by noticing that the difference $\psi = \phi_1 - \phi_1$ of two solutions ϕ_1 and ϕ_2 satisfies

$$\psi(x) = -d_F(x)\int_0^M \frac{\psi(x) - \psi(x')}{|F(x) - F(x')|}\, h^*(x', x)\, dx'. \tag{4.32}$$

Part (ii) and (iii) follow from part (ii) and (iii) of Lemma 4.3. Note that at points x where F and k are continuous, $\phi_F'(x)$ is given by:

$$\begin{aligned}
&\phi_F'(x)\left\{1 + d_F(x)\int_0^M \frac{1}{|F(x) - F(x')|}\, h^*(x', x)\, dx'\right\}\\
= \;& d_F'(x)\left\{k(x) - \int_0^M \frac{\phi_F(x) - \phi_F(x')}{|F(x) - F(x')|}\, h^*(x', x)\, dx'\right\}\\
&+ d_F(x)k'(x) - d_F(x)\int_0^M \frac{\phi_F(x) - \phi_F(x')}{|F(x) - F(x')|}\frac{\partial}{\partial x} h^*(x', x)\, dx',
\end{aligned} \tag{4.33}$$

and that the jump of ϕ_F at a point of jump x of k where F is continuous is given by

$$\{k(x) - k(x-)\} \Big/ \left\{d_F(x)^{-1} + \int_0^M \frac{h(x', x)}{|F(x) - F(x')|}\, dx'\right\}. \tag{4.34}$$

if $d_F(x) > 0$, see (4.28). If $d_F(x) = 0$, we also have $\phi_F(x) = 0$ and there is no jump in this case, even if k has a jump at x. □

The following lemma gives a (crucial) uniform $\sqrt{n}$-convergence property for integrals with respect to the indicator functions $1_{[0,t)}$. The proof makes essential use of the Fenchel duality theory, developed in chapter 2, see the inequality (4.38) below. We really need the Fenchel *inequalities* here, and not the *equalities* of the self-consistency equations.

Lemma 4.4 *Let $\hat{F}_n$ be the NPMLE, and let conditions (S1) to (S3) be satisfied. Then*

$$\sup_{t\in(0,M)} \sqrt{n}\int_0^t \{\hat{F}_n(x)-F_0(x)\}\,dx = \mathcal{O}_p(1). \tag{4.35}$$

Proof: For convenience of notation, I will denote the (relevant) ordered observation times by T_i instead of $T_{(i)}$. By Corollary 4.2 the equation

$$\phi(x) = d_F(x)\left\{1_{[0,t)}(x) - \int_0^M \frac{\phi(x)-\phi(x')}{|F(x)-F(x')|}\,h(x',x)\,dx'\right\},$$

has a right-continuous solution $\phi=\phi_{t,F}$ for each $t\in[0,M]$, if $F\in\mathcal{F}_\delta$. Defining $\tilde{\theta}_{t,F}$ by

$$\tilde{\theta}_{t,F}(u,v,\delta,\gamma) = -\delta_1\frac{\phi_{t,F}(u)}{F(u)} - \delta_2\frac{\phi_{t,F}(v)-\phi_{t,F}(u)}{F(v)-F(u)} + \delta_3\frac{\phi_{t,F}(v)}{1-F(v)},$$

we get

$$\int \tilde{\theta}_{t,F}\,dQ_{F_0} = \int_0^t \{F(x)-F_0(x)\}\,dx. \tag{4.36}$$

This is perhaps most easily seen by straightforward substitution:

$$\begin{aligned}
\int_0^t \{F(x)-F_0(x)\}\,dx &= \int_0^M 1_{[0,t)}(x)\{F(x)-F_0(x)\}\,dx\\
&= \iint_{(0,M)^2}\{F(x)-F_0(x)\}\frac{\phi_{t,F}(x)-\phi_{t,F}(x')}{|F(x)-F(x')|}\,h^*(x',x)\,dx'dx\\
&\quad+\int_0^M\{F(x)-F_0(x)\}\left\{\frac{h_1(x)}{F(x)}+\frac{h_2(x)}{1-F(x)}\right\}\phi_{t,F}(x)\,dx\\
&= \iint_{x<y}\left\{-\frac{\phi_{t,F}(x)}{F(x)}F_0(x) - \frac{\phi_{t,F}(y)-\phi_{t,F}(x)}{F(y)-F(x)}\{F_0(y)-F_0(x)\}\right.\\
&\qquad\left.+\frac{\phi_{t,F}(y)}{1-F(x)}\{1-F_0(y)\}\,h(x,y)\right\}dxdy\\
&= \int\tilde{\theta}_{t,F}\,dQ_{F_0}.
\end{aligned}$$

Since, with probability tending to 1, the NPMLE $\hat{F}_n\in\mathcal{F}_\delta$, as $n\to\infty$, we may assume $\hat{F}_n\in\mathcal{F}_\delta$. Define $\bar{\phi}_{t,\hat{F}_n}$ by

$$\bar{\phi}_{t,\hat{F}_n}(x) = \begin{cases}\phi_{t,\hat{F}_n}(T_i) &, \text{ if } x\in[T_i,T_{i+1}) \text{ and } [T_i,T_{i+1}) \text{ does not contain } t,\\ \phi_{t,\hat{F}_n}(t-) &, \text{ if } x\in[T_i,t) \text{ and } [T_i,T_{i+1}) \text{ contains } t,\\ \phi_{t,\hat{F}_n}(t) &, \text{ if } x\in[t,T_{i+1}) \text{ and } [T_i,T_{i+1}) \text{ contains } t.\end{cases} \tag{4.37}$$

Then $\bar{\phi}_{t,\hat{F}_n}$ is constant on the same intervals as $\hat{F}_n$, except possibly on the interval containing t. Note that the notation $\bar{\phi}_{t,\hat{F}_n}$ does *not* involve an inverse scale here (as in Lemma 4.3).

If $d_F(t) = 0$, $\bar{\phi}_{t,\hat{F}_n}$ has no jump at t. Otherwise the jump is given by

$$-1 \Big/ \left\{ d_F(x)^{-1} + \int_0^M \frac{h^*(x', x)}{|F(x) - F(x')|}\, dx' \right\}.$$

see (4.34). Hence $\phi_{t,\hat{F}_n}$ jumps down at t. We note here that it also follows from the structure of the integral equation that $\phi_{t,\hat{F}_n} \geq 0$.

Defining $\bar{\theta}_{t,\hat{F}_n}$ by replacing $\phi_{t,\hat{F}_n}$ by $\bar{\phi}_{t,\hat{F}_n}$ in $\tilde{\theta}_{t,\hat{F}_n}$, we get

$$\begin{aligned}
&\int \bar{\theta}_{t,\hat{F}_n}\, dQ_n \\
= &-\int \left\{ \frac{\delta_1}{\hat{F}_n(t')} - \frac{\delta_2}{\hat{F}_n(u) - \hat{F}_n(t')} \right\} \bar{\phi}_{\hat{F}_n}(t')\, dQ_n(t', u, \delta_1, \delta_2) \\
&-\int \left\{ \frac{\delta_2}{\hat{F}_n(u) - \hat{F}_n(t')} - \frac{1 - \delta_1 - \delta_2}{1 - \hat{F}_n(u)} \right\} \bar{\phi}_{\hat{F}_n}(u)\, dQ_n(t', u, \delta_1, \delta_2) \\
= &-\phi_{\hat{F}_n}(t-) \int_{t' \in [T_i, t)} dW_{\hat{F}_n}(t') - \phi_{\hat{F}_n}(t) \int_{t' \in [t, T_{i+1})} dW_{\hat{F}_n}(t') \leq 0 \quad (4.38)
\end{aligned}$$

where the process $W_{\hat{F}_n}$ is defined by (4.9). In (4.38), we use

$$-\phi_{t,\hat{F}_n}(t-) < -\phi_{t,\hat{F}_n}(t),$$

and the relation (see GW, p. 47)

$$\int \{F(t') - \hat{F}_n(t')\}\, dW_{\hat{F}_n}(t') \leq 0,$$

where F is chosen in such a way that $F(t') - \hat{F}_n(t')$ is proportional to $-\bar{\phi}_{\hat{F}_n}(t')$ on $[T_i, T_{i+1})$ and is zero on the other intervals. Using Corollary 6.1 in GESKUS AND GROENEBOOM (1996B) (or using the theory in chapter 2) it is seen that the integrals involving the remaining intervals in (4.38) are zero.

Moreover, by a proof which is entirely similar to the proof of Lemma 2.2 in GESKUS AND GROENEBOOM (1996B), it can be shown that

$$\left| \int \{\bar{\theta}_{t,\hat{F}_n} - \tilde{\theta}_{t,\hat{F}_n}\}\, dQ_{F_0} \right| \leq c\{\|\hat{F}_n - F_0\|_{H_1}^2 + \|\hat{F}_n - F_0\|_{H_2}^2\},$$

where $c > 0$ is independent of t, and where $\|\hat{F}_n - F_0\|_{H_i}^2$ is defined by

$$\|\hat{F}_n - F_0\|_{H_i}^2 = \int \{\hat{F}_n(x) - F_0(x)\}^2\, dH_i(x),\ i = 1, 2,$$

for the marginal dfs H_i of H. This implies, by Corollary 2.2 in GESKUS AND GROENEBOOM (1996B):

$$\sup_{t\in[0,M]}\left|\int\{\bar\theta_{t,\hat F_n}-\tilde\theta_{t,\hat F_n}\}\,dQ_{F_0}\right|=O_p(n^{-2/3}),\tag{4.39}$$

Let the function ψ_n be defined by

$$\psi_n(t)=\int_0^t\{\hat F_n(y)-F_0(y)\}\,dy,\ t\in[0,M].$$

If $t\mapsto\hat F_n(T_i)-F_0(t)$ is of constant sign on (T_i,T_{i+1}) we must have

$$\sup_{t\in(T_i,T_{i+1})}|\psi_n(t)|\le\max\{|\psi_n(T_i)|,|\psi_n(T_{i+1})|\},\tag{4.40}$$

since the function ψ_n is then either increasing or decreasing on $[T_i,T_{i+1})$.

If, on the other hand, $\hat F_n(T_i)=F_0(s)$ for an $s\in(T_i,T_{i+1})$, the function ψ_n first increases on the interval $[T_i,s]$ and then decreases on the interval $[s,T_{i+1})$. Hence we get in that case, if $t\in(T_i,T_{i+1})$:

$$\begin{aligned}&\min\{\psi_n(T_i),\psi_n(T_{i+1})\}\\&\le\psi_n(t)=\int\tilde\theta_{t,\hat F_n}\,dQ_{F_0}=\int\bar\theta_{t,\hat F_n}\,dQ_{F_0}+O_p(n^{-2/3})\\&=\int\bar\theta_{t,\hat F_n}\,d(Q_{F_0}-Q_n)+\int\bar\theta_{t,\hat F_n}\,dQ_n+O_p(n^{-2/3})\\&\le\int\bar\theta_{t,\hat F_n}\,d(Q_{F_0}-Q_n)+O_p(n^{-2/3}),\end{aligned}\tag{4.41}$$

where we use (4.39), and where we use (4.38) in the last step.

Noting that $\psi_n(T_i)$ can be written

$$\begin{aligned}\psi_n(T_i)&=\int\tilde\theta_{T_i,\hat F_n}\,dQ_{F_0}=\int\bar\theta_{T_i,\hat F_n}\,dQ_{F_0}+O_p(n^{-2/3})\\&=\int\bar\theta_{T_i,\hat F_n}\,d(Q_{F_0}-Q_n)+O_p(n^{-2/3}),\end{aligned}$$

where the remainder term $O_p(n^{-2/3})$ is uniform in T_i, we find that the statement of the lemma holds if

$$\left|\sup_{t\in[0,M]}\int\bar\theta_{t,\hat F_n}\,d(Q_{F_0}-Q_n)\right|=O_p(n^{-1/2}).\tag{4.42}$$

But (4.42) easily follows from the fact that the random entropy $H(\epsilon,\mathcal K,Q_n)$ of the set of functions

$$\mathcal K=\{\bar\theta_{t,F}-\tilde\theta_{t,F_0}:t\in[0,M],\ F\in\mathcal F_\delta\}$$

satisfies

$$\int_0^\epsilon H(u,\mathcal{K},Q_n)^{1/2}\,du = \mathcal{O}_p(\epsilon^{1/2}),\ \epsilon > 0,$$

see (the proof of) Lemma 2.3 in GESKUS AND GROENEBOOM (1996B). □

We now get the following corollary.

Corollary 4.3 *Let the conditions (S1) to (S3) be satisfied and let $\mathcal{G}$ be a set of right-continuous functions $g : [0, M] \to \mathbb{R}$ which are of uniformly bounded variation and satisfy the same conditions as the function k in Lemma 4.3. Then we have*

$$\sup_{g\in\mathcal{G}} \left| \int_0^M g(x)\{\hat{F}_n(x) - F_0(x)\}\,dx \right| = \mathcal{O}_p(n^{-1/2}).$$

Proof: By Corollary 4.2 the equation

$$\phi(x) = d_F(x)\left\{ g(x) - \int_0^x \frac{\phi(x) - \phi(x')}{|F(x) - F(x')|}\, h^*(x',x)\,dx' \right\}, \tag{4.43}$$

has a right-continuous solution $\phi = \phi_{g,F}$ for each $g \in \mathcal{G}$, if $F \in \mathcal{F}_\delta$, and $\delta > 0$ is sufficiently small. Defining $\tilde{\theta}_{g,F}$ by

$$\tilde{\theta}_{g,F}(u,v,\delta,\gamma) = -\delta_1 \frac{\phi_{g,F}(u)}{F(u)} - \delta_2 \frac{\phi_{g,F}(v) - \phi_{g,F}(u)}{F(v) - F(u)} + \delta_3 \frac{\phi_{g,F}(v)}{1 - F(v)}, \tag{4.44}$$

we get

$$\int \tilde{\theta}_{g,F}\, dQ_{F_0} = \int_0^M g(x)\{F(x) - F_0(x)\}\,dx. \tag{4.45}$$

This is seen by the substitution argument leading to (4.36) in the proof of Lemma 4.4.

If g has no points of jump, or if the points of jump of g coincide with those of $\hat{F}_n$, we can write

$$\int \tilde{\theta}_{\hat{F}_n}\, dQ_{F_0} = \int \tilde{\theta}_{\hat{F}_n}\, d(Q_{F_0} - Q_n) \tag{4.46}$$

If g has a jump at a point x where $\hat{F}_n$ is continuous, we can remove the singularity by adding to g the function $\{g(x) - g(x-)\}1_{[0,x)}$. Hence, by adding a finite number of functions of this type, we can change g into a function $\tilde{g}$, which only has jumps at locations of jumps of $\hat{F}_n$. So we can write $g = g_{1,n} + g_{2,n}$, where $g_{2,n}$ is a finite linear combination of indicator functions, and where $g_{1,n}$ only can have jumps at the points of jump of $\hat{F}_n$. The sum of the absolute values of the weights in the linear combination forming $g_{2,n}$ has a fixed upper bound, since these weights consist of sizes of jumps $g(x) - g(x-)$, for $g \in \mathcal{G}$.

It follows from Lemma 4.4 that

$$\sup_{g\in\mathcal{G}}\left|\int_0^M g_{2,n}(x)\{\hat{F}_n(x)-F_0(x)\}\,dx\right| = \mathcal{O}_p(n^{-1/2}).$$

For $g_{1,n}$ we get, as in the proof of Lemma 4.4,

$$\int \tilde{\theta}_{g_{1,n},\hat{F}_n}\,dQ_{F_0} = \int \bar{\theta}_{g_{1,n},\hat{F}_n}\,dQ_{F_0} + \mathcal{O}_p(n^{-2/3}).$$

Since

$$\int \bar{\theta}_{g_{1,n},\hat{F}_n}\,dQ_{F_0} = \int \bar{\theta}_{g_{1,n},\hat{F}_n}\,d(Q_{F_0}-Q_n),$$

and the random entropy $H(\epsilon,\mathcal{K},Q_n)$ of the set of functions

$$\mathcal{K} = \{\bar{\theta}_{g,F} - \bar{\theta}_{g,F_0} : g\in\mathcal{G},\, F\in\mathcal{F}_\delta\}$$

satisfies

$$\int_0^\epsilon H(u,\mathcal{K},Q_n)^{1/2}\,du = \mathcal{O}_p(\epsilon^{1/2}),\ \epsilon>0,$$

(essentially because of the uniformly bounded variation of the functions $\phi_{g,F}$ and F), the result follows. □

Remark. It is possible to strengthen the preceding result by proving that the process

$$g \mapsto \sqrt{n}\int_0^M g(x)\{\hat{F}_n(x)-F_0(x)\}\,dx,\ g\in\mathcal{G},$$

converges weakly to a Gaussian process with covariance function

$$C(g_1,g_2) = \int \tilde{\theta}_{g_1,F_0}\tilde{\theta}_{g_2,F_0}\,dQ_{F_0} - \int \tilde{\theta}_{g_1,F_0}\,dQ_{F_0}\int \tilde{\theta}_{g_2,F_0}\,dQ_{F_0},$$

but for reasons of space we omit the proof of this.

Another corollary to the preceding results is a bound on the supremum distance between $\hat{F}_n$ and F_0.

Corollary 4.4 *Let the conditions (S1) to (S3) be satisfied. Then*

$$\sup_{x\in[0,M]}|\hat{F}_n(x)-F_0(x)| = \mathcal{O}_p(n^{-1/4}). \tag{4.47}$$

Proof: By the monotonicity of $\hat{F}_n$, the boundedness of the density f_0, and Lemma 4.4 we have, if $x + n^{-1/4} < M$:

$$n^{1/4}\{\hat{F}_n(x) - F_0(x)\} \le n^{1/2} \int_x^{x+n^{-1/4}} \hat{F}_n(t)\,dt - n^{-1/4}F_0(x)$$

$$\le n^{1/2}\left|\int_x^{x+n^{-1/4}} \{\hat{F}_n(t) - F_0(t)\}\,dt\right| + n^{1/2}\int_x^{x+n^{-1/4}} \{F_0(t) - F_0(x)\}\,dt$$

$$\le k_1, \tag{4.48}$$

for some constant $k_1 > 0$, independent of x. Likewise, if $x > n^{-1/4}$,

$$n^{1/4}\{\hat{F}_n(x) - F_0(x)\} \ge n^{1/2} \int_{x-n^{-1/4}}^{x} \hat{F}_n(t)\,dt - n^{-1/4}F_0(x)$$

$$\ge -n^{1/2}\left|\int_{x-n^{-1/4}}^{x} \{\hat{F}_n(t) - F_0(t)\}\,dt\right| - n^{1/2}\int_{x-n^{-1/4}}^{x} \{F_0(x) - F_0(t)\}\,dt$$

$$\ge -k_2,$$

for some constant $k_2 > 0$, independent of x.

For $x \in [0, n^{-1/4}]$ we get from (4.48):

$$n^{1/4}\{\hat{F}_n(x) - F_0(x)\} \le k_1,$$

but since we also have:

$$n^{1/4}\{\hat{F}_n(x) - F_0(x)\} \ge -n^{1/4}F_0(n^{-1/4}) \ge -c,$$

for some $c > 0$, the result also holds if $x \in [0, n^{-1/4}]$. A similar argument is used if $x \in [M - n^{-1/4}, M]$. □

To be able to apply the argument on page 96 of GW, we also need the following lemma.

Lemma 4.5 *Let conditions (S1) to (S3) be satisfied and let J_n be a non-degenerate (open, closed or half-closed) interval, with left an right endpoints $\tau_{n,1}$ and $\tau_{n,2}$, respectively, satifying $\tau_{n,i} - t_0 = O_p(n^{-1/4})$, $i = 1, 2$, where $t_0 \in (0, M)$. Then*

$$\int_{u\in J_n}\left\{\frac{\delta_1}{\hat{F}_n(u)} - \frac{\delta_2}{\hat{F}_n(v) - \hat{F}_n(u)}\right\} dQ_n$$

$$+ \int_{v\in J_n}\left\{\frac{\delta_2}{\hat{F}_n(v) - \hat{F}_n(u)} - \frac{\delta_3}{1 - \hat{F}_n(v)}\right\} dQ_n$$

$$= \int_{u\in J_n}\left\{\frac{\delta_1}{\hat{F}_n(u)} - \frac{\delta_2}{F_0(v) - \hat{F}_n(u)}\right\} dQ_n$$

$$+ \int_{v\in J_n}\left\{\frac{\delta_2}{\hat{F}_n(v) - F_0(u)} - \frac{\delta_3}{1 - \hat{F}_n(v)}\right\} dQ_n + o_p(n^{-2/3}), \tag{4.49}$$

Note that in Lemma 4.5 the "off-diagonal" values of $\hat F_n$ are replaced by the corresponding values of F_0, and that we have to show that the remainder term is $o_p(n^{-2/3})$.

Proof of Lemma 4.5: Consider

$$\int_{u\in J_n}\left\{\frac{\delta_2}{\hat F_n(v)-\hat F_n(u)}-\frac{\delta_2}{F_0(v)-\hat F_n(u)}\right\}dQ_n$$
$$=\int_{u\in J_n}\frac{\delta_2(F_0(v)-\hat F_n(v))}{\{\hat F_n(v)-\hat F_n(u)\}\{F_0(v)-\hat F_n(u)\}}\,dQ_n.$$

We have, by Corollary 4.4, using the strict separation hypothesis,

$$\int_{u\in J_n}\left\{\frac{\delta_2(F_0(v)-\hat F_n(v))}{\{\hat F_n(v)-\hat F_n(u)\}\{F_0(v)-\hat F_n(u)\}}-\frac{\delta_2(F_0(v)-\hat F_n(v))}{\{F_0(v)-F_0(u)\}^2}\right\}dQ_n$$
$$=\mathcal{O}_p(n^{-1/2})\int 1_{J_n}(u)\,dH_n(u,v)=\mathcal{O}_p(n^{-3/4})=o_p(n^{-2/3}).$$

Now write

$$\int_{u\in J_n}\frac{\delta_2(F_0(v)-\hat F_n(v))}{\{F_0(v)-F_0(u)\}^2}\,dQ_n$$
$$=\int_{u\in J_n}\frac{F_0(v)-\hat F_n(v)}{F_0(v)-F_0(u)}\,dH(u,v)$$
$$+\int_{u\in J_n}\frac{F_0(v)-\hat F_n(v)}{\{F_0(v)-F_0(u)\}^2}\,\delta_2\,d(Q_n-Q_{F_0}). \tag{4.50}$$

To the first term on the right of (4.50) we can apply smooth functional theory! Integrating u over J_n, and using the facts that $\sup_{x\in[0,M]}|\hat F_n(x)-F_0(x)|=\mathcal{O}_p(n^{-1/4})$ and the length of J_n is $\mathcal{O}_p(n^{-1/4})$, we get

$$\int_{u\in J_n}\frac{F_0(v)-\hat F_n(v)}{F_0(v)-F_0(u)}\,dH(u,v)$$
$$=\mathcal{O}_p(n^{-1/4})\int g(v)\{F_0(v)-\hat F_n(v)\}\,dv+\mathcal{O}_p(n^{-3/4}),$$

where $g(v)=h(t_0,v)/\{F_0(v)-F_0(t_0)\}$. Since g clearly belongs to the class of functions $\mathcal{G}$ of Corollary 4.3, we get from this corollary:

$$\int_0^M g(v)\{F_0(v)-\hat F_n(v)\}\,du=\mathcal{O}_p(n^{-1/2}).$$

Hence

$$\int_{u\in J_n}\frac{F_0(v)-\hat F_n(v)}{F_0(v)-F_0(u)}\,dH(u,v)$$

$$= \mathcal{O}_p(n^{-1/4}) \int_0^M g(v)\{F_0(v) - \hat{F}_n(v)\}\,dv + o_p(n^{-2/3})$$
$$= \mathcal{O}_p(n^{-3/4}) + o_p(n^{-2/3}) = o_p(n^{-2/3}).$$

For the second term on the right of (4.50), we have to consider the process

$$F \mapsto Z_n(F) = \int_{u \in J_n} \frac{F_0(v) - F(v)}{\{F_0(v) - F_0(u)\}^2}\,\delta_2\,d(\mathbb{Q}_n - Q_{F_0}),$$

where $F \in \mathcal{F}_\delta$, and we have to show: $Z_n(\hat{F}_n) = o_p(n^{-2/3})$. We have

$$\sup_{F \in \mathcal{F}_{Kn^{-1/4}}} |Z_n(F)| = \mathcal{O}_p(n^{-11/16}), \tag{4.51}$$

for each $K > 0$. This is seen as follows. The function

$$E_n : (u, v, \delta_1, \delta_2) \mapsto 1_{[t_0 - Kn^{-1/4}, t_0 + Kn^{-1/4}]}(u) K n^{-1/4},$$

is an envelope for the class $\mathcal{A}_n$ of functions $a_{I,F}$, defined by

$$a_{I,F}(u, v, \delta_1, \delta_2) = 1_I(u)(F(v) - F_0(v)),$$
$$I \subset [t_0 - Kn^{-1/4}, t_0 + Kn^{-1/4}],\ F \in \mathcal{F}_{Kn^{-1/4}}.$$

For these functions we define a pseudo-distance

$$d(a_{I_1,F}, a_{I_2,G}) = \left\{\int \{a_{I_1,F} - a_{I_2,G}\}^2 \{F_0(v) - F_0(u)\}^{-4}\,\delta_2\,dQ_{F_0}\right\}^{1/2}.$$

Let $N_n(\epsilon)$ be the minimum number of ϵ-brackets, needed to cover $\mathcal{A}_n$ for the pseudo-distance d, and let, for $\delta > 0$, $J_{n,\|E_n\|}(\delta)$ denote the bracketing entropy integral

$$\int_0^\delta \sqrt{1 + \log N_n(\epsilon \|E_n\|)}\,d\epsilon,$$

where $\|E_n\|$ is defined by

$$\|E_n\| = Kn^{-1/4}\left\{\int 1_{[t_0 - Kn^{-1/4}, t_0 + Kn^{-1/4}]}(u)\{F_0(v) - F_0(u)\}^{-3}\,dQ_{F_0}\right\}^{1/2}.$$

Then we get, by (a slightly generalized version of) the bracketing maximal inequality 2.14.2 on p. 240 of VAN DER VAART AND WELLNER(1996) and Markov's inequality,

$$\sup_{F \in \mathcal{F}_{Kn^{-1/4}}} n^{1/2}|Z_n(F)| = \mathcal{O}_p(J_{n,\|E_n\|}(1)\|E_n\|) = \mathcal{O}_p(\|E_n\|^{1/2}),$$

noting that, for $a_{I_1,F}, a_{I_2,G} \in \mathcal{A}_n$, we have by the triangle inequality:

$$
\begin{aligned}
& d(a_{I_1,F}, a_{I_2,G}) \\
& \le \left\{ \int 1_{I_1}(u)\{F(v) - G(v)\}^2\{F_0(v) - F_0(u)\}^{-3}\, dQ_{F_0} \right\}^{1/2} \\
& \quad + \left\{ \int \left\{1_{I_1}(u) - 1_{I_2}(u)\right\}^2 \{G(v) - F_0(v)\}^2\{F_0(v) - F_0(u)\}^{-3}\, dQ_{F_0} \right\}^{1/2} \\
& \le \left\{ \int \{F(v) - G(v)\}^2\{F_0(v) - F_0(u)\}^{-3}\, dQ_{F_0} \right\}^{1/2} \\
& \quad + Kn^{-1/4} \left\{ \int \left\{1_{I_1}(u) - 1_{I_2}(u)\right\}^2 \{F_0(v) - F_0(u)\}^{-3}\, dQ_{F_0} \right\}^{1/2}
\end{aligned}
$$

and hence

$$
J_{n,\|E_n\|}(1)\|E_n\| = \|E_n\| \int_0^1 \sqrt{1 + \log N_n(\epsilon\|E_n\|)}\, d\epsilon = \mathcal{O}\big(\|E_n\|^{1/2}\big) = \mathcal{O}\big(n^{-3/16}\big).
$$

The inequality (4.51) follows.

Since, by Corollary 4.4,

$$
\liminf_{n\to\infty} Pr\{\hat{F}_n \in \mathcal{F}_{Kn^{-1/4}}\} > 1 - \epsilon,
$$

for suitably chosen $K = K(\epsilon)$, we get

$$
Z_n(\hat{F}_n) = \mathcal{O}_p(n^{-11/16}) = o_p(n^{-2/3}).
$$

So we have proved

$$
\int_{u\in J_n} \left\{ \frac{\delta_2}{\hat{F}_n(v) - \hat{F}_n(u)} - \frac{\delta_2}{F_0(v) - \hat{F}_n(u)} \right\} dQ_n = o_p(n^{-2/3}).
$$

An entirely similar argument shows

$$
\int_{v\in J_n} \left\{ \frac{\delta_2}{\hat{F}_n(v) - \hat{F}_n(u)} - \frac{\delta_2}{F_0(v) - \hat{F}_n(u)} \right\} dQ_n = o_p(n^{-2/3}).
$$

The result of the lemma now follows. □

The following lemma shows that the distance between $\hat{F}_n(t)$ and $F_0(t_0)$ is of order $\mathcal{O}_p(n^{-1/3})$, for t in an interval $[t_0 - cn^{-1/3}, t_0 + cn^{-1/3}]$, where $t_0 \in (0, M)$. It is similar to Lemma 5.4 in GW.

Lemma 4.6 *Let the conditions (S1) to (S3) be satisfied and let F_0 be continuous with a bounded derivative f_0 on $(0,M)$, satisfying*

$$f_0(x) \geq c_0 > 0,\ x \in (0,M),$$

for some constant $c_0 > 0$. Then we have at each point $t_0 \in (0,M)$ and for each $c > 0$:

$$\sup_{t\in[-c,c]} \left|\hat{F}_n(t_0 + n^{-1/3}t) - F_0(t_0)\right| = \mathcal{O}_p(n^{-1/3}).$$

Proof: We follow the argument on page 96 of GW. We first show that for each $\epsilon > 0$ there exists an $K = K(\epsilon)$ such that

$$Pr\{\hat{F}_n(t_0 - Kn^{-1/3}) \geq F_0(t_0)\} < \epsilon, \tag{4.52}$$

for all large n.

Let τ_n be the last jump time of $\hat{F}_n$ before $t_0 - Kn^{-1/3}$, i.e.,

$$\tau_n = \max\{t \in [0,M] : t \leq t_0 - Kn^{-1/3},\ \hat{F}_n(t-) \neq \hat{F}_n(t)\}.$$

Since, by Corollary 4.4,

$$\sup_{x\in[0,M]} |\hat{F}_n(x) - F_0(x)| = \mathcal{O}_p(n^{-1/4}),$$

we also have: $t_0 - \tau_n = \mathcal{O}_p(n^{-1/4})$.

By Proposition 2.1 in chapter 2 we get:

$$\int_{u\in[\tau_n,t)} \left\{\frac{\delta_1}{\hat{F}_n(u)} - \frac{\delta_2}{\hat{F}_n(v) - \hat{F}_n(u)}\right\} dQ_n + \int_{v\in[\tau_n,t)} \left\{\frac{\delta_2}{\hat{F}_n(v) - \hat{F}_n(u)} - \frac{\delta_3}{1 - \hat{F}_n(v)}\right\} dQ_n \geq 0. \tag{4.53}$$

for each $t > \tau_n$. Hence, if $\hat{F}_n(t_0 - Kn^{-1/3}) \geq F_0(t_0)$ we have for each $t > \tau_n$:

$$\begin{aligned}
&\int_{u\in[\tau_n,t)} \left\{\frac{\delta_1}{F_0(t_0)} - \frac{\delta_2}{\hat{F}_n(v) - F_0(t_0)}\right\} dQ_n \\
&\qquad + \int_{v\in[\tau_n,t)} \left\{\frac{\delta_2}{F_0(t_0) - \hat{F}_n(u)} - \frac{\delta_3}{1 - F_0(t_0)}\right\} dQ_n \\
&\geq \int_{u\in[\tau_n,t)} \left\{\frac{\delta_1}{\hat{F}_n(u)} - \frac{\delta_2}{\hat{F}_n(v) - \hat{F}_n(u)}\right\} dQ_n \\
&\qquad + \int_{v\in[\tau_n,t)} \left\{\frac{\delta_2}{\hat{F}_n(v) - \hat{F}_n(u)} - \frac{\delta_3}{1 - \hat{F}_n(v)}\right\} dQ_n \geq 0.
\end{aligned} \tag{4.54}$$

But, by Lemma 4.5,

$$\int_{u\in[\tau_n,t_0-\frac{1}{2}Kn^{-1/3})} \left\{\frac{\delta_1}{F_0(t_0)} - \frac{\delta_2}{\hat{F}_n(v) - F_0(t_0)}\right\} dQ_n$$

$$
+\int_{v\in[\tau_n,t_0-\frac12 Kn^{-1/3})}\left\{\frac{\delta_2}{F_0(t_0)-\hat F_n(u)}-\frac{\delta_3}{1-F_0(t_0)}\right\}dQ_n
$$

$$
=\int_{u\in[\tau_n,t_0-\frac12 Kn^{-1/3})}\left\{\frac{\delta_1}{F_0(t_0)}-\frac{\delta_2}{F_0(v)-F_0(t_0)}\right\}dQ_n
$$

$$
+\int_{v\in[\tau_n,t_0-\frac12 Kn^{-1/3})}\left\{\frac{\delta_2}{F_0(t_0)-F_0(u)}-\frac{\delta_3}{1-F_0(t_0)}\right\}dQ_n+o_p(n^{-2/3})
$$

$$
=\int_{u\in[\tau_n,t_0-\frac12 Kn^{-1/3})}\left\{\frac{F_0(u)}{F_0(t_0)}-\frac{F_0(v)-F_0(u)}{F_0(v)-F_0(t_0)}\right\}dH(u,v)
$$

$$
+\int_{v\in[\tau_n,t_0-\frac12 Kn^{-1/3})}\left\{\frac{F_0(v)-F_0(u)}{F_0(t_0)-F_0(u)}-\frac{1-F_0(v)}{1-F_0(t_0)}\right\}dH(u,v)
$$

$$
+\int_{u\in[\tau_n,t_0-\frac12 Kn^{-1/3})}\left\{\frac{\delta_1}{F_0(t_0)}-\frac{\delta_2}{F_0(v)-F_0(t_0)}\right\}d(Q_n-Q_{F_0})
$$

$$
+\int_{v\in[\tau_n,t_0-\frac12 Kn^{-1/3})}\left\{\frac{\delta_2}{F_0(t_0)-F_0(u)}-\frac{\delta_3}{1-F_0(t_0)}\right\}d(Q_n-Q_{F_0})
$$

$$
+o_p(n^{-2/3}). \tag{4.55}
$$

By the same arguments as used in Lemma 4.1 in KIM AND POLLARD (1990) (see also p. 94 of GW for an application of this argument) we get

$$
\int_{u\in[\tau_n,t_0-\frac12 Kn^{-1/3})}\left\{\frac{\delta_1}{F_0(t_0)}-\frac{\delta_2}{F_0(v)-F_0(t_0)}\right\}d(Q_n-Q_{F_0})
$$

$$
+\int_{v\in[\tau_n,t_0-\frac12 Kn^{-1/3})}\left\{\frac{\delta_2}{F_0(t_0)-F_0(u)}-\frac{\delta_3}{1-F_0(t_0)}\right\}d(Q_n-Q_{F_0})
$$

$$
= o_p((t_0-\tfrac12 Kn^{-1/3}-\tau_n)^2)+n^{-2/3}A_n, \tag{4.56}
$$

where $A_n=O_p(1)$. On the other hand we have, using the conditions on F_0 and the fact that the length of the interval $[\tau_n,t_0-\frac12 Kn^{-1/3})$ is (at most) $O_p(n^{-1/4})$:

$$
\int_{u\in[\tau_n,t_0-\frac12 Kn^{-1/3})}\left\{\frac{F_0(u)}{F_0(t_0)}-\frac{F_0(v)-F_0(u)}{F_0(v)-F_0(t_0)}\right\}dH(u,v)
$$

$$
+\int_{v\in[\tau_n,t_0-\frac12 Kn^{-1/3})}\left\{\frac{F_0(v)-F_0(u)}{F_0(t_0)-F_0(u)}-\frac{1-F_0(v)}{1-F_0(t_0)}\right\}dH(u,v)
$$

$$
\le -c_1(t_0-\tfrac12 Kn^{-1/3}-\tau_n)^2\{1+o_p(1)\}, \tag{4.57}
$$

for a constant $c_1>0$. It now follows from (4.55), (4.56) and (4.57), that, for all large n,

$$
\int_{u\in[\tau_n,t_0-\frac12 Kn^{-1/3})}\left\{\frac{\delta_1}{F_0(t_0)}-\frac{\delta_2}{\hat F_n(v)-F_0(t_0)}\right\}dQ_n
$$

$$
+\int_{v\in[\tau_n,t_0-\frac12 Kn^{-1/3})}\left\{\frac{\delta_2}{F_0(t_0)-\hat F_n(u)}-\frac{\delta_3}{1-F_0(t_0)}\right\}dQ_n
$$

$$< 0, \tag{4.58}$$

with probability bigger than $1-\epsilon$ for suitably chosen $K = K(\epsilon)$. Note here that, by definition, $\tau_n \le t_0 - Kn^{-1/3}$. Since, by (4.54), the event (4.58) is contained in the event

$$\hat{F}_n(t_0 - Kn^{-1/3}) < F_0(t_0),$$

(the logic law of contraposition!) (4.52) follows.

In a similar way one can show that for each $\epsilon > 0$ there exists an $K = K(\epsilon)$ such that

$$Pr\{\hat{F}_n(t_0 + Kn^{-1/3}) \le F_0(t_0)\} < \epsilon, \tag{4.59}$$

for all large n. Hence for each $\epsilon > 0$ there exists an $K = K(\epsilon)$ such that

$$Pr\{\hat{F}_n(t_0 - Kn^{-1/3}) < F_0(t_0) < \hat{F}_n(t_0 + Kn^{-1/3})\} > 1 - 2\epsilon,$$

for all large n, implying that $\hat{F}_n$ has a jump in the interval $[t_0 - Kn^{-1/3}, t_0 + Kn^{-1/3}]$ with probability bigger than $> 1 - 2\epsilon$.

The remaining part of the proof is similar and proceeds along the lines of the proof of Lemma 5.4 in GW. □

We can now prove the main result.

Theorem 4.4 *Let the conditions (S1) to (S3) be satisfied and let F_0 be continuous with a bounded derivative f_0 on $[0, M]$, satisfying*

$$f_0(x) \ge c_0 > 0,\ x \in (0, M),$$

for some constant $c_0 > 0$. Then we have at each point $t_0 \in (0, M)$:

$$n^{1/3}\{2a(t_0)/f_0(t_0)\}^{1/3}\{\hat{F}_n(t_0) - F_0(t_0)\} \xrightarrow{\mathcal{D}} 2Z,$$

where a and Z are defined as in Theorem 4.3. Hence $\hat{F}_n$ has the same asymptotic distribution as the "toy estimator" $F_n^{(1)}$ of Theorem 4.3.

Proof: After a preliminary reduction, we may assume that the conditions of Proposition 4.1 are satisfied, that is, if $T_{(1)}$ to $T_{(m)}$ are the ordered observation times, we may assume that $T_{(1)}$ corresponds to an observation time U_i such that $\Delta_{1i} = 1$, and that the largest observation time $T_{(m)}$ corresponds to an observation V_i such that $\Delta_{3i} = 1$. Then $\hat{F}_n(t_0)$ is the left derivative of the convex minorant at $G_{\hat{F}_n}(t_0)$ of the self-induced cumulative sum diagram formed by the points

$$P_j = \left(G_{\hat{F}_n}(T_{(j)}), V_{\hat{F}_n}(T_{(j)})\right), j = 1, \ldots, m,$$

and $P_0 = (0, 0)$. By Lemma 4.6 we have for each $c > 0$:

$$\sup_{t \in [t_0 - cn^{-1/3}, t_0 + cn^{-1/3}]} |\hat{F}_n(t) - F_0(t_0)| = \mathcal{O}_p(n^{-1/3}),$$

and the distance between jumps of $\hat{F}_n$ is therefore also of order $\mathcal{O}_p(n^{-1/3})$.

Let J_n be an interval of the form $(t_0, t_0 + tn^{-1/3}]$. Using Lemma 4.5 and Taylor expansion, we get

$$\begin{aligned}
&\int_{u\in J_n}\left\{\frac{\delta_1}{\hat{F}_n(u)} - \frac{\delta_2}{\hat{F}_n(v)-\hat{F}_n(u)}\right\} dQ_n \\
&\qquad + \int_{v\in J_n}\left\{\frac{\delta_2}{\hat{F}_n(v)-\hat{F}_n(u)} - \frac{\delta_3}{1-\hat{F}_n(v)}\right\} dQ_n \\
&= \int_{u\in J_n}\left\{\frac{\delta_1}{\hat{F}_n(u)} - \frac{\delta_2}{F_0(v)-\hat{F}_n(u)}\right\} dQ_n \\
&\qquad + \int_{v\in J_n}\left\{\frac{\delta_2}{\hat{F}_n(v)-F_0(u)} - \frac{\delta_3}{1-F_0(v)}\right\} dQ_n + o_p(n^{-2/3}) \\
&= \int_{u\in J_n}\left\{\frac{\delta_1}{F_0(u)} - \frac{\delta_2}{F_0(v)-F_0(u)}\right\} dQ_n \qquad (4.60)\\
&\qquad + \int_{v\in J_n}\left\{\frac{\delta_2}{F_0(v)-F_0(u)} - \frac{\delta_3}{1-F_0(v)}\right\} dQ_n \\
&\qquad + \int_{t\in J_n}\{F_0(t)-\hat{F}_n(t)\}\, dG_{F_0}(t) + o_p(n^{-2/3}) \\
&= W_{F_0}(t_0+tn^{-1/3}) - W_{F_0}(t_0) \qquad (4.61)\\
&\qquad + \int_{t\in J_n}\{F_0(t)-\hat{F}_n(t)\}\, dG_{F_0}(t) + o_p(n^{-2/3}), \qquad (4.62)
\end{aligned}$$

where the remainder term is uniform in t, for $t \geq 0$ in a compact interval. Taylor expansion also yields:

$$\int_{t\in J_n}\{\hat{F}_n(t)-F_0(t_0)\}\, dG_{\hat{F}_n}(t) = \int_{t\in J_n}\{\hat{F}_n(t)-F_0(t_0)\}\, dG_{F_0}(t) + o_p(n^{-2/3}), \qquad (4.63)$$

where again the remainder term is uniform in t, for $t \geq 0$ in a compact interval.

Hence, using the notation of Corollary 4.1, we get

$$V_{\hat{F}_n}^{(0)}(t_0+tn^{-1/3}) - V_{\hat{F}_n}^{(0)}(t_0) = V_{F_0}^{(0)}(t_0+tn^{-1/3}) - V_{F_0}^{(0)}(t_0) + o_p(n^{-2/3}).$$

We also have

$$G_{\hat{F}_n}(t_0+tn^{-1/3}) - G_{\hat{F}_n}(t_0) = G_{F_0}(t_0+tn^{-1/3}) - G_{F_0}(t_0) + \mathcal{O}_p(n^{-2/3}),$$

uniformly for $t \geq 0$ in a compact interval. Similarly we get

$$V_{\hat{F}_n}^{(0)}(t_0) - V_{\hat{F}_n}^{(0)}(t_0-tn^{-1/3}) = V_{F_0}^{(0)}(t_0) - V_{F_0}^{(0)}(t_0-tn^{-1/3}) + o_p(n^{-2/3}).$$

and

$$G_{\hat{F}_n}(t_0) - G_{\hat{F}_n}(t_0-tn^{-1/3}) = G_{F_0}(t_0) - G_{F_0}(t_0-tn^{-1/3}) + \mathcal{O}_p(n^{-2/3}),$$

uniformly for $t \geq 0$ in a compact interval.

Since, for each $c > 0$,

$$\sup_{t\in[t_0-cn^{-1/3},t_0+cn^{-1/3}]} |\hat{F}_n(t) - F_0(t_0)| = O_p(n^{-1/3}),$$

the distribution of the process $t \mapsto n^{1/3}\{\hat{F}_n(t_0 + tn^{-1/3}) - F_0(t_0)\}$ is determined by the locations of minima in t of the process

$$(t,a) \mapsto n^{2/3}\{V_{\hat{F}_n}^{(0)}(t_0+tn^{-1/3}) - V_{\hat{F}_n}^{(0)}(t_0)\} - n^{1/3}a\{G_{\hat{F}_n}(t_0+tn^{-1/3}) - G_{\hat{F}_n}(t_0)\},$$

where the parameters t and a vary over bounded intervals. Iain Johnstone offered the nice expression "switching relation" for this, while I was giving a Stanford summer course on these matters in 1990. For more details on the "switching relation", see, e.g., pp. 91 to 95 in GW and examples 3.2.14 and 3.2.15 in VAN DER VAART AND WELLNER(1996).

By the preceding equations, we have

$$\begin{aligned}
& n^{2/3}\{V_{\hat{F}_n}^{(0)}(t_0 + tn^{-1/3}) - V_{\hat{F}_n}^{(0)}(t_0)\} - n^{1/3}a\{G_{\hat{F}_n}(t_0 + tn^{-1/3}) - G_{\hat{F}_n}(t_0)\} \\
= \; & n^{2/3}\{V_{F_0}^{(0)}(t_0 + tn^{-1/3}) - V_{F_0}^{(0)}(t_0)\} - n^{1/3}a\{G_{F_0}(t_0 + tn^{-1/3}) - G_{F_0}(t_0)\} \\
& + o_p(1),
\end{aligned} \tag{4.64}$$

uniformly for t and a in bounded intervals.

Since similarly the distribution of the process

$$t \mapsto n^{1/3}\{F_n^{(1)}(t_0 + tn^{-1/3}) - F_0(t_0)\}$$

is determined by the locations of minima in t of the process

$$\begin{aligned}
(t,a) \mapsto n^{2/3}&\{V_{F_0}^{(0)}(t_0 + tn^{-1/3}) - V_{F_0}^{(0)}(t_0)\} \\
& -n^{1/3}a\{G_{F_0}(t_0 + tn^{-1/3}) - G_{F_0}(t_0)\} + o_p(1),
\end{aligned}$$

where the parameters t and a vary over bounded intervals, and where $F_n^{(1)}$ is the "toy estimator" of Theorem 4.3, the result now follows from (4.64). □

Bibliography

ANDERSEN, P.K., BORGAN, O, GILL, R.D. AND KEIDING, N. (1993). *Statistical models based on counting processes*, Springer, New York.

ARAGÓN, J. AND EBERLY, D. (1992). *On convergence of convex minorant algorithms for distribution estimation with interval censored data*, J. of Comp. and Graph. Statist., vol. 1, 129-140.

AYER, M., BRUNK, H.D., EWING, G.M., REID, W.T., SILVERMAN, E. (1955). *An empirical distribution function for sampling with incomplete information*, Ann. Math. Statist., vol. 26, 641–647.

BALL, K. AND PAJOR, A. (1990). *The entropy of convex bodies with "few" extreme points*, Proceedings of the 1989 Conference in Banach Spaces at Strob. Austria. Cambridge Univ. Press.

BAKKER, D. (1988). *Nonparametric maximum likelihood estimation of the distribution function of interval censored observations*, Unpublished master's thesis, University of Amsterdam.

BERMAN A. AND PLEMMONS, R.J. (1979). *Nonnegative Matrices in the Mathematical Sciences*, Academic Press, New York.

BICKEL, P.J., KLAASSEN, C.A.J., RITOV, Y. AND WELLNER, J.A. (1993). *Efficient and adaptive estimation for semiparametric models*, John Hopkins University Press, Baltimore.

BIRMAN, M.Š. AND SOLOMJAK, A. (1967) . *Piecewise-polynomial approximations of functions of the classes* W_p^α, Mat. Sb. , vol. 73, 295-317.

COSSLETT, S.R. (1983). *Distribution-free maximum likelihood estimator of the binary choice model*, Econometrica, vol. 51, 765-782.

COSSLETT, S.R. (1987). *Efficiency bounds for distribution-free estimators of the binary choice and censored regression models*, Econometrica, vol. 55, 559-585.

COX, D.R. (1972). *Regression models and life-tables (with discussion)*, Journal of the Royal Statistical Society, series B, vol. 34, 187-220.

GESKUS, R.B. (1992). *Efficient estimation of the mean for interval censoring case II*, Technical Report 92-83, Delft University of Technology.
ftp://ftp.twi.tudelft.nl/pub/publications/tech-reports/1992

GESKUS, R.B. AND GROENEBOOM, P. (1996a). *Asymptotically optimal estimation of smooth functionals for interval censoring, part 1.* Statistica Neerlandica, vol. 50, 69-88.

GESKUS, R.B. AND GROENEBOOM, P. (1996b). *Asymptotically optimal estimation of smooth functionals for interval censoring, part 2.* to appear in Statistica Neerlandica.

GESKUS, R.B. AND GROENEBOOM, P. (1996c). *Asymptotically optimal estimation of smooth functionals for interval censoring, case 2*, Technical Report 96-36, Delft University of Technology.
ftp://ftp.twi.tudelft.nl/pub/publications/tech-reports/1996

GILL, R.D. (1994). *Lectures in survival analysis*, in: Lectures on Probability theory, Ecole d'Eté de Saint-Flour XXII, 1992, Ed. P. Bernard. Lecture Notes in Mathematics, 1581, 115-241. Springer Verlag, Berlin.

GILL, R.D. AND LEVIT, B.Y. (1995). *Applications of the van Trees inequality: a Bayesian Cramér-Rao bound*, Bernoulli vol. 1, 59-79.

GOLUBEV, G.K. AND LEVITT, B.Y. (1996). *Asymptotically efficient estimation in the Wicksell problem.* Preprint 970, Department of Mathematics, University of Utrecht.

GROENEBOOM, P. (1987). *Asymptotics for interval censored observations.* Technical Report 87-18, Department of Mathematics, University of Amsterdam.

GROENEBOOM, P. (1989). *Brownian motion with a parabolic drift and Airy functions*, Probability Theory and Related Fields., vol. 79, p. 327-368.

GROENEBOOM, P. (1991). *Nonparametric maximum likelihood estimators for interval censoring and deconvolution.* Technical Report 378, Department of Statistics, Stanford University.

GROENEBOOM, P. AND WELLNER, J.A. (1992). *Information bounds and nonparametric maximum likelihood estimation*, Birkhäuser Verlag.

GROENEBOOM, P. (1995). *Nonparametric estimators for interval censoring*, in: Analysis of Censored Data, IMS Lecture Notes-Monograph Series, Vol 27, pp. 105-128. Editors: H. L. Koul and J. V. Deshpande. Hayword.

GROENEBOOM, P. AND JONGBLOED, G. (1995). *Isotonic estimation and rates of convergence in Wicksell's problem*, Ann. Statist., vol. 23, p. 1518-1542.

HAMPEL, F.R. (1987). *Design, modelling, and analysis of some biological datasets.*, in: Design, data and analysis, by some friends of Cuthbert Daniel, pp. 111-115. Editor: C.L. Mallows. Wiley, New York.

HUANG, J. (1996). *Efficient estimation for the proportional hazards model with interval censoring.*, Ann. Statist. 24, 540 - 568.

HUANG, J. AND WELLNER, J.A. (1995a). *Asymptotic normality of the NPMLE of linear functionals for interval censored data, case 1*, Statistica Neerlandica, vol. 49, 153-163.

HUANG, J. AND WELLNER, J.A. (1995b). *Efficient estimation for the proportional hazards model with "case 2" interval censoring*, Technical Report 290, Department of Statistics, University of Washington, submitted.

HUANG, J. AND WELLNER, J.A. (1996). *Interval censored survival data: a review of recent progress*, Proceedings of the First Seattle Symposium in Biostatistics: Survival Analysis. Lecture notes in Statistics, Springer Verlag, Berlin (To appear).

IBRAGIMOV, I.A. AND HAS'MINSKII, R.Z. (1981). *Statistical Estimation: Asymptotic Theory*, Springer Verlag. New York.

JONGBLOED, G. (1995a). *Three statistical inverse problems*, Ph. D. dissertation, Delft University.

JONGBLOED, G. (1995b). *The iterative convex minorant algorithm for nonparametric estimation*, Technical Report 95-105, Delft University of Technology, submitted.

JONGBLOED, G. (1994). *Exponential deconvolution: two asymptotically equivalent estimators*, Technical Report 94-17, Delft University of Technology. To appear in Statistica Neerlandica.
`ftp://ftp.twi.tudelft.nl/pub/publications/tech-reports/1994`

KIM, J, AND POLLARD, D. (1990). *Cube root aymptotics*, Ann. Statist., vol. 18, p.191-219.

KRESS, R. (1989). *Linear integral equations*, Applied Mathematical Sciences vol. 82, Springer Verlag, New York.

LEHMANN E.L. (1983). *Theory of point estimation*, Wiley, New York.

MAMMEN, E. (1991). *Nonparametric regression under qualitative smoothness assumptions*, Ann. Statist., vol. 19, 741-759.

MANSKI, C.F. (1985). *Semi-parametric analysis of discrete response: Asymptotic properties of the maximum score operator*, J. Econometrics, vol. 27, 313-333.

POLLARD, D. (1984). Convergence of stochastic processes. Springer-Verlag.

ROBERTSON,T., WRIGHT, F.T. AND DYKSTRA, R.L. (1988), *Order Restricted Statistical Inference.* Wiley, New York.

ROCKAFELLAR, R.T. (1970). Convex analysis. Princeton University Press.

SHEEHY, A. AND WELLNER, J.A. (1992). *Uniform Donsker classes of functions*, Ann. Prob.,vol 20, p. 1983-2030.

SZEGÖ, G. (1978). *Orthogonal Polynomials*, Colloquium Publications 23, A.M.S., Providence, Rhode Island.

TSAI, W.I. AND CROWLEY, J. (1985), *A large sample study of generalized maximum likelihood estimators from incomplete data via self-consistency*, Ann. Statist, vol. 13, 1317–1334.

TSAI, W.I. AND CROWLEY, J. (1990), *Correction on: A large sample study of generalized maximum likelihood estimators from incomplete data via self-consistency*, Ann. Statist., vol. 18, 470.

TURNBULL, B.W. (1974), *Nonparametric estimation of a survivorship function with doubly censored data*, J. Amer. Statist. Assoc., vol. 69, 169–173.

TÜRNBULL, B.W. (1976), *The empirical distribution function with arbitrarily grouped censored and truncated data*, J.R. Statist. Soc. B, vol. 38, 290–295.

TURNBULL, B. W. AND MITCHELL, T. J. (1984), *Nonparametric estimation of the distribution of time to onset for specific diseases in survival/sacrifice experiments*, Biometrics, vol. 40, 41–50.

VAN DE GEER, S. (1996). *Rates of convergence for the maximum likelihood estimator in mixture models*, Journal of Nonparametric Statistics, vol. 6, 293-310.

VAN DER VAART, A.W. (1991). *On differentiable functionals*, Ann. Statist., vol. 19, p.178-204.

VAN DER VAART, A.W. AND WELLNER, J.A. (1996). *Weak convergence and empirical processes.* Springer, New York.

VAN EEDEN, C. (1956). *Maximum likelihood estimation of ordered probabilities*, Indag. Mathematicae.,vol 18, p. 444-455.

VAN ES, A.J. (1991). *Aspects of Nonparametric Density Estimation.* CWI Tract 77, Centre for Mathematics and Computer Science, Amsterdam (revision of the 1988 Ph. D. thesis of the author at the University of Amsterdam).

VAN ES, A.J., JONGBLOED, G. AND VAN ZUIJLEN, M.C.A. (1995b). *Nonparametric deconvolution for decreasing kernels*, submitted for publication. Technical Report 95-77, Delft University of Technology.
`ftp://ftp.twi.tudelft.nl/TWI/publications/tech-reports/1995`

VAN ES, A.J.AND VAN ZUIJLEN, M.C.A. (1996). *Convex minorant estimators of distributions in nonparametric deconvolution problems*, Scandinavian journal of Statistics, vol. 23, 85-104.

WATSON, G.S. (1971). *Estimating functionals of particle size distributions*, Biometrika, vol. 58, 483-490.

WICKSELL, S.D. (1925). *The corpuscle problem*, Biometrika, vol. 17, 84-99.

WELLNER, J.A. (1995). *Interval censoring case 2: an exploration of alternative hypotheses*, in: Analysis of Censored Data, IMS Lecture Notes-Monograph Series, Vol 27, pp. 271-291. Editors: H. L. Koul and J. V. Deshpande. Hayword.

ISOPERIMETRY AND GAUSSIAN ANALYSIS

Michel LEDOUX

Twenty years after his celebrated St-Flour course

on regularity of Gaussian processes,

I dedicate these notes to my teacher X. Fernique

Table of contents

In memory of A. Ehrhard

The Gaussian isoperimetric inequality, and its related concentration phenomenon, is one of the most important properties of Gaussian measures. These notes aim to present, in a concise and selfcontained form, the fundamental results on Gaussian processes and measures based on the isoperimetric tool. In particular, our exposition will include, from this modern point of view, some of the by now classical aspects such as integrability and tail behavior of Gaussian seminorms, large deviations or regularity of Gaussian sample paths. We will also concentrate on some of the more recent aspects of the theory which deal with small ball probabilities. Actually, the Gaussian concentration inequality will be the opportunity to develop some functional analytic ideas around the concentration of measure phenomenon. In particular, we will see how simple semigroup tools and the geometry of abstract Markov generator may be used to study concentration and isoperimetric inequalities. We investigate in this context some of the deep connections between isoperimetric inequalities and functional inequalities of Sobolev type. We also survey recent work on concentration inequalities in product spaces. Actually, although the main theme is Gaussian isoperimetry and analysis, many ideas and results have a much broader range of applications. We will try to indicate some of the related fields of interest.

The Gaussian isoperimetric and concentration inequalities were developed most vigorously in the study of the functional analytic aspects of probability theory (probability in Banach spaces and its relation to geometry and the local theory of Banach spaces) through the contributions of A. Badrikian, C. Borell, S. Chevet, A. Ehrhard, X. Fernique, H. J. Landau and L. A. Shepp, B. Maurey, V. D. Milman, G. Pisier, V. N. Sudakov and B. S. Tsirel'son, M. Talagrand among others. In particular, the new proof by V. D. Milman of Dvoretzky's theorem on spherical sections of convex bodies started the development of the concentration ideas and of their applications in geometry and probability in Banach spaces. Actually, most of the tools and inspiration come from analysis rather than probability. From this analytical point of view, emphasis is put on inequalities in finite dimension as well as on the fundamental Gaussian measurable structure consisting of the product measure on $\mathbb{R}^{\mathbb{N}}$ when each coordinate is endowed with the standard Gaussian measure. It is no surprise therefore that most of the results, developed in the seventies and eighties, often do not seem familiar to true probabilists, and even analysts on Wiener spaces. The aim of this course is to try to advertise these powerful and useful ideas to the probability community although all the results presented here are known and already appeared elsewhere. In particular, M. Talagrand's ideas and contributions, that strongly influenced the author's comprehension of the subject, take an important part in this exposition.

After a short introduction on isoperimetry, where we present the classical isoperimetric inequality, the isoperimetric inequality on spheres and the Gaussian isoperimetric inequality, our first task, in Chapter 2, will be to develop the concentration of measure phenomenon from a functional analytic point of view based on semigroup theory. In particular we show how the Gaussian concentration inequality may easily be obtained from the commutation property of the Ornstein-Uhlenbeck semigroup. In the last chapter, we further investigate the deep connections between isoperimetric and functional inequalities (Sobolev inequalities, hypercontractivity, heat kernel estimates...). We follow in this matter the ideas of N. Varopoulos in his functional approach to isoperimetric inequalities and heat kernel bounds on groups and manifolds. In Chapter 3, we will survey the remarkable recent isoperimetric and concentration inequalities for product measures of M. Talagrand. This section aims to demonstrate the power of abstract concentration arguments and induction techniques in this setting. These deep ideas appear of potential use in a number of problems in probability and applied probability. In Chapter 4, we present, from the concentration viewpoint, the classical integrability properties and tail behaviors of norms of Gaussian measures or random vectors as well as their large deviations. We also show how the isoperimetric and concentration ideas allow a nontopological approach to large deviations of Gaussian measures. The next chapter deals with the corresponding questions for Wiener chaos as remarkably investigated by C. Borell in the late seventies and early eighties. In Chapter 6, we provide a complete treatment of regularity of Gaussian processes based on the results of R. M. Dudley, X. Fernique, V. N. Sudakov and M. Talagrand. In particular, we present the recent short proof of M. Talagrand, based on concentration, of the necessity of the majorizing measure condition for bounded or continuous Gaussian processes. Chapter 7 is devoted to some of the recent aspects of the study of Gaussian measures, namely small ball probabilities. We also investigate in this chapter some correlation and conditional inequalities for norms of Gaussian measures (which have been applied recently to the support of a diffusion theorem and the Freidlin-Wentzell large deviation principle for stronger topologies on Wiener space). Finally, and as announced, we come back in Chapter 8 to a semigroup approach of the Gaussian isoperimetric inequality based on hypercontractivity. Most chapters are completed with short notes for further reading. We also tried to appropriately complete the list of references although we did not put emphasis on historical details and comments.

I sincerely thank the organizers of the École d'Été de St-Flour for their invitation to present this course. My warmest thanks to Ph. Barbe, M. Capitaine, M. A. Lifshits and W. Stolz for a careful reading of the early version of these notes and to C. Borell and S. Kwapień for several helful comments and indications. Many thanks to P. Baldi, S. Chevet, Ch. Léonard, A. Millet and J. Wellner for their comments, remarks and corrections during the school and to all the participants for their interest in this course.

St-Flour, Toulouse 1994 Michel Ledoux

1. SOME ISOPERIMETRIC INEQUALITIES

In this first chapter, we present the basic isoperimetric inequalities which form the geometric background of this study. Although we will not directly be concerned with true isoperimetric problems and description of extremal sets later on, these inequalities are at the basis of the concentration inequalities of the next chapter on which most results of these notes will be based. We introduce the isoperimetric ideas with the classical isoperimetric inequality on $\mathbb{R}^n$ but the main result will actually be the isoperimetric property on spheres and its limit version, the Gaussian isoperimetric inequality. More on isoperimetry may be found e.g. in the book [B-Z] as well as in the survey paper [Os] and the references therein.

The classical isoperimetric inequality in $\mathbb{R}^n$ (see e.g. [B-Z], [Ha], [Os]...), which at least in dimension 2 and for convex sets may be considered as one of the oldest mathematical statements (cf. [Os]), asserts that among all compact sets A in $\mathbb{R}^n$ with smooth boundary ∂A and with fixed volume, Euclidean balls are the ones with the minimal surface measure. In other words, whenever $\mathrm{vol}_n(A) = \mathrm{vol}_n(B)$ where B is a ball (and $n > 1$),

$$\mathrm{vol}_{n-1}(\partial A) \geq \mathrm{vol}_{n-1}(\partial B). \tag{1.1}$$

There is an equivalent, although less familiar, formulation of this result in terms of isoperimetric neighborhoods or enlargements which in particular avoids surface measures and boundary considerations; namely, if A_r denotes the (closed) Euclidean neighborhood of A of order $r \geq 0$, and if B is as before a ball with the same volume as A, then, for every $r \geq 0$,

$$\mathrm{vol}_n(A_r) \geq \mathrm{vol}_n(B_r). \tag{1.2}$$

Note that A_r is simply the Minkowski sum $A + B(0,r)$ of A and of the (closed) Euclidean ball $B(0,r)$ with center the origin and radius r. The equivalence between (1.1) and (1.2) follows from the Minkowski content formula

$$\mathrm{vol}_{n-1}(\partial A) = \liminf_{r \to 0} \frac{1}{r}\left[\mathrm{vol}_n(A_r) - \mathrm{vol}_n(A)\right]$$

(whenever the boundary ∂A of A is regular enough). Actually, if we take the latter as the definition of $\mathrm{vol}_{n-1}(\partial A)$, it is not too difficult to see that (1.1) and (1.2) are equivalent for every Borel set A (see Chapter 8 for a related result). The simplest proof of this isoperimetric inequality goes through the Brunn-Minkowski inequality which states that if A and B are two compact sets in $\mathbb{R}^n$, then

$$\mathrm{vol}_n(A+B)^{1/n} \geq \mathrm{vol}_n(A)^{1/n} + \mathrm{vol}_n(B)^{1/n}. \tag{1.3}$$

To deduce the isoperimetric inequality (1.2) from the Brunn-Minkowski inequality (1.3), let $r_0 > 0$ be such that $\mathrm{vol}_n(A) = \mathrm{vol}_n(B(0,r_0))$. Then, by (1.3),

$$\begin{aligned}
\mathrm{vol}_n(A_r)^{1/n} = \mathrm{vol}_n\big(A + B(0,r)\big)^{1/n} &\geq \mathrm{vol}_n(A)^{1/n} + \mathrm{vol}_n\big(B(0,r)\big)^{1/n} \\
&= \mathrm{vol}_n\big(B(0,r_0)\big)^{1/n} + \mathrm{vol}_n\big(B(0,r)\big)^{1/n} \\
&= (r_0 + r)\mathrm{vol}_n\big(B(0,1)\big)^{1/n} \\
&= \mathrm{vol}_n\big(B(0,r_0+r)\big)^{1/n} = \mathrm{vol}_n\big(B(0,r_0)_r\big)^{1/n}.
\end{aligned}$$

As an illustration of the methods, let us briefly sketch the proof of the Brunn-Minkowski inequality (1.3) following [Ha] (for an alternate simple proof, see [Pi3]). By a simple approximation procedure, we may assume that each of A and B is a union of finitely many disjoint sets, each of which is a product of intervals with edges parallel to the coordinate axes. The proof is by induction on the total number p of these rectangular boxes in A and B. If $p = 2$, that is if A and B are products of intervals with sides of length $(a_i)_{1\leq i\leq n}$ and $(b_i)_{1\leq i\leq n}$ respectively, then

$$\begin{aligned}
\frac{\mathrm{vol}_n(A)^{1/n} + \mathrm{vol}_n(B)^{1/n}}{\mathrm{vol}_n(A+B)^{1/n}} &= \prod_{i=1}^n \left(\frac{a_i}{a_i+b_i}\right)^{1/n} + \prod_{i=1}^n \left(\frac{b_i}{a_i+b_i}\right)^{1/n} \\
&\leq \frac{1}{n}\sum_{i=1}^n \frac{a_i}{a_i+b_i} + \frac{1}{n}\sum_{i=1}^n \frac{b_i}{a_i+b_i} = 1
\end{aligned}$$

where we have used the inequality between geometric and arithmetic means. Now, assume that A and B consist of a total of $p > 2$ products of intervals and that (1.3) holds for all sets A' and B' which are composed of a total of at most $p-1$ rectangular boxes. We may and do assume that the number of rectangular boxes in A is at least 2. Parallel shifts of A and B do not change the volume of A, B or $A+B$. Take then a shift of A with the property that one of the coordinate hyperplanes divides A in such a way that there is at least one rectangular box in A on each side of this hyperplane. Therefore A is the union of A' and A'' where A' and A'' are disjoint unions of a number of rectangular boxes strictly smaller than the number in A. Now shift B parallel to the coordinate axes in such a manner that the same hyperplane divides B into B' and B'' with

$$\frac{\mathrm{vol}_n(B')}{\mathrm{vol}_n(B)} = \frac{\mathrm{vol}_n(A')}{\mathrm{vol}_n(A)} = \lambda.$$

Each of B' and B'' has at most the same number of products of intervals as B has. Now, by the induction hypothesis,

$$\begin{aligned}
&\mathrm{vol}_n(A+B)\\
&\quad \geq \mathrm{vol}_n(A'+B') + \mathrm{vol}_n(A''+B'')\\
&\quad \geq \left[\mathrm{vol}_n(A')^{1/n} + \mathrm{vol}_n(B')^{1/n}\right]^n + \left[\mathrm{vol}_n(A'')^{1/n} + \mathrm{vol}_n(B'')^{1/n}\right]^n\\
&\quad = \lambda\left[\mathrm{vol}_n(A)^{1/n} + \mathrm{vol}_n(B)^{1/n}\right]^n + (1-\lambda)\left[\mathrm{vol}_n(A)^{1/n} + \mathrm{vol}_n(B)^{1/n}\right]^n\\
&\quad = \left[\mathrm{vol}_n(A)^{1/n} + \mathrm{vol}_n(B)^{1/n}\right]^n
\end{aligned}$$

which is the result. Note that, by concavity, (1.3) implies (is actually equivalent to the fact) that, for every λ in $[0,1]$,

$$\mathrm{vol}_n\big(\lambda A + (1-\lambda)B\big) \geq \left[\lambda \mathrm{vol}_n(A)^{1/n} + (1-\lambda)\mathrm{vol}_n(B)^{1/n}\right]^n \geq \mathrm{vol}_n(A)^\lambda \mathrm{vol}_n(B)^{1-\lambda}.$$

In the probabilistic applications, it is the isoperimetric inequality on spheres, rather than the classical isoperimetric inequality, which is of fundamental importance. The use of the isoperimetric inequality on spheres in analysis and probability goes back to the new proof, by V. D. Milman [Mi1], [Mi3], of the famous Dvoretzky theorem on spherical sections of convex bodies [Dv]. Since then, it has been used extensively in the local theory of Banach spaces (see [F-L-M], [Mi-S], [Pi3]...) and in probability theory via its Gaussian version (see below). The purpose of this course is actually to present a complete account on the Gaussian isoperimetric inequality and its probabilistic applications. For the applications to Banach space theory, we refer to [Mi-S], [Pi1], [Pi3].

Very much as (1.1), the isoperimetric inequality on spheres expresses that among all subsets with fixed volume on a sphere, geodesic balls (caps) achieve the minimal surface measure. This inequality has been established independently by E. Schmidt [Sch] and P. Lévy [Lé] in the late forties (but apparently for sets with smooth boundaries). Schmidt's proof is based on the classical isoperimetric rearrangement or symmetrization techniques due to J. Steiner (see [F-L-M] for a complete proof along these lines, perhaps the first in this generality). A nice two-point symmetrization technique may also be used (see [Be2]). Lévy's argument, which applies to more general types of surfaces, uses the modern tools of minimal hypersurfaces and integral currents. His proof has been generalized to Riemannian manifolds with positive Ricci curvature by M. Gromov [Gro], [Mi-S], [G-H-L]. Let M be a compact connected Riemannian manifold of dimension N (≥ 2), and let d be its Riemannian metric and μ its normalized Riemannian measure. Denote by $R(M)$ the infimum of the Ricci tensor $\mathrm{Ric}(\cdot,\cdot)$ of M over all unit tangent vectors. Recall that if S_ρ^N is the sphere of radius $\rho > 0$ in $\mathbb{R}^{N+1}$, $R(S_\rho^N) = (N-1)/\rho^2$ (see [G-H-L]). We denote below by σ_ρ^N the normalized rotation invariant measure on S_ρ^N. If A is a subset of M, we let as before $A_r = \{x \in M; d(x,A) \leq r\}$, $r \geq 0$.

Theorem 1.1. *Assume that $R(M) = R > 0$ and let S_ρ^N be the manifold of constant curvature equal to R (i.e. ρ is such that $R(S_\rho^N) = (N-1)/\rho^2 = R$). Let A be* *measurable in M and let B be a geodesic ball, or cap, of S_ρ^N such that $\mu(A) \geq \sigma_\rho^N(B)$. Then, for every $r \geq 0$,*

$$\mu(A_r) \geq \sigma_\rho^N(B_r). \tag{1.4}$$

Theorem 1.1 of course applies to the sphere S_ρ^N itself. Equality in (1.4) occurs only if M is a sphere and A a cap on this sphere. Notice furthermore that Theorem 1.1 applied to sets the diameter of which tends to zero contains the classical isoperimetric inequality in Euclidean space. We refer to [Gro], [Mi-S] or [G-H-L] for the proof of Theorem 1.1.

Theorem 1.1 is of particular interest in probability theory via its limit version which gives rise to the Gaussian isoperimetric inequality, our tool of fundamental importance in this course. The Gaussian isoperimetric inequality may indeed be considered as the limit of the isoperimetric inequality on the spheres S_ρ^N when the dimension N and the radius ρ both tend to infinity in the geometric ($R(S_\rho^N) = (N-1)/\rho^2$) and probabilistic ratio $\rho^2 = N$. It has indeed been known for some time that the measures $\sigma_{\sqrt{N}}^N$ on $S_{\sqrt{N}}^N$, projected on a fixed subspace $\mathbb{R}^n$, converge when N goes to infinity to the canonical Gaussian measure on $\mathbb{R}^n$. To be more precise, denote by $\Pi_{N+1,n}$, $N \geq n$, the projection from $\mathbb{R}^{N+1}$ onto $\mathbb{R}^n$. Let γ_n be the canonical Gaussian measure on $\mathbb{R}^n$ with density $\varphi_n(x) = (2\pi)^{-n/2}\exp(-|x|^2/2)$ with respect to Lebesgue measure (where $|x|$ is the Euclidean norm of $x \in \mathbb{R}^n$).

Lemma 1.2. *For every Borel set A in $\mathbb{R}^n$,*

$$\lim_{N\to\infty} \sigma_{\sqrt{N}}^N(\Pi_{N+1,n}^{-1}(A) \cap S_{\sqrt{N}}^N) = \gamma_n(A).$$

Lemma 1.2 is commonly known as Poincaré's lemma [MK] although it does not seem to be due to H. Poincaré (cf. [D-F]). The convergence is better than only weak convergence of the sequence of measures $\Pi_{N+1,n}(\sigma_{\sqrt{N}}^N)$ to γ_n. Simple analytic or probabilistic proofs of Lemma 1.2 may be found in the literature ([Eh1], [Gal], [Fe5], [D-F]...). The following elegant proof was kindly communicated to us by J. Rosiński.

Proof. Let $(g_i)_{i\geq 1}$ be a sequence of independent standard normal random variables. For every integer $N \geq 1$, set $R_N^2 = g_1^2 + \cdots + g_N^2$. Now, $(\sqrt{N}/R_{N+1})\cdot(g_1,\ldots,g_{N+1})$ is equal in distribution to $\sigma_{\sqrt{N}}^N$, and thus $(\sqrt{N}/R_{N+1})\cdot(g_1,\ldots,g_n)$ is equal in distribution to $\Pi_{N+1,n}(\sigma_{\sqrt{N}}^N)$ $(N \geq n)$. Since $R_N^2/N \to 1$ almost surely by the strong law of large numbers, we already get the weak convergence result. Lemma 1.2 is however stronger since convergence is claimed for every Borel set. In order to get the full conclusion, notice that R_n^2, $R_{N+1}^2 - R_n^2$ and $(g_1,\ldots,g_n)/R_n$ are independent. Therefore R_n^2/R_{N+1}^2 is independent of $(g_1,\ldots,g_n)/R_n$ and has beta distribution β with parameters $n/2$, $(N+1-n)/2$. Now,

$$\begin{aligned}\sigma_{\sqrt{N}}^N(\Pi_{N+1,n}^{-1}(A)\cap S_{\sqrt{N}}^N) &= \mathbb{P}\Big\{\frac{\sqrt{N}}{R_{N+1}}(g_1,\ldots,g_n)\in A\Big\}\\ &= \mathbb{P}\Big\{\Big(N\frac{R_n^2}{R_{N+1}^2}\Big)^{1/2}\cdot\frac{1}{R_n}(g_1,\ldots,g_n)\in A\Big\}.\end{aligned}$$

Therefore,

$$\begin{aligned}
&\sigma^N_{\sqrt{N}}\big(\Pi^{-1}_{N+1,n}(A)\cap S^N_{\sqrt{N}}\big)\\
&\quad=\beta\big(\tfrac{n}{2},\tfrac{N+1-n}{2}\big)^{-1}\int_{S_1^{n-1}}\int_0^1 I_A(\sqrt{Nt}x)t^{\frac{n}{2}-1}(1-t)^{\frac{N+1-n}{2}-1}d\sigma_1^{n-1}(x)dt\\
&\quad=\beta\big(\tfrac{n}{2},\tfrac{N+1-n}{2}\big)^{-1}\frac{2}{N^{n/2}}\int_{S_1^{n-1}}\int_0^{\sqrt{N}} I_A(ux)u^{n-1}\Big(1-\frac{u^2}{N}\Big)^{\frac{N+1-n}{2}-1}d\sigma_1^{n-1}(x)du
\end{aligned}$$

by the change of variables $u=\sqrt{Nt}$. Letting $N\to\infty$, the last integral converges by the dominated convergence theorem to

$$\frac{2}{2^{n/2}\Gamma(n/2)}\int_{S_1^{n-1}}\int_0^\infty I_A(ux)u^{n-1}e^{-u^2/2}d\sigma_1^{n-1}(x)du$$

which is precisely $\gamma_n(A)$ in polar coordinates. The proof of Lemma 1.2 is thus complete. This proof is easily modified to actually yield uniform convergence of densities on compacts sets ([Eh1], [Gal], [Fe5]) and in the variation metric [D-F]. □

As we have seen, caps are the extremal sets of the isoperimetric problem on spheres. Now, a cap may be regarded as the intersection of a sphere and a half-space, and, by Poincaré's limit, caps will thus converge to half-spaces. There are therefore strong indications that half-spaces will be the extremal sets of the isoperimetric problem for Gaussian measures. A half-space H in $\mathbb{R}^n$ is defined as

$$H=\{x\in\mathbb{R}^n;\langle x,u\rangle\le a\}$$

for some real number a and some unit vector u in $\mathbb{R}^n$. The isoperimetric inequality for the canonical Gaussian measure γ_n in $\mathbb{R}^n$ may then be stated as follows. If A is a set in $\mathbb{R}^n$, A_r denotes below its Euclidean neighborhood of order $r\ge 0$.

Theorem 1.3. *Let A be a Borel set in $\mathbb{R}^n$ and let H be a half-space such that $\gamma_n(A)\ge\gamma_n(H)$. Then, for every $r\ge 0$,*

$$\gamma_n(A_r)\ge\gamma_n(H_r).$$

Since γ_n is both rotation invariant and a product measure, the measure of a half-space is actually computed in dimension one. Denote by Φ the distribution function of γ_1, that is

$$\Phi(t)=\int_{-\infty}^t e^{-x^2/2}\frac{dx}{\sqrt{2\pi}},\quad t\in\mathbb{R}.$$

Then, if $H=\{x\in\mathbb{R}^n;\langle x,u\rangle\le a\}$, $\gamma_n(H)=\Phi(a)$, and Theorem 1.3 expresses equivalently that when $\gamma_n(A)\ge\Phi(a)$, then

$$\gamma_n(A_r)\ge\Phi(a+r)\tag{1.5}$$

for every $r \geq 0$. In other words, if Φ^{-1} is the inverse function of Φ, for every Borel set A in $\mathbb{R}^n$ and every $r \geq 0$,

$$\Phi^{-1}\big(\gamma_n(A_r)\big) \geq \Phi^{-1}\big(\gamma_n(A)\big) + r. \tag{1.6}$$

The Gaussian isoperimetric inequality is thus essentially dimension free, a characteristic feature of the Gaussian setting.

Proof of Theorem 1.3. We prove (1.5) and use the isoperimetric inequality on spheres (Theorem 1.1) and Lemma 1.2. We may assume that $a = \Phi^{-1}(\gamma_n(A)) > -\infty$. Let then $b \in (-\infty, a)$. Since $\gamma_n(A) > \Phi(b) = \gamma_1((-\infty, b])$, by Lemma 1.2, for every N ($\geq n$) large enough,

$$\sigma^N_{\sqrt{N}}\big(\Pi^{-1}_{N+1,n}(A) \cap S^N_{\sqrt{N}}\big) > \sigma^N_{\sqrt{N}}\big(\Pi^{-1}_{N+1,1}(]-\infty, b]) \cap S^N_{\sqrt{N}}\big). \tag{1.7}$$

It is easy to see that $\Pi^{-1}_{N+1,n}(A_r) \cap S^N_{\sqrt{N}} \supset \big(\Pi^{-1}_{N+1,n}(A) \cap S^N_{\sqrt{N}}\big)_r$ where the neighborhood of order r on the right hand side is understood with respect to the geodesic distance on $S^N_{\sqrt{N}}$. Since $\Pi^{-1}_{N+1,1}((-\infty, b]) \cap S^N_{\sqrt{N}}$ is a cap on $S^N_{\sqrt{N}}$, by (1.7) and the isoperimetric inequality on spheres (Theorem 1.1),

$$\begin{aligned}\sigma^N_{\sqrt{N}}\big(\Pi^{-1}_{N+1,n}(A_r) \cap S^N_{\sqrt{N}}\big) &\geq \sigma^N_{\sqrt{N}}\big((\Pi^{-1}_{N+1,n}(A) \cap S^N_{\sqrt{N}})_r\big)\\ &\geq \sigma^N_{\sqrt{N}}\big((\Pi^{-1}_{N+1,1}((-\infty, b]) \cap S^N_{\sqrt{N}})_r\big).\end{aligned}$$

Now, $\big(\Pi^{-1}_{N+1,1}((-\infty, b]) \cap S^N_{\sqrt{N}}\big)_r = \Pi^{-1}_{N+1,1}((-\infty, b + r(N)]) \cap S^N_{\sqrt{N}}$ where (for N large)

$$r(N) = \sqrt{N}\cos\big[\arccos(b/\sqrt{N}) - r/\sqrt{N}\big] - b.$$

Since $\lim r(N) = r$, by Lemma 1.2 again, $\gamma_n(A_r) \geq \Phi(b+r)$. Since $b < a$ is arbitrary, the conclusion follows. □

Theorem 1.3 is due independently to C. Borell [Bo2] and to V. N. Sudakov and B. S. Tsirel'son [S-T] with the same proof based on the isoperimetric inequality on spheres and Poincaré's limit. A. Ehrhard [Eh2] (see also [Eh3], [Eh5]) gave a different proof using an intrinsic Gaussian symmetrization procedure similar to the Steiner symmetrization used by E. Schmidt in his proof of Theorem 1.1. In any case, Ehrhard's proof or the proof of isoperimetry on spheres are rather delicate, as it is usually the case with isoperimetric inequalities and the description of their extremal sets.

With this same Gaussian symmetrization tool, A. Ehrhard [Eh2] established furthermore a Brunn-Minkowski type inequality for γ_n, however only for convex sets. More precisely, he showed that whenever A and B are convex sets in $\mathbb{R}^n$, for every $\lambda \in [0,1]$,

$$\Phi^{-1}\big(\gamma_n(\lambda A + (1-\lambda)B)\big) \geq \lambda\Phi^{-1}\big(\gamma_n(A)\big) + (1-\lambda)\Phi^{-1}\big(\gamma_n(B)\big). \tag{1.8}$$

It might be worthwhile noting that if we apply this inequality to B the Euclidean ball with center the origin and radius $r/(1-\lambda)$ and let λ tend to one, we recover inequality (1.6) (for A convex). However, it is still an open problem to know whether (1.8) holds true for every Borel sets A and B and not only convex sets*. It would improve upon the more classical logconcavity of Gaussian measures (cf. [Bo1]) that states that, for every Borel sets A and B, and every $\lambda \in [0,1]$,

$$\gamma_n(\lambda A + (1-\lambda)B) \geq \gamma_n(A)^\lambda \gamma_n(B)^{1-\lambda}. \tag{1.9}$$

As another inequality of interest, let us note that if A is a Borel set with $\gamma_n(A) = \Phi(a)$, and if $h \in \mathbb{R}^n$,

$$\gamma_n(A+h) \leq \Phi(a + |h|). \tag{1.10}$$

By rotational invariance, we may assume that $h = re_1$, $r = |h|$, where e_1 is the first unit vector on $\mathbb{R}^n$, changing A into some new set A' with $\gamma_n(A) = \gamma_n(A') = \Phi(a)$. Then, by the translation formula for γ_n,

$$\begin{aligned} e^{r^2/2}\gamma_n(A'+h) &= \int_{A'} e^{-rx_1} d\gamma_n(x) \\ &\leq \int_{A'\cap\{x_1 \leq a\}} e^{-rx_1} d\gamma_n(x) + e^{-ra}\gamma_n\big(A' \cap \{x_1 > a\}\big). \end{aligned}$$

Since $\gamma_n(A' \cap \{x_1 > a\}) = \gamma_n((A')^c \cap \{x_1 \leq a\})$ where $(A')^c$ is the complement of A',

$$\begin{aligned} e^{r^2/2}\gamma_n(A'+h) &\leq \int_{A'\cap\{x_1 \leq a\}} e^{-rx_1} d\gamma_n(x) + e^{-ra}\gamma_n\big((A')^c \cap \{x_1 \leq a\}\big) \\ &\leq \int_{\{x_1 \leq a\}} e^{-rx_1} d\gamma_n(x) \\ &= e^{r^2/2}\gamma_n(x; x_1 \leq a + r) = e^{r^2/2}\Phi(a+r). \end{aligned}$$

The claim (1.10) follows.

Notes for further reading. Very recently, S. Bobkov [Bob2] gave a remarkable new simple proof of the isoperimetric inequality for Gaussian measures based on a sharp two point isoperimetric inequality (inspired by [Ta11]) and the central limit theorem. This proof is similar in nature to Gross' original proof [Gr3] of the logarithmic Sobolev inequality for Gaussian measures and does not use any kind of isoperimetric symmetrization or rearrangement (cf. also Chapter 8). In addition to the preceding open problem (1.8), the following conjecture is still open. Is it true that for every symmetric closed convex set A in $\mathbb{R}^n$,

$$\gamma_n(\lambda A) \geq \gamma_n(\lambda S) \tag{1.11}$$

---

* During the school, R. Latała [La] proved that (1.8) holds when only one of the two sets A and B is convex. Thus, due to the preceding comment, the Brunn-Minkowski principle generalizes to the Gaussian setting.

for each $\lambda > 1$, where S is a symmetric strip such that $\gamma_n(A) = \gamma_n(S)$? This conjecture has been known since an unpublished preprint by L. Shepp on the existence of strong exponential moments of Gaussian measures (cf. Chapter 4 and [L-S]). Recent work of S. Kwapień and J. Sawa [K-S] shows that the conjecture is true under the additional assumption that A is sufficiently symmetric (A is an ellipsoid for example). Examples of isoperimetric processes in probability theory are presented in [Bo11], [Bo12].

2. THE CONCENTRATION OF MEASURE PHENOMENON

In this section, we present the concentration of measure phenomenon which was most vigourously put forward by V. D. Milman in the local theory of Banach spaces (cf. [Mi2], [Mi3], [Mi-S]). Isoperimetry is concerned with infinitesimal neighborhoods and surface areas and with extremal sets. The concentration of measure phenomenon rather concerns the behavior of "large" isoperimetric neighborhoods. Although of basic isoperimetric inspiration, the concentration of measure phenomenon is a milder property that may be shown, as we will see, to be satisfied in a large number of settings, sometimes rather far from the geometrical frame of isoperimetry. It roughly states that if a set $A \subset X$ has measure at least one half, "most" of the points in X are "close" to A. The main task is to make precise the meaning of the words "most" and "close" in the examples of interest. Moreover, new tools may be used to establish concentration inequalities. In particular, we will present in this chapter simple semigroup and probabilistic proofs of both the concentration inequalities on spheres and in Gauss space. In chapter 8, we further develop the functional approach and try to reach with these tools the full isoperimetric statements.

As we mentioned it at the end of the preceding chapter, isoperimetric inequalities and description of their extremal sets are often rather delicate, if not unknown. However, in almost all the applications presented here, the Gaussian isoperimetric inequality is only used in the form of the corresponding concentration inequality. Since the latter will be established here in an elementary way, it can be freely used in the applications.

In the setting of Theorem 1.1, if A is a set on M with sufficiently large measure, for example if $\mu(A) \geq \frac{1}{2}$, then, by the explicit expression of the measure of a cap, we get that, for every $r \geq 0$

$$\mu(A_r) \geq 1 - \exp\left(-R\frac{r^2}{2}\right), \tag{2.1}$$

that is a Gaussian bound, only depending on R, on the complement of the neighborhood of order r of A, uniformly in those A's such that $\mu(A) \geq \frac{1}{2}$. More precisely, if $\mu(A) \geq \frac{1}{2}$, for "most" x's in M, there exists y in A within distance $1/\sqrt{R}$ of x. Of course, the ratio $1/\sqrt{R}$ is in general much smaller than the diameter of the manifold (see below the example of S_1^N). Equivalently, let f be a Lipschitz map on M and let m be a median of f for μ (i.e. $\mu(f \geq m) \geq \frac{1}{2}$ and $\mu(f \leq m) \geq \frac{1}{2}$). If we apply (2.1) to the set $A = \{f \leq m\}$, it easily follows that, for every $r \geq 0$,

$$\mu(f \geq m + r) \leq \exp\left(-\frac{R r^2}{2\|f\|_{\mathrm{Lip}}^2}\right).$$

Together with the corresponding inequality for $A = \{f \geq m\}$, for every $r \geq 0$,

$$\mu(|f - m| \geq r) \leq 2\exp\left(-\frac{R r^2}{2\|f\|_{\mathrm{Lip}}^2}\right). \tag{2.2}$$

Thus, f is concentrated around some mean value with a large probability depending on some exponential of the ratio $R/\|f\|_{\mathrm{Lip}}^2$. This property has taken the name of concentration of measure phenomenon (cf. [G-M], [Mi-S]).

The preceding bounds are of particular interest for families of probability measures such as for example the measures σ_1^N on the unit spheres S_1^N as N tends to infinity for which (2.2) becomes (since $R(S_1^N) = N - 1$),

$$\sigma_1^N(|f - m| \geq r) \leq 2\exp\left(-\frac{(N-1)r^2}{2\|f\|_{\mathrm{Lip}}^2}\right).$$

Think thus of the dimension N to be large. Of course, if $\|f\|_{\mathrm{Lip}} \leq 1$, for every x, y in S_1^N, $|f(x) - f(y)| \leq \pi$. But the preceding concentration inequality tells us that, already for r of the order of $1/\sqrt{N}$, $|f(x) - m| \leq r$ on a large set (in the sense of the measure) of x's. It is then from the interplay, in this inequality, between N large, r of the order of $1/\sqrt{N}$ and the respective values of m and $\|f\|_{\mathrm{Lip}}$ for f the gauge of a convex body that V. D. Milman draws the information in order to choose at random the Euclidean sections of the convex body and to prove in this way Dvoretzky's theorem (see [Mi1], [Mi3], [Mi-S]).

Another (this time noncompact) concentration example is of course the Gaussian measure γ_n on $\mathbb{R}^n$ (the canonical Gaussian measure on $\mathbb{R}^n$ with density with respect to Lebesgue measure $(2\pi)^{-n/2}\exp(-|x|^2/2)$). If $\gamma_n(A) \geq \frac{1}{2}$, we may take $a = 0$ in (1.5) and thus, for every $r \geq 0$,

$$\gamma_n(A_r) \geq \Phi(r) \geq 1 - \frac{1}{2}\mathrm{e}^{-r^2/2}. \tag{2.3}$$

Let f be a Lipschitz function on $\mathbb{R}^n$ with Lipschitz (semi-) norm

$$\|f\|_{\mathrm{Lip}} = \sup_{x \neq y} \frac{|f(x) - f(y)|}{|x - y|}$$

(where $|\cdot|$ is the Euclidean norm on $\mathbb{R}^n$) and let m be a median of f for γ_n. As efore, it follows from (2.3) that for every $r \geq 0$,

$$\gamma_n(|f - m| \geq r) \leq 2\big(1 - \Phi(r/\|f\|_{\mathrm{Lip}})\big) \leq \exp\left(-\frac{r^2}{2\|f\|_{\mathrm{Lip}}^2}\right). \tag{2.4}$$

Thus, for r of the order of $\|f\|_{\mathrm{Lip}}$, $|f-m|\le r$ on "most" of the space. The word "most" is described here by a Gaussian bound.

Isoperimetric and concentration inequalities involve both a measure and a metric structure (to define the isoperimetric neighborhoods or enlargements). On an abstract (Polish) metric space (X,d) equipped with a probability measure μ, or a family of probability measures, the concentration of measure phenomenon may be described via the concentration function

$$\alpha(r)=\alpha\big((X,d;\mu);r\big)=\sup\big\{1-\mu(A_r);\ A\subset X,\mu(A)\ge\tfrac{1}{2}\big\},\quad r\ge 0.$$

It is a remarkable property that this concentration function may be controlled in a rather large number of cases, and very often by a Gaussian decay as above. Isoperimetric tools are one of the most important and powerful arguments used to establish concentration inequalities. However, since we are concerned here with enlargements A_r for (relatively) large values of r rather than infinitesimal values, the study of the concentration phenomenon can be quite different from the study of isoperimetric inequalities, both in establishing new concentration inequalities and in applying them. Indeed, the framework of concentration inequalities is less restrictive than the isoperimetric setting as we will see for example in the next chapter, due mainly to the fact that we are not looking here for the extremal sets.

New tools to establish concentration inequalities were thus developed. For example, M. Gromov and V. D. Milman [G-M] showed that if X is a compact Riemannian manifold, for every $r\ge 0$,

$$\alpha(r)\le C\exp\left(-c\sqrt{\lambda_1}\,r\right)$$

(with $C=\frac{3}{4}$ and $c=\log(\frac{3}{2})$) where λ_1 is the first nontrivial eigenvalue of the Laplacian on X (see also [A-M-S] for a similar result in an abstract setting). In case $R(X)>0$, this is however weaker than (2.1). They also developed in this paper [G-M] several topological applications of concentration such as fixed point theorems. On the probabilistic side, some martingale inequalities put forward by B. Maurey [Ma1] have been used in the local theory of Banach spaces in extensions of Dvoretzky's theorem (cf. [Ma2], [Mi-S], [Pi1]). The main idea consists in writing, for a well-behaved function f, the difference $f-\mathbb{E}(f)$ as a sum of martingale differences $d_i=\mathbb{E}(f|\mathcal{F}_i)-\mathbb{E}(f|\mathcal{F}_{i-1})$ where $(\mathcal{F}_i)_i$ is some (finite) filtration. The classical arguments on sums of independent random variables then show in the same way that if $\sum_i\|d_i\|_\infty^2\le 1$, for every $r\ge 0$,

$$\mathbb{P}\big\{|f-\mathbb{E}(f)|\ge r\big\}\le 2e^{-r^2/2}\tag{2.5}$$

([Azu], [Ma1]). As a corollary, one can deduce from this result the concentration of Haar measure μ on $\{0,1\}^n$ equipped with the Hamming metric as

$$\alpha(r)\le C\exp\left(-\frac{r^2}{Cn}\right)$$

for some numerical constant $C>0$. This property may be established from the corresponding isoperimetric inequality ([Har], [W-W]), but V. D. Milman et G. Schechtman [Mi-S] deduce it from inequality (2.5) (see Chapter 3).

Our first aim here will be to give a simple proof of the concentration inequality (2.4). The proof is based on the Hermite or Ornstein-Uhlenbeck semigroup $(P_t)_{t\geq 0}$ defined, for every well-behaved function f on $\mathbb{R}^n$, by (Mehler's formula)

$$P_t f(x) = \int_{\mathbb{R}^n} f\big(e^{-t}x + (1-e^{-2t})^{1/2}y\big)\, d\gamma_n(y), \quad x \in \mathbb{R}^n, \quad t \geq 0,$$

and more precisely on its commutation property

$$\nabla P_t f = e^{-t} P_t(\nabla f). \tag{2.6}$$

The generator L of $(P_t)_{t\geq 0}$ is given by $Lf(x) = \Delta f(x) - \langle x, \nabla f(x)\rangle$, f smooth enough on $\mathbb{R}^n$, and we have the integration by parts formula

$$\int f(-Lg)d\gamma_n = \int \langle \nabla f, \nabla g\rangle d\gamma_n$$

for all smooth functions f and g on $\mathbb{R}^n$.

Proposition 2.1. *Let f be a Lipschitz function on $\mathbb{R}^n$ with $\|f\|_{\mathrm{Lip}} \leq 1$ and $\int f d\gamma_n = 0$. Then, for every real number λ,*

$$\int e^{\lambda f} d\gamma_n \leq e^{\lambda^2/2}. \tag{2.7}$$

Before turning to the proof of this proposition, let us briefly indicate how to deduce a concentration inequality from (2.7). Let f be any Lipschitz function on $\mathbb{R}^n$. As a consequence of (2.7), for every real number λ,

$$\int \exp\big(\lambda(f - \textstyle\int f d\gamma_n)\big) d\gamma_n \leq \exp\Big(\frac{1}{2}\lambda^2 \|f\|^2_{\mathrm{Lip}}\Big).$$

By Chebyshev's inequality, for every λ and $r \geq 0$,

$$\gamma_n\big(f \geq \textstyle\int f d\gamma_n + r\big) \leq \exp\Big(-\lambda r + \frac{1}{2}\lambda^2\|f\|^2_{\mathrm{Lip}}\Big)$$

and, optimizing in λ,

$$\gamma_n\big(f \geq \textstyle\int f d\gamma_n + r\big) \leq \exp\Big(-\frac{r^2}{2\|f\|^2_{\mathrm{Lip}}}\Big). \tag{2.8}$$

Applying (2.8) to both f and $-f$, we get a concentration inequality similar to (2.4) (around the mean rather than the median)

$$\gamma_n\big(\big|f - \textstyle\int f d\gamma_n\big| \geq r\big) \leq 2\exp\Big(-\frac{r^2}{2\|f\|^2_{\mathrm{Lip}}}\Big). \tag{2.9}$$

Two parameters are thus entering those concentration inequalities, the median or the mean of a function f with respect to γ_n and its Lipschitz norm. These can be very different. For example, if f is the Euclidean norm on $\mathbb{R}^n$, the median or the mean of f are of the order of $\sqrt{n}$ while $\|f\|_{\mathrm{Lip}} = 1$. This is one of the main features of these inequalities (cf. [L-T2]) and is at the basis of the Gaussian proofs of Dvoretzky's theorem (see [Pi1], [Pi3]).

Proof of Proposition 2.1. Let f be smooth enough on $\mathbb{R}^n$ with mean zero and $\|f\|_{\mathrm{Lip}} \leq 1$. For λ fixed, set $G(t) = \int \exp(\lambda P_t f) d\gamma_n$, $t \geq 0$. Since $\|f\|_{\mathrm{Lip}} \leq 1$, it follows from (2.6) that $|\nabla(P_s f)|^2 \leq \mathrm{e}^{-2s}$ almost everywhere for every s. Since $\int f d\gamma_n = 0$, $G(\infty) = 1$. Hence, for every $t \geq 0$,

$$\begin{aligned} G(t) = 1 - \int_t^\infty G'(s)\, ds &= 1 - \lambda \int_t^\infty \Big(\int L(P_s f) \exp(\lambda P_s f) d\gamma_n \Big) ds \\ &= 1 + \lambda^2 \int_t^\infty \Big(\int |\nabla(P_s f)|^2 \exp(\lambda P_s f) d\gamma_n \Big) ds \\ &\leq 1 + \lambda^2 \int_t^\infty \mathrm{e}^{-2s} G(s) ds \end{aligned}$$

where we used integration by parts in the space variable. Let $H(t)$ be the logarithm of the right hand side of this inequality. Then the preceding inequality tells us that $-H'(t) \leq \lambda^2 \mathrm{e}^{-2t}$ for every $t \geq 0$. Therefore

$$\log G(0) \leq H(0) = -\int_0^\infty H'(t)\, dt \leq \frac{\lambda^2}{2}$$

which is the claim of the proposition, at least for a smooth function f. The general case follows from a standard approximation, by considering for example $P_\varepsilon f$ instead of f and by letting then ε tend to zero. The proof is complete. □

Inequalities (2.8) and (2.9) will be our key argument in the study of integrability properties and tail behavior of Gaussian random vectors, as well as in the various applications throughout these notes. While the concentration inequalities (2.4) of isoperimetric nature may of course be used equivalently, we would like to emphasize here the simple proof of Proposition 2.1 from which (2.8) and (2.9) follow. Proposition 2.1 is due to I. A. Ibragimov, V. N. Sudakov and B. S. Tsirel'son [I-S-T] (see also B. Maurey [Pi1, p. 181]). Their argument is actually of more probabilistic nature. For every smooth enough function f on $\mathbb{R}^n$, write

$$f(W(1)) - \mathbb{E} f(W(1)) = \int_0^1 \langle \nabla T_{1-t} f(W(t)) dW(t) \rangle$$

where $(W(t))_{t \geq 0}$ is Brownian motion on $\mathbb{R}^n$ starting at the origin and where $(T_t)_{t \geq 0}$ denotes its associated semigroup (the heat semigroup), with the probabilistic normalization. Note then that the above stochastic integral has the same distribution as $\beta(\tau)$ where $(\beta(t))_{t \geq 0}$ is a one-dimensional Brownian motion and where

$\tau = \int_0^1 |\nabla T_{1-t} f(W(t))|^2 \, dt$. Therefore, for every Lipschitz function f such that $\|f\|_{\mathrm{Lip}} \le 1$, $\tau \le 1$ almost surely so that, for all $r \ge 0$,

$$\mathbb{P}\{f(W(1)) - \mathbb{E}f(W(1)) \ge r\} \le \mathbb{P}\{\max_{0 \le t \le 1} \beta(t) \ge r\}$$
$$= 2\int_r^\infty \mathrm{e}^{-x^2/2} \frac{dx}{\sqrt{2\pi}} \le \mathrm{e}^{-r^2/2}.$$

Since $W(1)$ has distribution γ_n, this is thus simply (2.8).

Proposition 2.1 and its proof may actually be extended to a larger setting to yield for example a simple proof of the concentration (2.2) (up to some numerical constants) on spheres or on Riemannian manifolds M with positive curvature $R(M)$. The proof uses the heat semigroup on M and Bochner's formula. It is inspired by the work of D. Bakry and M. Émery [B-É] (cf. also [D-S]) on hypercontractivity and logarithmic Sobolev inequalities. We will come back to this observation in Chapter 8. We establish the following fact (cf. [Led4]).

Proposition 2.2. *Let M be a compact Riemannian manifold of dimension N (≥ 2) and with $R(M) = R > 0$. Let f be a Lipschitz function on M with $\|f\|_{\mathrm{Lip}} \le 1$ and assume that $\int f d\mu = 0$. Then, for every real number λ,*

$$\int \mathrm{e}^{\lambda f} d\mu \le \mathrm{e}^{\lambda^2/2R}.$$

Proof. Let ∇ be the gradient on M and Δ be the Laplace-Beltrami operator. By Bochner's formula (see e.g. [G-H-L]), for every smooth function f on M, pointwise

$$\frac{1}{2}\Delta(|\nabla f|^2) - \langle \nabla f, \nabla(\Delta f)\rangle = \mathrm{Ric}(\nabla f, \nabla f) + \|\mathrm{Hess}(f)\|_2^2.$$

In particular,

$$\frac{1}{2}\Delta(|\nabla f|^2) - \langle \nabla f, \nabla(\Delta f)\rangle \ge R|\nabla f|^2 + \frac{1}{N}(\Delta f)^2. \tag{2.10}$$

The dimensional term in this inequality will actually not be used and we will only be concerned here with the inequality

$$\frac{1}{2}\Delta(|\nabla f|^2) - \langle \nabla f, \nabla(\Delta f)\rangle \ge R|\nabla f|^2. \tag{2.11}$$

Now, consider the heat semigroup $P_t = \mathrm{e}^{t\Delta}$, $t \ge 0$, and let f be smooth on M. Let further $s > 0$ be fixed and set, for every $t \le s$, $F(t) = P_t(|\nabla(P_{s-t}f)|^2)$. It is an immediate consequence of (2.11) applied to $P_{s-t}f$ that $F'(t) \ge 2RF(t)$, $t \le s$. Hence, the function $\mathrm{e}^{-2Rt}F(t)$ is increasing on the interval $[0,s]$ and we have thus that, for every $s \ge 0$,

$$|\nabla(P_s f)|^2 \le \mathrm{e}^{-2Rs} P_s(|\nabla f|^2). \tag{2.12}$$

This relation, which is actually equivalent to (2.11), expresses a commutation property between the respective actions of the semigroup and the gradient (cf. (2.6)). It is the only property which is used in the proof itself which is now exactly as the proof of Proposition 2.1. Let f be smooth on M with $\|f\|_{\mathrm{Lip}} \leq 1$ and $\int f d\mu = 0$. For λ fixed, set $G(t) = \int \exp(\lambda P_t f) d\mu$, $t \geq 0$. Since $\|f\|_{\mathrm{Lip}} \leq 1$, it follows from (2.12) that $|\nabla(P_s f)|^2 \leq \mathrm{e}^{-2Rs}$ almost everywhere for every s. Since $\int f d\mu = 0$, $G(\infty) = 1$. Hence, for every $t \geq 0$,

$$\begin{aligned} G(t) = 1 - \int_t^\infty G'(s)\,ds &= 1 - \lambda \int_t^\infty \left(\int \Delta(P_s f) \exp(\lambda P_s f) d\mu \right) ds \\ &= 1 + \lambda^2 \int_t^\infty \left(\int \left|\nabla(P_s f)\right|^2 \exp(\lambda P_s f) d\mu \right) ds \\ &\leq 1 + \lambda^2 \int_t^\infty \mathrm{e}^{-2Rs}\, G(s) ds \end{aligned}$$

where we used integration by parts in the space variable. The proof is completed as in Proposition 2.1. □

The commutation formula $\nabla P_t f = \mathrm{e}^{-t} P_t(\nabla f)$ of the Ornstein-Uhlenbeck semigroup expresses equivalently a Bochner formula for the second order generator L of infinite dimension ($N = \infty$) and constant curvature 1 (limit of $R(S^N_{\sqrt{N}})$ when N goes to infinity) of the type (2.10) or (2.11)

$$\frac{1}{2} L(|\nabla f|^2) - \langle \nabla f, \nabla(Lf) \rangle \geq (Lf)^2.$$

The geometry of the Ornstein-Uhlenbeck generator is thus purely infinite dimensional, even on a finite dimensional state space (as the Gaussian isoperimetric inequality itself, cf. Chapter 1). The abstract consequences of these observations are at the origin of the study by D. Bakry and M. Émery of hypercontractive diffusions under curvature-dimension hypotheses [B-É], [Bak]. We will come back to this question in Chapter 8 and actually show, according to [A-M-S], that (2.7) can be deduced directly from hypercontractivity.

At this point, we realized that simple semigroup arguments may be used to establish concentration properties, however on Lipschitz functions rather than sets. It is not difficult however to deduce from Propositions 2.1 and 2.2 inequalities on sets very close to the inequalities which follow from isoperimetry (but still for "large" neighborhoods). We briefly indicate the procedure in the Gaussian setting.

Let A be a Borel set in $\mathbb{R}^n$ with canonical Gaussian measure $\gamma_n(A) > 0$. For every $u \geq 0$, let

$$f_{A,u}(x) = \min(d(x, A), u)$$

where $d(x, A)$ is the Euclidean distance from the point x to the set A. Clearly $\|f_{A,u}\|_{\mathrm{Lip}} \leq 1$ so that we may apply inequality (2.8) to this family of Lipschitz functions when u varies. Let $E_{A,u} = \int f_{A,u} d\gamma_n$. Inequality (2.8) applied to $f_{A,u}$ and $r = u - E_{A,u}$ yields

$$\gamma_n\big(x \in \mathbb{R}^n; \min(d(x, A), u) \geq u\big) \leq \exp\left(-\frac{1}{2}(u - E_{A,u})^2\right),$$

that is

$$\gamma_n(x; x \notin A_u) \le \exp\Big(-\frac{1}{2}(u - E_{A,u})^2\Big) \tag{2.13}$$

since $d(x,A) > u$ if and only if $x \notin A_u$. We have now to appropriately control the expectations $E_{A,u} = \int f_{A,u} d\gamma_n$, possibly only with u and the measure of A. A first bound is simply $E_{A,u} \le u\,\gamma(A^c)$ which already yields,

$$\gamma_n(x; x \notin A_u) \le \exp\Big(-\frac{1}{2}u^2\gamma_n(A)^2\Big)$$

for every $u \ge 0$. This inequality may already be compared to (2.3). However, if we use this estimate recursively, we get

$$E_{A,u} = \int_0^u \gamma_n\big(x; d(x,A) > t\big)dt \le \int_0^u \min\big(\gamma_n(A^c), e^{-t^2\gamma_n(A)^2/2}\big)dt.$$

If we let then $\delta(v)$ be the decreasing function on $(0,1]$ defined by

$$\delta(v) = \int_0^\infty \min\big(1 - v, e^{-t^2v^2/2}\big)dt, \tag{2.14}$$

we have $E_{A,u} \le \delta(\gamma_n(A))$ uniformly in u. Hence, from (2.13), for every $u \ge 0$,

$$\gamma_n(x; x \notin A_u) \le \exp\Big(-\frac{u^2}{2} + uE_{A,u}\Big) \le \exp\Big(-\frac{u^2}{2} + u\delta\big(\gamma_n(A)\big)\Big).$$

In conclusion, we obtained from Proposition 2.1 and inequality (2.8) that, for every $r \ge 0$,

$$\gamma_n(A_r) \ge 1 - \exp\Big(-\frac{r^2}{2} + r\delta\big(\gamma_n(A)\big)\Big). \tag{2.15}$$

This simple argument thus yields an inequality very similar in nature to the isoperimetric bound (2.3), with however the extra factor $r\delta(\gamma_n(A))$. (Using the preceding recursive argument, one may of course improve further and further this estimate.) Due to the fact that $\delta(\gamma_n(A)) \to 0$ as $\gamma_n(A) \to 1$, this result can be used exactly as the isoperimetric inequality in almost all the applications presented in these notes. We will come back to this in Chapter 4 for example, and we will always use (2.15) rather than isoperimetry in the applications.

We conclude this chapter with a proposition closely related to Proposition 2.1 and the proof of which is similar. It will be used in Chapter 4 in some large deviation statement for the Ornstein-Uhlenbeck process. We only consider the Gaussian setting.

Proposition 2.3. *Let f be a Lipschitz function on $\mathbb{R}^n$ with $\|f\|_{\mathrm{Lip}} \le 1$. Then, for every real number λ and every $t \ge 0$,*

$$\iint \exp\big(\lambda\big[f(e^{-t}x + (1-e^{-2t})^{1/2}y) - f(x)\big]\big)\,d\gamma_n(x)d\gamma_n(y) \le \exp\big(\lambda^2(1-e^{-t})\big).$$

Proposition 2.3 will be used for the small values of the time t. When $t \to \infty$, it is somewhat weaker than Proposition 2.1. The stochastic version of this proposition is inspired from the forward and backward martingales of T. Lyons and W. Zheng [L-Z] (see [Tak], [Fa]).

Proof. The left hand side of the inequality of Proposition 2.3 may be rewritten as

$$G(t) = \int e^{-\lambda f} P_t(e^{\lambda f})\,d\gamma_n.$$

Let λ be fixed and f be smooth enough. For every $t \ge 0$,

$$\begin{aligned} G(t) = 1 + \int_0^t G'(s)\,ds &= 1 + \int_0^t \Big(\int e^{-\lambda f} L P_s(e^{\lambda f})\,d\gamma_n\Big)\,ds \\ &= 1 + \lambda^2 \int_0^t e^{-s}\Big(\int e^{-\lambda f}\langle \nabla f, P_s(e^{\lambda f}\nabla f)\rangle d\gamma_n\Big) \\ &\le 1 + \lambda^2 \int_0^t e^{-s} G(s)ds \end{aligned}$$

since $|\nabla f| \le 1$ almost everywhere. Let $H(t) = \log(1 + \lambda^2 \int_0^t e^{-s} G(s)ds)$, $t \ge 0$. We just showed that $H'(t) \le \lambda^2 e^{-t}$ for every $t \ge 0$. Hence,

$$H(t) = \int_0^t H'(s)\,ds \le \lambda^2 \int_0^t e^{-s}\,ds = \lambda^2(1-e^{-t})$$

and the proof is complete. □

If A and B are subsets of $\mathbb{R}^n$, and if $t \ge 0$, set

$$K_t(A,B) = \int_A P_t(I_B)\,d\gamma_n \quad \Big(= \int_B P_t(I_A)\,d\gamma_n\Big)$$

where I_A is the indicator function of the set A. Assume that $d(A,B) > r > 0$ (for the Euclidean distance on $\mathbb{R}^n$). In particular, $B \subset (A_r)^c$ so that

$$K_t(A,B) \le K_t\big(A,(A_r)^c\big).$$

Apply then Proposition 2.3 to the Lipschitz map $f(x) = d(x,A)$. For every $t \ge 0$ and every $\lambda \ge 0$,

$$K_t\big(A,(A_r)^c\big) = \int_A P_t\big(I_{(A_r)^c}\big)d\gamma_n \le e^{-\lambda r}\int_A e^{-\lambda f} P_t(e^{\lambda f})\,d\gamma_n$$

since $I_{(A_r)^c} \leq e^{-\lambda r} e^{\lambda f}$. Hence

$$K_t(A,(A_r)^c) \leq e^{-\lambda r} e^{\lambda^2(1-e^{-t})}.$$

Optimizing in λ yields

$$K_t(A,B) \leq K_t(A,(A_r)^c) \leq \exp\left(-\frac{r^2}{4(1-e^{-t})}\right). \tag{2.16}$$

Formula (2.16) will thus be used in Chapter 4 in applications to large deviations for the Ornstein-Uhlenbeck process.

3. ISOPERIMETRIC AND CONCENTRATION INEQUALITIES FOR PRODUCT MEASURES

In this chapter, we present several isoperimetric and concentration inequalities for product measures due to M. Talagrand. On the basis of the product structure of the canonical Gaussian measure γ_n and various open problems on sums of independent vector valued random variables, M. Talagrand developed in the past years new inequalities in products of probability spaces by defining several different notions of isoperimetric enlargement in this setting. These results appear as a striking illustration of the power of abstract concentration ideas which can be developed far beyond the framework of the classical geometrical isoperimetric inequalities. One of the main applications of his powerful techniques and results concerns tail behaviors and limit properties of sums of independent Banach space valued random variables. It partly motivated the writing of the book [L-T2] and we thus refer the interested reader to this reference for this kind of applications. New applications concern geometric probabilities, percolation, statistical mechanics... We will concentrate here on some of the theoretical inequalities and their relations with the Gaussian isoperimetric inequality, as well as on some recent and new aspects of the work of M. Talagrand [Ta16]. We actually refer to [Ta16] for complete proofs and details of some of the main results we present here. The reader that is interested first in the applications of the Gaussian isoperimetric and concentration inequalities may skip this chapter and come back to it after Chapter 7.

One first example studied by M. Talagrand is uniform measure on $\{0,1\}^{\mathbb{N}}$. For this example, he established a concentration inequality independent of the dimension [Ta3]. More importantly, he developed a new powerful scheme of proof based on induction on the number of coordinates. This technique allowed him to investigate isoperimetric and concentration inequalities in abstract product spaces.

Let (Ω, Σ, μ) be a (fixed but arbitrary) probability space and let P be the product measure $\mu^{\otimes n}$ on Ω^n. A point x in Ω^n has coordinates $x = (x_1, \ldots, x_n)$. (In the results which we present, one should notice that one does not increase the generality with arbitrary products spaces $\left(\Pi_{i=1}^n \Omega_i, \bigotimes_{i=1}^n \mu_i\right)$. Since the crucial inequalities will not depend on n, we need simply towork on products of $(\widetilde{\Omega}, \widetilde{\mu}) = \left(\Pi_{i=1}^n \Omega_i, \bigotimes_{i=1}^n \mu_i\right)$

with itself and consider the coordinate map

$$\widetilde{x} = (\widetilde{x}_1, \ldots, \widetilde{x}_n) \in \widetilde{\Omega}^n \to (\widetilde{x}_i)_i \in \Omega_i, \quad \widetilde{x}_i = ((\widetilde{x}_i)_1, \ldots, (\widetilde{x}_i)_n) \in \widetilde{\Omega},$$

that only depends on the i-th factor.)

The Hamming distance on Ω^n is defined by

$$d(x,y) = \mathrm{Card}\{1 \leq i \leq n;\, x_i \neq y_i\}.$$

The concentration function α of Ω^n for d satisfies, for every product probability P,

$$\alpha(r) \leq C \exp\left(-\frac{r^2}{Cn}\right), \quad r \geq 0, \tag{3.1}$$

where $C > 0$ is numerical. In particular, if $P(A) \geq \frac{1}{2}$, for most of the elements x in Ω^n, there exists $y \in A$ within distance $\sqrt{n}$ of x. On the two point space, this may be shown to follow from the corresponding isoperimetric inequality [Har], [W-W]. A proof using the martingale inequality (2.4) is given in [Mi-S] (see also [MD] for a version with a better constant). As we will see later on, one can actually give an elementary proof of this result by induction on n. It might be important for the sequel to briefly indicate the procedure. If A is a subset of Ω^n and $x \in \Omega^n$, denote by $\varphi_A^1(x)$ the Hamming distance from x to A thus defined by

$$\varphi_A^1(x) = \inf\big\{k \geq 0;\, \exists\, y \in A,\, \mathrm{Card}\{1 \leq i \leq n;\, x_i \neq y_i\} \leq k\big\}.$$

(Although this is nothing more than $d(x,A)$, this notation will be of better use in the subsequent developments.) M. Talagrand's approach [Ta3], [Ta16] then consists in showing that, for every $\lambda > 0$ and every product probability P,

$$\int e^{\lambda \varphi_A^1}\, dP \leq \frac{1}{P(A)}\, e^{n\lambda^2/4}. \tag{3.2}$$

In particular, by Chebyshev's inequality, for every integer k,

$$P(\varphi_A^1 \geq k) \leq \frac{1}{P(A)}\, e^{-k^2/n},$$

that is the concentration (3.1). The same proof actually applies to all the Hamming metrics

$$d_a(x,y) = \sum_{i=1}^n a_i I_{\{x_i \neq y_i\}}, \quad a = (a_1, \ldots, a_n) \in \mathbb{R}_+^n,$$

with $|a|^2 = \sum_{i=1}^n a_i^2$ instead of n in the right hand side of (3.2). One can improve this result by studying functions of the probability of A in (3.2) such as $P(A)^{-\gamma}$. Optimizing in $\gamma > 0$, it is then shown in [Ta16] that for $k \geq \left(\frac{n}{2} \log \frac{1}{P(A)}\right)^{1/2}$,

$$P(\varphi_A^1 \geq k) \leq \exp\left(-\frac{2}{n}\left[k - \left(\frac{n}{2} \log \frac{1}{P(A)}\right)^{1/2}\right]^2\right),$$

which is close to the best exponent $-2k^2/n$ ([MD]).

Note that various measurability questions arise on φ_A^1. These are actually completely unessential and will be ignored in what follows (start for example with a finite probability space Ω).

The previous definition of φ_A^1 allows one to investigate various and very different ways to measure the "distance" of a point x to a set A. In particular, this need not anymore be metric. The functional φ_A^1 controls a point x in Ω^n by a single point in A. Besides this function, M. Talagrand defines two new main controls, or enlargement functions: one using several points in A, and one using a convex hull procedure. In each case, a Gaussian concentration will be proved.

The convex hull control is defined with the metric $\varphi_A^c(x) = \sup_{|a|=1} d_a(x, A)$. However, this definition somewhat hides the convexity properties of the functional φ_A^c which will be needed in its investigation. For a subset $A \subset \Omega^n$ and $x \in \Omega^n$, let

$$U_A(x) = \big\{ s = (s_i)_{1 \le i \le n} \in \{0,1\}^n;\ \exists y \in A \text{ such that } y_i = x_i \text{ if } s_i = 0 \big\}.$$

(One can use instead the collection of the indicator functions $I_{\{x_i \ne y_i\}}$, $y \in A$.) Denote by $V_A(x)$ the convex hull of $U_A(x)$ as a subset of $\mathbb{R}^n$. Note that $0 \in V_A(x)$ if and only if $x \in A$. One may then measure the distance from x to A by the Euclidean distance $d(0, V_A(x))$ from 0 to $V_A(x)$. It is easily seen that $d(0, V_A(x)) = \varphi_A^c(x)$. If $d(0, V_A(x)) \le r$, there exists z in $V_A(x)$ with $|z| \le r$. Let $a \in \mathbb{R}_+^n$ with $|a| = 1$. Then

$$\inf_{y \in V_A(x)} \langle a, y \rangle \le \langle a, z \rangle \le |z| \le r.$$

Since

$$\inf_{y \in V_A(x)} \langle a, y \rangle = \inf_{s \in U_A(x)} \langle a, s \rangle = d_a(x, A),$$

$\varphi_A^c(x) \le r$. The converse, that is not needed below, follows from Hahn-Banach's theorem.

The functional $\varphi_A^c(x)$ is a kind of uniform control in the Hamming metrics d_a, $|a| = 1$. The next theorem [Ta6], [Ta16] extends the concentration (3.2) to this uniformity.

Theorem 3.1. *For every subset A of Ω^n, and every product probability P,*

$$\int \exp\Big(\frac{1}{4}\,(\varphi_A^c)^2\Big)\, dP \le \frac{1}{P(A)}\,.$$

In particular, for every $r \ge 0$,

$$P(\varphi_A^c \ge r) \le \frac{1}{P(A)}\, \mathrm{e}^{-r^2/4}.$$

To briefly describe the general scheme of proofs by induction on the number of coordinates, we present the proof of Theorem 3.1. The main difficulty in this type of statements is to find the adapted recurrence hypothesis expressed here by the exponential integral inequalities.

Proof. The case $n = 1$ is easy. To go from n to $n+1$, let A be a subset of Ω^{n+1} and let B be its projection on Ω^n. Let furthermore, for $\omega \in \Omega$, $A(\omega)$ be the section of A along ω. If $x \in \Omega^n$ and $\omega \in \Omega$, set $z = (x, \omega)$. The key observation is the following: if $s \in U_{A(\omega)}(x)$, then $(s, 0) \in U_A(z)$, and if $t \in U_B(x)$, then $(t, 1) \in U_A(z)$. It follows that if $u \in V_{A(\omega)}(x)$, $v \in V_B(x)$ and $0 \leq \lambda \leq 1$, then $(\lambda u + (1-\lambda)v, 1-\lambda) \in V_A(z)$. By definition of φ_A^c and convexity of the square function,

$$\varphi_A^c(z)^2 \leq (1-\lambda)^2 + \left|\lambda u + (1-\lambda)v\right|^2 \leq (1-\lambda)^2 + \lambda|u|^2 + (1-\lambda)|v|^2.$$

Hence,

$$\varphi_A^c(z)^2 \leq (1-\lambda)^2 + \lambda\varphi_{A(\omega)}^c(x)^2 + (1-\lambda)\varphi_B^c(x)^2.$$

Now, by Hölder's inequality and the induction hypothesis, for every ω in Ω,

$$\begin{aligned}
&\int_{\Omega^n} \exp\Big(\frac{1}{4}\left(\varphi_A^c(x,\omega)\right)^2\Big)dP(x) \\
&\qquad \leq e^{(1-\lambda)^2/4}\Big(\int_{\Omega^n} \exp\Big(\frac{1}{4}\left(\varphi_{A(\omega)}^c\right)^2\Big)dP\Big)^{\lambda}\Big(\int_{\Omega^n} \exp\Big(\frac{1}{4}\left(\varphi_B^c\right)^2\Big)dP\Big)^{1-\lambda} \\
&\qquad \leq e^{(1-\lambda)^2/4}\Big(\frac{1}{P(A(\omega))}\Big)^{\lambda}\Big(\frac{1}{P(B)}\Big)^{1-\lambda}
\end{aligned}$$

that is,

$$\int_{\Omega^n} \exp\Big(\frac{1}{4}\left(\varphi_A^c(x,\omega)\right)^2\Big)dP(x) \leq \frac{1}{P(B)}\, e^{(1-\lambda)^2/4}\Big(\frac{P(A(\omega))}{P(B)}\Big)^{-\lambda}.$$

Optimize now in λ (cf. [Ta16]) to get that

$$\int_{\Omega^n} \exp\Big(\frac{1}{4}\left(\varphi_A^c(x,\omega)\right)^2\Big)dP(x) \leq \frac{1}{P(B)}\Big(2 - \frac{P(A(\omega))}{P(B)}\Big).$$

To conclude, integrate in ω, and, by Fubini's theorem,

$$\int_{\Omega^{n+1}} \exp\Big(\frac{1}{4}\left(\varphi_A^c(x,\omega)\right)^2\Big)dP(x)d\mu(\omega) \leq \frac{1}{P(B)}\Big(2 - \frac{P\otimes\mu(A)}{P(B)}\Big) \leq \frac{1}{P\otimes\mu(A)}$$

since $u(2-u) \leq 1$ for every real number u. Theorem 3.1 is established. □

It is easy to check that if $\Omega = [0,1]$ and if d_A is the Euclidean distance to the convex hull $\mathrm{Conv}(A)$ of A, then $d_A \leq \varphi_A^c$. Let then f be a convex function on $[0,1]^n$ such that $\|f\|_{\mathrm{Lip}} \leq 1$, and let m be a median of f for P and $A = \{f \leq m\}$. Since f is convex, $f \leq m$ on $\mathrm{Conv}(A)$. Using that $\|f\|_{\mathrm{Lip}} \leq 1$, we see that $f(x) < m + r$ for every x such that $d_A(x) < r$, $r \geq 0$. Hence, by Theorem 3.1, $P(f \geq m+r) \leq 2e^{-r^2/4}$.

On the other hand, let $B = \{f \le m - r\}$. As above, $d_B(x) < r$ implies $f(x) < m$. By definition of the median, it follows from Theorem 3.1 again that

$$1 - \frac{1}{P(B)}\, e^{-r^2/4} \le P(d_B < r) \le P(f < m) \le \frac{1}{2}.$$

Hence $P(f \le m - r) \le 2e^{-r^2/4}$. Therefore

$$P(|f - m| \ge r) \le 4\,e^{-r^2/4} \tag{3.3}$$

for every $r \ge 0$ and every probability measure μ on $[0,1]$. The numerical constant 4 in the exponent may be improved to get close to the best possible value 2. This concentration inequality (3.3) is very similar to the Gaussian concentration (2.4) or (2.9), with however f convex. It applies to norms of vector valued sums $\sum_i \varphi_i e_i$ with coefficients e_i in a Banach space E where the φ_i's are independent real valued uniformly bounded random variables on some probability space $(\Omega, \mathcal{A}, \mathbb{P})$. This applies in particular to independent symmetric Bernoulli (or Rademacher) random variables and (3.3) allows us in particular to recover and improve the pioneer inequalities by J.-P. Kahane [Ka1], [Ka2] (cf. also Chapter 4). More precisely, if $\|\varphi_i\|_\infty \le 1$ for all i's and if $S = \sum_i \varphi_i e_i$ is almost surely convergent in E, for every $r \ge 0$,

$$\mathbb{P}\big(\big|\|S\| - m\big| \ge r\big) \le 4\,e^{-r^2/16\sigma^2} \tag{3.4}$$

where m is a median of $\|S\|$ and where

$$\sigma = \sup_{\xi \in E^*, \|\xi\| \le 1} \Big(\sum_i \langle \xi, e_i\rangle^2\Big)^{1/2}.$$

Typically, the martingale inequality (2.5) would only yield a similar inequality but with σ replaced by the larger quantity $\sum_i \|e_i\|^2$. This result is the exact analogue of what we will obtain on Gaussian series in the next chapter through isoperimetric and concentration inequalities. It shows how powerful the preceding induction techniques can be. In particular, we may integrate by parts (3.4) to see that $\mathbb{E}\exp(\alpha\|S\|^2) < \infty$ for every α (cf. [L-T2]). Furthermore, for every $p > 0$,

$$\big(\mathbb{E}\|S\|^p\big)^{1/p} \le m + C_p\sigma \tag{3.5}$$

where C_p is of the order of $\sqrt{p}$ as $p \to \infty$.

When the φ_i's are Rademacher random variables, by the classical Khintchine inequalities, one easily sees that $\sigma \le 2\sqrt{2}m'$ for every m' such that $\mathbb{P}\{\|S\| \ge m'\} \le \frac{1}{8}$ (see [L-T2], p. 99). Since $m \le m' \le (8\mathbb{E}\|S\|^q)^{1/q}$ for every $0 < q < \infty$, we also deduce from (3.5) the moment equivalences for $\|S\|$: for every $0 < p, q < \infty$, there exists $C_{p,q} > 0$ only depending on p and q such that

$$\big(\mathbb{E}\|S\|^p\big)^{1/p} \le C_{p,q}\big(\mathbb{E}\|S\|^q\big)^{1/q}. \tag{3.6}$$

By the classical central limit theorem, these inequalities imply the corresponding ones for Gaussian averages (see (4.5)). In the case of the two point space, the method of proof by induction on the dimension is very similar to hypercontractivity techniques [Bon], [Gr3], [Bel] (which, in particular, also show (3.6) [Bo6]). Some further connections on the basis of this observation are developed in [Ta11] in analogy with the Gaussian example (Chapter 8). See also [Bo8]. It was recently shown in [L-O] that $C_{2,1} = \sqrt{2}$ as in the real case [Sz].

We turn to the control by a finite number of points. If q is an integer ≥ 2 and if $A^1, \ldots, A^q$ are subsets of Ω^n, let, for every $x = (x_1, \ldots, x_n)$ in Ω^n,

$$\varphi^q(x) = \varphi^q_{A^1,\ldots,A^q}(x) = \inf\big\{k \geq 0;\ \exists\, y^1 \in A^1, \ldots, \exists\, y^q \in A^q \text{ such that } \operatorname{Card}\{1 \leq i \leq n;\ x_i \notin \{y_i^1, \ldots, y_i^q\}\} \leq k\big\}.$$

(We agree that $\varphi^q = \infty$ if one of the A_i's is empty.) If, for every $i = 1, \ldots, n$, $A^i = A$ for some $A \subset \Omega^n$, $\varphi^q(x) \leq k$ means that the coordinates of x may be copied, at the exception of k of them, by the coordinates of q elements in A. Using again a proof by induction on the number of coordinates, M. Talagrand [Ta16] established the following result.

Theorem 3.2. *Under the previous notations,*

$$\int q^{\varphi^q(x)}\, dP(x) \leq \prod_{i=1}^{q} \frac{1}{P(A^i)}\,.$$

In particular, for every integer k,

$$P(\varphi^q \geq k) \leq q^{-k} \prod_{i=1}^{q} \frac{1}{P(A^i)}\,.$$

Proof. One first observes that if g is a function on Ω such that $\frac{1}{q} \leq g \leq 1$, then

$$\int \frac{1}{g}\, d\mu \left(\int g\, d\mu\right)^q \leq 1. \tag{3.7}$$

Since $\log u \leq u - 1$, it suffices to show that

$$\int \frac{1}{g}\, d\mu + q \int g\, d\mu = \int \Big(\frac{1}{g} + qg\Big) d\mu \leq q + 1.$$

But this is obvious since $\frac{1}{u} + qu \leq q + 1$ for $\frac{1}{q} \leq u \leq 1$.

Let g_i, $i = 1, \ldots, q$, be functions on Ω such that $0 \leq g_i \leq 1$. Applying (3.7) to g given by $\frac{1}{g} = \min(q, \min_{1 \leq i \leq q} \frac{1}{g_i})$ yields

$$\int \min\Big(q, \min_{1 \leq i \leq q} \frac{1}{g_i}\Big) d\mu \left(\prod_{i=1}^{q} \int g_i\, d\mu\right) \leq 1 \tag{3.8}$$

since $g_i \le g$ for every $i = 1, \dots, q$.

We prove the theorem by induction over n. If $n = 1$, the result follows from (3.8) by taking $g_i = I_{A^i}$. Assume Theorem 3.2 has been proved for n and let us prove it for $n+1$. Consider sets $A^1, \dots, A^q$ of Ω^{n+1}. For $\omega \in \Omega$, consider $A^i(\omega)$, $i = 1, \dots, q$, as well as the projections B^i of A^i on Ω^n, $i = 1, \dots, q$. Note that if we set $g_i = P(A^i(\omega))/P(B^i)$ in (3.8), we get by Fubini's theorem that

$$\int \min\left(q \prod_{i=1}^{q} \frac{1}{P(B^i)}, \min_{1 \le j \le q} \prod_{i=1}^{q} \frac{1}{P(C^{ij})} \right) d\mu \le \prod_{i=1}^{q} \frac{1}{P \otimes \mu(A^i)} \tag{3.9}$$

where $C^{ij} = B^i$ if $i \ne j$ and $C^{ii} = A^i(\omega)$. The basic observation is now the following: for $(x, \omega) \in \Omega^n \times \Omega$,

$$\varphi^q_{A^1,\dots,A^q}(x,\omega) \le 1 + \varphi^q_{B^1,\dots,B^q}(x)$$

and, for every $1 \le j \le q$,

$$\varphi^q_{A^1,\dots,A^q}(x,\omega) \le \varphi^q_{C^{1j},\dots,C^{qj}}(x).$$

It follows that

$$\begin{aligned}
&\int_{\Omega^{n+1}} q^{\varphi^q_{A^1,\dots,A^q}(x,\omega)} dP(x) d\mu(\omega) \\
&\quad \le \int_{\Omega^{n+1}} \min\left(q\, q^{\varphi^q_{B^1,\dots,B^q}(x)}, \min_{1 \le j \le q} q^{\varphi^q_{C^{1j},\dots,C^{qj}}(x)}\right) dP(x) d\mu(\omega) \\
&\quad \le \int_{\Omega} \min\left(q \int_{\Omega^n} q^{\varphi^q_{B^1,\dots,B^q}(x)} dP(x), \min_{1 \le j \le q} \int_{\Omega^n} q^{\varphi^q_{C^{1j},\dots,C^{qj}}(x)} dP(x) \right) d\mu(\omega) \\
&\quad \le \int_{\Omega} \min\left(q \prod_{i=1}^{q} \frac{1}{P(B^i)}, \min_{1 \le j \le q} \prod_{i=1}^{q} \frac{1}{P(C^{ij})} \right) d\mu(\omega)
\end{aligned}$$

by the recurrence hypothesis. The conclusion follows from (3.9). □

In the applications, q is usually fixed, for example equal to 2. Theorem 3.2 then shows how to control, with a fixed subset A, arbitrary samples with an exponential decay of the probability in the number of coordinates which are neglected. Theorem 3.2 was first proved by delicate rearrangement techniques (of isoperimetric flavor) in [Ta5]. It allowed M. Talagrand to solve a number of open questions in probability in Banach spaces (and may be considered at the origin of the subsequent abstract developments, see [Ta5], [Ta16], [L-T2]). To briefly illustrate how Theorem 3.2 is used in the applications, let us consider a sum $S = X_1 + \cdots + X_n$ of independent nonnegative random variables on some probability space $(\Omega, \mathcal{A}, \mathbb{P})$. In the preceding language, we may simply equip $[0, \infty)^n$ with the product P of the laws of the X_i's. Let $A = \{\sum_{i=1}^n x_i \le m\}$ where m is such that, for example, $P(A) \ge \frac{1}{2}$. Let $\varphi^q = \varphi^q_{A,\dots,A}$. If $x \in \{\varphi^q \le k\}$, there exist $y^1, \dots, y^q$ in A such that $\mathrm{Card}\, I \le k$ where $I = \{1 \le i \le n;\ x_i \notin \{y^1_i, \dots, y^q_i\}\}$. Take then a partition $(J_j)_{1 \le j \le q}$ of $\{1, \dots, n\} \setminus I$ such that $x_i = y^j_i$ if $i \in J_j$. Then,

$$\sum_{i \notin I} x_i = \sum_{j=1}^{q} \sum_{i \in J_j} y^j_i \le \sum_{j=1}^{q} \sum_{i=1}^{n} y^j_i \le qm$$

where we are using a crucial monotonicity property since the coordinates are non-negative. It follows that

$$\sum_{i=1}^{n} x_i \leq qm + \sum_{i=1}^{k} x_i^*$$

where $\{x_1^*, \ldots, x_n^*\}$ is the nonincreasing rearrangement of the sample $\{x_1, \ldots, x_n\}$. Hence, according to Theorem 3.2, for every integers $k, q \geq 1$, and every $t \geq 0$,

$$\mathbb{P}\{S \geq qm + t\} \leq 2^q q^{-(k+1)} + \mathbb{P}\left\{\sum_{i=1}^{k} X_i^* \geq t\right\}. \tag{3.10}$$

Let $\mathcal{F}$ be a family of n-tuples $\alpha = (\alpha_i)_{1 \leq i \leq n}$, $\alpha_i \geq 0$. It is plain that the preceding argument leading to (3.10) applies in the same way to

$$S = \sup_{\alpha \in \mathcal{F}} \sum_{i=1}^{n} \alpha_i X_i$$

to yield

$$\mathbb{P}\{S \geq qm + t\} \leq 2^q q^{-(k+1)} + \mathbb{P}\left\{\sigma \sum_{i=1}^{k} X_i^* \geq t\right\}$$

where $\sigma = \sup\{\alpha_i ; 1 \leq i \leq n, \alpha \in \mathcal{F}\}$.

Now, in probability in Banach spaces or in the study of empirical processes, one does not usually deal with nonnegative summands. One general situation is the following (cf. [L-T2], [Ta13] for the notations and further details). Let $X_1, \ldots, X_n$ be independent random variables taking values in some space $\mathcal{S}$ and consider say a countable family $\mathcal{F}$ of (measurable) real valued functions on $\mathcal{S}$. We are interested in bounds on the tail of

$$\left\| \sum_{i=1}^{n} f(X_i) \right\|_{\mathcal{F}} = \sup_{f \in \mathcal{F}} \left| \sum_{i=1}^{n} f(X_i) \right|.$$

If $\mathbb{E} f(X_i) = 0$ for every $1 \leq i \leq n$ and every $f \in \mathcal{F}$, standard symmetrization techniques (cf. [L-T2]) reduce to the investigation of

$$\left\| \sum_{i=1}^{n} \varepsilon_i f(X_i) \right\|_{\mathcal{F}}$$

where $(\varepsilon_i)_{1 \leq i \leq n}$ are independent symmetric Bernoulli random variables independent of the X_i's. Although the isoperimetric approach applies in the same way, we may not use directly here the crucial monotonicity property on the coordinates. We turn over this difficulty via a symmetrization procedure with Rademacher random variables which was developed first in the study of the law of the iterated logarithm [L-T1]. One writes

$$\left\| \sum_{i=1}^{n} \varepsilon_i f(X_i) \right\|_{\mathcal{F}} = \left(\left\| \sum_{i=1}^{n} \varepsilon_i f(X_i) \right\|_{\mathcal{F}} - \mathbb{E}_\varepsilon \left\| \sum_{i=1}^{n} \varepsilon_i f(X_i) \right\|_{\mathcal{F}} \right) + \mathbb{E}_\varepsilon \left\| \sum_{i=1}^{n} \varepsilon_i f(X_i) \right\|_{\mathcal{F}}$$

where $\mathbb{E}_\varepsilon$ is partial integration with respect to the Bernoulli variables $\varepsilon_1,\ldots,\varepsilon_n$. Now, on $\mathbb{E}_\varepsilon\|\sum_{i=1}^n \varepsilon_i f(X_i)\|_{\mathcal{F}}$ the monotonicity property is satisfied, since, by Jensen's inequality and independence, for every subset $I\subset\{1,\ldots,n\}$,

$$\mathbb{E}_\varepsilon\Big\|\sum_{i\in I}\varepsilon_i f(X_i)\Big\|_{\mathcal{F}}\le \mathbb{E}_\varepsilon\Big\|\sum_{i=1}^n\varepsilon_i f(X_i)\Big\|_{\mathcal{F}}.$$

Therefore, the isoperimetric method may be used efficiently on this term. The remainder term

$$\Big\|\sum_{i=1}^n\varepsilon_i f(X_i)\Big\|_{\mathcal{F}}-\mathbb{E}_\varepsilon\Big\|\sum_{i=1}^n\varepsilon_i f(X_i)\Big\|_{\mathcal{F}}$$

is bounded, conditionally on the X_i's, with the deviation inequality (3.4) by a Gaussian tail involving

$$\sup_{f\in\mathcal{F}}\sum_{i=1}^n f(X_i)^2$$

which will again satisfy this monotonicity property. The proper details are presented in [L-T2], p. 166-169. Combining the arguments yields the inequality, for nonnegative integers k,q and real numbers $s,t\ge 0$,

$$\mathbb{P}\bigg\{\Big\|\sum_{i=1}^n\varepsilon_i f(X_i)\Big\|_{\mathcal{F}}\ge 8qM+s+t\bigg\}$$
$$\le 2^q q^{-k}+\mathbb{P}\bigg\{\sum_{i=1}^k\|f(X_i)\|_{\mathcal{F}}^*\ge s\bigg\}+2\exp\Big(-\frac{t^2}{128qm^2}\Big)$$

where

$$M=\mathbb{E}\Big\|\sum_{i=1}^n\varepsilon_i f(X_i)\Big\|_{\mathcal{F}},\quad m=\mathbb{E}\bigg(\sup_{f\in\mathcal{F}}\Big(\sum_{i=1}^n f(X_i)^2\Big)^{1/2}\bigg)$$

and where $(\|f(X_i)\|_{\mathcal{F}}^*)_{1\le i\le n}$ denotes the nonincreasing rearrangement of the sample $(\|f(X_i)\|_{\mathcal{F}})_{1\le i\le n}$. (Of course, if the functions f of $\mathcal{F}$ are such that $|f|\le 1$, one may choose for example $s=k$.) We find again in this type of inequalities the basic parameters of concentration inequalities of Gaussian type.

This approach to bounds on sums of independent Banach space valued random variables (or empirical processes) is today one of the main successful tools in the study of integrability and limit properties of these sums. The results which may be obtained with this isoperimetric technique are rather sharp and often improve even the scalar case. The range of applications appears to be much broader than what can be obtained for example from the martingale inequalities (2.5). We refer to the monograph [L-T2] for a complete exposition of these applications in the context of probability in Banach spaces.

In his recent developments, M. Talagrand further analyzes the control functionals φ^1, φ^q and φ^c and extends their potential use and interest by a new concept of penalty. Indeed, in the functional φ^1 for example, the coordinates of x which differ from the coordinates of a point in A are accounted for one. One may therefore imagine a more precise measure of this control with some adapted weight. Let, for a nonnegative function h on $\Omega \times \Omega$ such that $h(\omega, \omega) = 0$, and for $A \subset \Omega^n$, $x \in \Omega^n$,

$$\varphi_A^{1,h}(x) = \inf\left\{ \sum_{i=1}^n h(x_i, y_i);\ y \in A \right\}.$$

When $h(\omega, \omega') = 1$ if $\omega \neq \omega'$, we simply recover the Hamming metric φ_A^1. The new functional $\varphi_A^{1,h}$ thus puts a variable penalty $h(x,y)$ on the coordinates of x and y which differ.

Provided with these functionals, one may therefore take again the preceding study and obtain, by the same method of proof based on induction on the number of coordinates, several new and important concentration inequalities. The first result resembles Bernstein's classical exponential bound. Denote by $\|h\|_2$ and $\|h\|_\infty$ respectively the L^2 and L^∞-norms of h with respect to $\mu \otimes \mu$.

Theorem 3.3. *For each subset A of Ω^n and every product probability P, and every $r \geq 0$,*

$$P(\varphi_A^{1,h} \geq r) \leq \frac{1}{P(A)} \exp\left(- \min\left(\frac{r^2}{8n\|h\|_2^2}, \frac{r}{2\|h\|_\infty} \right)\right).$$

To better analyze the conditions on the penalty function h, set, for $B \subset \Omega$ and $\omega \in \Omega$,

$$h(\omega, B) = \inf\{ h(\omega, \omega');\ \omega' \in B\}.$$

Assume that for all $B \subset \Omega$,

$$\int \mathrm{e}^{2h(\omega, B)}\, d\mu(\omega) \leq \frac{\mathrm{e}}{\mu(B)}.$$

A typical statement of [Ta16] is then that, for every $0 \leq \lambda \leq 1$,

$$\int \mathrm{e}^{\lambda \varphi_A^{1,h}}\, dP \leq \frac{1}{P(A)}\, \mathrm{e}^{Cn\lambda^2} \tag{3.11}$$

where $C > 0$ is a numerical constant. With respect to (3.2), we easily see how successful (3.11) can be for an appropriate choice of the penalty function h. One may also prove extensions where, as we already mentioned it, the probability of A is replaced by more complicated functions of this probability (related of course to h.) The penalty or interacting functions h which are used in such a result are of various types. For example, on $\mathbb{R}$, one may take $h(\omega, \omega') = |\omega - \omega'|$ or $h(\omega, \omega') = (\omega - \omega')^+$. One of the striking observations by M. Talagrand is the dissymmetric behavior of the two variables of h, that is on the point x that we would like to control and the point y in the fixed set A. For example, if h only depends on the first coordinate, then it should be bounded; if it only depends on the second coordinate, only weak integrability properties (with respect to μ) are required.

These extensions can also be performed on the functionals φ^q and φ^c, the latter being probably the most interesting for the applications. For a nonnegative penalty function h as before, let, for $A \subset \Omega^n$ and $x \in \Omega$,

$$U_A(x) = \big\{ s = (s_i)_{1 \le i \le n} \in \mathbb{R}_+^n; \exists\, y \in A \text{ such that } s_i \ge h(x_i, y_i) \text{ for every } i = 1, \ldots, n \big\}.$$

Denote by $V_A(x)$ the convex hull of $U_A(x)$. To measure the "distance" from 0 to $V_A(x)$, let us consider a function ψ on $\mathbb{R}$ with $\psi(0) = 0$ and such that $\psi(t) \le t^2$ if $t \le 1$ and $\psi(t) \ge t$ if $t \ge 1$. Then, let

$$\varphi_A^{c,h,\psi}(x) = \inf \bigg\{ \sum_{i=1}^n \psi(s_i);\ s = (s_i)_{1 \le i \le n} \in V_A(x) \bigg\}.$$

The metric φ^c thus simply corresponds to $h(\omega, \omega') = 1$ if $\omega \ne \omega'$ and $\psi(t) = t^2$. Again by induction on the dimension, M. Talagrand [Ta16] then establishes a general form of Theorem 3.2. He shows that, for some constant $\alpha > 0$,

$$\int \exp\big(\alpha\, \varphi_A^{c,h,\psi}\big) \le \exp\big(\theta(P(A))\big)$$

under various conditions connecting μ, h and ψ to the function θ of the probability of A. The proof is of course more involved due to the level of generality.

This abstract study of isoperimetry and concentration in product spaces is motivated by the large number of applications, both in theoretical and more applied probabilistic topics proposed today by M. Talagrand [Ta16]. Most often, the preceding inequalities allow one to establish a concentration inequality once an appropriate mean or median is known. To briefly present such an example of application, let us deal with first passage time in percolation theory. Let $G = (V, \mathcal{E})$ be a graph with vertices V and edges $\mathcal{E}$. Let, on some probability space $(\Omega, \mathcal{A}, \mathbb{P})$, $(X_e)_{e \in \mathcal{E}}$ be a family of nonnegative independent and identically distributed random variables with the same distribution as X. X_e represents the passage time through the edge e. Let $\mathcal{T}$ be a family of (finite) subsets of $\mathcal{E}$, and, for $T \in \mathcal{T}$, set $X_T = \sum_{e \in T} X_e$. If T is made of contiguous edges, X_T represents the passage time through the path T. Set $Z_{\mathcal{T}} = \inf_{T \in \mathcal{T}} X_T$ and $D = \sup_{T \in \mathcal{T}} \mathrm{Card}(T)$, and let m be a median of $Z_{\mathcal{T}}$. As a corollary of his penalty theorems, M. Talagrand [Ta16] proved the following result.

Theorem 3.4. *There exists a numerical constant $c > 0$ such that, if $\mathbb{E}(e^{cX}) \le 2$, for every $r \ge 0$,*

$$\mathbb{P}\big(|Z_{\mathcal{T}} - m| \ge r\big) \le \exp\bigg(-c \min\Big(\frac{r^2}{D}, r\Big)\bigg).$$

When V is $\mathbb{Z}^2$ and $\mathcal{E}$ the edges connecting two adjacent points, and when $\mathcal{T} = \mathcal{T}_n$ is the set of all selfavoiding paths connecting the origin to the point $(0, n)$,

H. Kesten [Ke] showed that, when $0 \leq X \leq 1$ almost surely and $\mathbb{P}(X = 0) < \frac{1}{2}$ (percolation), one may reduce, in Z_T, to paths with length less than some multiple of n. Together with this result, Theorem 3.4 indicates that

$$\mathbb{P}\left(|Z_{T_n} - m| \geq r\right) \leq 5\exp\left(-\frac{r^2}{Cn}\right)$$

for every $r \leq n/C$ where $C > 0$ is a constant independent of n. This result strengthens the previous estimate by H. Kesten [Ke] which was of the order of $r/C\sqrt{n}$ in the exponent and the proof of which was based on martingale inequalities.

Let us mention to conclude a further application of these methods to spin glasses. Consider a sequence $(\varepsilon_i)_{i\in\mathbb{N}}$ of independent symmetric random variables taking values ± 1. Each ε_i represents the spin of particule i. Consider then interactions H_{ij}, $i < j$, between spins. For some parameter $\beta > 0$ (that plays the role of the inverse of the temperature), the so-called partition function is defined by

$$Z_n = Z_n(\beta) = \mathbb{E}_\varepsilon\left(\exp\left(\frac{\beta}{\sqrt{n}}\sum_{1\leq i<j\leq n} H_{ij}\varepsilon_i\varepsilon_j\right)\right), \quad n \geq 2,$$

where $\mathbb{E}_\varepsilon$ is integration with respect to the ε_i's. In the model we study, the interactions H_{ij} are random and the H_{ij}'s will be assumed independent and identically distributed. We assume that, for every $i < j$,

$$\mathbb{E}(H_{ij}) = \mathbb{E}(H_{ij}^3) = 0, \quad \mathbb{E}(H_{ij}^2) = 1,$$

and

$$\mathbb{E}\big(\exp(\pm H_{ij})\big) \leq 2$$

(for normalization purposes). The typical example is of course the example of a standard Gaussian sequence. In this case, it was shown in [A-L-R] and [C-N] that for $\beta < 1$, the sequence

$$\log Z_n - \frac{\beta^2 n}{4}, \quad n \geq 2,$$

converges in distribution to a (nonstandard) centered Gaussian variable. Of equal interest, but of rather different nature, is a concentration result of $\log Z_n$ around $\frac{\beta^2 n}{4}$ for n fixed, that M. Talagrand deduces from its penalty theorems [Ta16].

Theorem 3.5. *There is a numerical constant $C > 1$ such that for $0 \leq r \leq n/C$ and $\beta < 1$,*

$$\mathbb{P}\left\{\left|\log Z_n - \frac{\beta^2 n}{4}\right| \geq C\left(r + \left(\log\frac{C}{1-\beta^2}\right)^{1/2}\right)\sqrt{n}\right\} \leq 4e^{-r^2}.$$

In particular,

$$-\frac{C}{\sqrt{n}}\left(\log\frac{C}{1-\beta^2}\right)^{1/2} \leq \frac{1}{n}\mathbb{E}(\log Z_n) - \frac{\beta^2}{4} \leq \frac{C}{n}.$$

In case the interactions H_{ij}, $i < j$, are independent standard Gaussian, Theorem 3.5 immediately follows from the Gaussian concentration inequalities. Let indeed, on $\mathbb{R}^k$, $k = (n(n-1)/2$,

$$f(x) = \log \mathbb{E}_\varepsilon\left(\exp\left(\frac{\beta}{\sqrt{n}} \sum_{1\le i<j\le n} x_{ij}\varepsilon_i\varepsilon_j\right)\right), \quad x = (x_{ij})_{1\le i<j\le n}.$$

It is easily seen that $\|f\|_{\mathrm{Lip}} \le \beta\sqrt{(n-1)/2}$ so that, by (2.9), for every $r \ge 0$,

$$\mathbb{P}\big\{|\log Z_n - \mathbb{E}(\log Z_n)| \ge r\big\} \le 2\exp\left(-\frac{r^2}{\beta^2(n-1)}\right). \tag{3.12}$$

Now $\mathbb{E}(\log Z_n) \le \log \mathbb{E}(Z_n) = \beta^2(n-1)/4$. Conversely, it may easily be shown (cf. [Ta16]) that

$$\mathbb{E}(Z_n^2) = \big(\mathbb{E}(Z_n)\big)^2 e^{-\beta^2/2}\mathbb{E}\left(\exp\left(\frac{\beta^2}{2n}\left(\sum_{i=1}^n \varepsilon_i\right)^2\right)\right). \tag{3.13}$$

In particular (using the subgaussian inequality for sums of Rademacher random variables [L-T2], p. 90), if $\beta < 1$,

$$\mathbb{E}(Z_n^2) \le \frac{3}{1-\beta^2}\big(\mathbb{E}(Z_n)\big)^2.$$

Hence, by the Paley-Zygmund inequality ([L-T2], p. 92),

$$\mathbb{P}\left\{Z_n \ge \frac{1}{2}\,\mathbb{E}(Z_n)\right\} \ge \frac{\big(\mathbb{E}(Z_n)\big)^2}{4\mathbb{E}(Z_n^2)} \ge \frac{1-\beta^2}{12}.$$

Assume first that $r = \log(\frac{1}{2}\mathbb{E}(Z_n)) - \mathbb{E}(\log Z_n) > 0$. Then, by (3.12) applied to this r,

$$\begin{aligned}\frac{1-\beta^2}{12} &\le \mathbb{P}\left\{\log Z_n \ge \log\left(\frac{1}{2}\,\mathbb{E}(Z_n)\right)\right\} \\ &\le \mathbb{P}\big\{\log Z_n \ge \mathbb{E}(\log Z_n) + r\big\} \le 2\exp\left(-\frac{r^2}{\beta^2(n-1)}\right)\end{aligned}$$

so that

$$r \le \sqrt{n}\left(\log\frac{24}{1-\beta^2}\right)^{1/2}.$$

Hence, in any case,

$$\begin{aligned}\frac{\beta^2(n-1)}{4} \ge \mathbb{E}(\log Z_n) &\ge \log\left(\frac{1}{2}\,\mathbb{E}(Z_n)\right) - \sqrt{n}\left(\log\frac{24}{1-\beta^2}\right)^{1/2} \\ &\ge \frac{\beta^2(n-1)}{4} - 2\sqrt{n}\left(\log\frac{24}{1-\beta^2}\right)^{1/2}\end{aligned}$$

and the theorem follows in this case.

Note that, by (3.12) and the Borel-Cantelli lemma, for any $\beta > 0$,

$$\lim_{n\to\infty} \left| \frac{1}{n} \log Z_n - \frac{1}{n}\mathbb{E}(\log Z_n) \right| = 0$$

almost surely. In particular,

$$0 \leq \limsup_{n\to\infty} \frac{1}{n} \log Z_n \leq \frac{\beta^2}{4}$$

almost surely. This supports the conjecture that $\frac{1}{n}\log Z_n$ should converge in an appropriate sense for every $\beta > 0$ (cf. [A-L-R] and [Co] for precise bounds using different techniques).

As we have seen, the application to probability in Banach spaces is one main topic in which these isoperimetric and concentration inequalities for product measures prove all their strength and efficiency. Besides, M. Talagrand has thus shown how these tools may be used in a variety of problems (random subsequences, random graphs, percolation, geometric probability, spin glasses...). We refer the interested reader to his important contribution [Ta16].

Notes for further reading. As already mentioned, the interested reader may find in the book [L-T2] an extensive description of the application of the isoperimetric inequalities for product measures to probability in Banach spaces (integrability of the norm of sums of independent Banach space valued random variables, strong limit theorems such as laws of large numbers and laws of the iterated logarithm...). Sharper bounds for empirical processes using these methods, and based on Gaussian ideas, are obtained in [Ta13]. The recent paper [Ta16] produces new fields of potential interest for applications of these ideas. [Ta17] provides further sharpenings with approximations by very many points.

4. INTEGRABILITY AND LARGE DEVIATIONS OF GAUSSIAN MEASURES

In this chapter, we make use of the isoperimetric and concentration inequalities of Chapters 1 and 2 to study the integrability properties of functionals of a Gaussian measure as well as large deviation statements. In particular, we will only use in this study the concentration inequalities which were obtained by rather elementary arguments in Chapter 2 so that the results presented here actually proceed from a very simple scheme. We first establish the, by now classical, strong integrability theorems of norms of Gaussian measures. In a second part, we present, on the basis of the Gaussian isoperimetric and concentration inequalities, a large deviation theorem for Gaussian measures without topology. We conclude this chapter with a large deviation statement for the Ornstein-Uhlenbeck process.

A Gaussian measure μ on a real separable Banach space E equipped with its Borel σ-algebra $\mathcal{B}$ and with norm $\|\cdot\|$ is a Borel probability measure on $(E,\mathcal{B})$ such that the law of each continuous linear functional on E is Gaussian. Throughout this work, we only consider centered Gaussian measures or random variables. Although the study of the integrability properties may be developed in a single step from the isoperimetric or concentration inequalities of Chapters 1 and 2, we prefer to decompose the procedure, for pedagogical reasons, in two separate arguments.

Let thus μ be a centered Gaussian measure on $(E,\mathcal{B})$. We first claim that

$$\sigma = \sup_{\xi\in E^*, \|\xi\|\leq 1} \left(\int \langle \xi, x\rangle^2 d\mu(x)\right)^{1/2} < \infty. \tag{4.1}$$

Indeed, if we denote by j the injection map from E^* into $\mathrm{L}^2(\mu) = \mathrm{L}^2(E,\mathcal{B},\mu;\mathbb{R})$, $\|j\| = \sigma$ and j is bounded by the closed graph theorem. Alternatively, let $m > 0$ be such that $\mu(x; \|x\| \leq m) \geq \frac{1}{2}$. Then, for every element ξ in E^* with $\|\xi\| \leq 1$, $\mu(x; |\langle \xi, x\rangle| \leq m) \geq \frac{1}{2}$. Now, under μ, $\langle \xi, x\rangle$ is Gaussian with variance $\int \langle \xi, x\rangle^2 d\mu(x)$. Since $2[1-\Phi(\frac{1}{2})] > \frac{1}{2}$, it immediately follows that $(\int \langle \xi, x\rangle^2 d\mu(x))^{1/2} \leq 2m$.

Since E is separable, the norm $\|\cdot\|$ on E may be described as a supremum over a countable set $(\xi_n)_{n\geq 1}$ of elements of the unit ball of the dual space E^*, that is, for every x in E,

$$\|x\| = \sup_{n\geq 1} \langle \xi_n, x\rangle.$$

In particular, the norm $\|\cdot\|$ can freely be used as a measurable map on $(E, \mathcal{B})$. Let $\Xi = \{\xi_1, \ldots, \xi_n\}$ be a finite subset of $(\xi_n)_{n \geq 1}$. Denote by $\Gamma = M\,{}^t M$ the (semi-) positive definite covariance matrix of the Gaussian vector $(\langle \xi_1, x\rangle, \ldots, \langle \xi_n, x\rangle)$ on $\mathbb{R}^n$. This random vector has the same distribution as $M\Lambda$ where Λ is distributed according to the canonical Gaussian measure γ_n. Let then $f : \mathbb{R}^n \to \mathbb{R}$ be defined by

$$f(z) = \max_{1 \leq i \leq n} M(z)_i, \quad z = (z_1, \ldots, z_n) \in \mathbb{R}^n.$$

It is easily seen that the Lipschitz norm $\|f\|_{\mathrm{Lip}}$ of f is less than or equal to the norm $\|M\|$ of M as an operator from $\mathbb{R}^n$ equipped with the Euclidean norm into $\mathbb{R}^n$ with the supnorm, and that furthermore this operator norm $\|M\|$ is equal, by construction, to

$$\max_{1 \leq i \leq n} \left(\int \langle \xi_i, x\rangle^2 \, d\mu(x) \right)^{1/2} \leq \sigma.$$

Therefore, inequality (2.8) applied to this Lipschitz function f yields, for every $r \geq 0$,

$$\mu\left(x;\, \sup_{\xi \in \Xi} \langle \xi, x\rangle \geq \int \sup_{\xi \in \Xi} \langle \xi, x\rangle d\mu(x) + r\right) \leq \exp\left(-\frac{r^2}{2\sigma^2}\right). \tag{4.2}$$

The same inequality applied to $-f$ yields

$$\mu\left(x;\, \sup_{\xi \in \Xi} \langle \xi, x\rangle + r \leq \int \sup_{\xi \in \Xi} \langle \xi, x\rangle d\mu(x)\right) \leq \exp\left(-\frac{r^2}{2\sigma^2}\right). \tag{4.3}$$

Let then r_0 be large enough so that $\exp(-r_0^2/2\sigma^2) < \frac{1}{2}$. Let also m be large enough in order that $\mu(x; \|x\| \leq m) \geq \frac{1}{2}$. Intersecting this probability with the one in (4.3) for $r = r_0$, we see that

$$\int \sup_{\xi \in \Xi} \langle \xi, x\rangle \, d\mu(x) \leq r_0 + m.$$

Since m and r_0 have been chosen independently of Ξ, we already notice that

$$\int \|x\| \, d\mu(x) < \infty.$$

Now, one can use monotone convergence in (4.2) and thus one obtains that, for every $r \geq 0$,

$$\mu\big(x; \|x\| \geq \textstyle\int \|x\| d\mu(x) + r\big) \leq e^{-r^2/2\sigma^2}. \tag{4.4}$$

Note that an entirely similar result may be obtained exactly in the same way (even simpler) from the concentration inequality (2.4) around the median of a Lipschitz function. As an immediate consequence of (4.4), we may already state the basic theorem about the integrability properties of norms of Gaussian measures. The lower bound and necessity part easily follow from the scalar case. As we have seen in Chapter 2, the two parameters $\int \|x\| d\mu(x)$ and σ in inequality (4.4) may be very different so that this inequality is a much stronger result than the following well-known consequence.

Theorem 4.1. *Let μ be a centered Gaussian measure on a separable Banach space E with norm $\|\cdot\|$. Then*

$$\lim_{r\to\infty}\frac{1}{r^2}\log\mu(x;\|x\|\ge r) = -\frac{1}{2\sigma^2}\,.$$

In other words,

$$\int \exp(\alpha\|x\|^2)\,d\mu(x) < \infty \quad \textit{if and only if} \quad \alpha < \frac{1}{2\sigma^2}\,.$$

The question of the integrability (actually only the square integrability) of the norm of a Gaussian measure was first raised by L. Gross [Gr1], [Gr2]. In 1969, A. V. Skorohod [Sk] was able to show that $\int\exp(\alpha\|x\|)\,d\mu(x) < \infty$ (for every $\alpha > 0$) using the strong Markov property of Brownian motion. The existence of some $\alpha > 0$ for which $\int\exp(\alpha\|x\|^2)\,d\mu(x) < \infty$ was then established independently by X. Fernique [Fe2] and H. J. Landau and L. A. Shepp [L-S] (with a proof already isoperimetric in nature). It may also be shown to follow from Skorokod's early result. The best possible value for α was first obtained in [M-S]. Recently, S. Kwapień mentioned to me that J.-P. Kahane, back in 1964 [Ka1] (cf. [Ka2]), proved an inequality on norms of Rademacher series which, together with a simple central limit theorem argument, already implied that $\int\exp(\alpha\|x\|)\,d\mu(x) < \infty$ for every $\alpha > 0$.

From inequality (4.4), we may also mention the equivalence of all moments of norms of Gaussian measures: for every $0 < p, q < \infty$, there exists a constant $C_{p,q} > 0$ only depending on p and q such that

$$\left(\int \|x\|^p\,d\mu(x)\right)^{1/p} \le C_{p,q}\left(\int \|x\|^q\,d\mu(x)\right)^{1/q}. \tag{4.5}$$

For the proof, simply integrate by parts inequality (4.4) together with the fact that $\sigma \le C_q(\int\|x\|^q\,d\mu(x))^{1/q}$ for every $q > 0$ by the one-dimensional equivalence of Gaussian moments. This yields (4.5) for every $q \ge 1$. When $0 < q \le 1$, simply note that if $m = (2C_{2,1})^{2/q}\big(\int\|x\|^q d\mu(x)\big)^{1/q}$,

$$\begin{aligned}
\int \|x\|d\mu(x) &\le m + \mu(x;\|x\|\ge m)^{1/2}\left(\int \|x\|^2\,d\mu(x)\right)^{1/2}\\
&\le m + C_{2,1}\mu(x;\|x\|\ge m)^{1/2}\int\|x\|d\mu(x)\\
&\le 2(2C_{2,1})^{2/q}\left(\int\|x\|^q d\mu(x)\right)^{1/q}
\end{aligned}$$

since $C_{2,1}\mu(x;\|x\|\ge m)^{1/2} \le \frac{1}{2}$. Note that $C_{p,2}$ (for example) is of the order of $\sqrt{p}$ as p goes to infinity. We will come back to this remark in the last chapter where we will relate (4.5) to hypercontractivity. It is conjectured that $C_{2,1} = \sqrt{\pi/2}$ (that is, the constant of the real case). S. Szarek recently noticed that if conjecture (1.11) holds, then the best possible $C_{p,q}$ are given by the real case.

The preceding integrability properties may also be applied in the context of almost surely bounded Gaussian processes. Let $X = (X_t)_{t \in T}$ be a centered Gaussian process indexed by a set T on some probability space $(\Omega, \mathcal{A}, \mathbb{P})$ such that $\sup_{t\in T} X_t(\omega) < \infty$ for almost all ω in Ω (or $\sup_{t\in T} |X_t(\omega)| < \infty$, which, by symmetry, is equivalent to the preceding, at least if the process is separable). Then, the same proof as above shows in particular that

$$\sup\big\{\mathbb{E}\big(\sup_{t\in U} X_t\big); U \text{ finite in } T\big\} < \infty.$$

We will actually take this as the definition of an almost surely bounded Gaussian process in Chapter 6. Under a separability assumption on the process, one can actually formulate the analogue of Theorem 4.1 in this context. Assume there exists a countable subset S of T such that the set $\{\omega; \sup_{t\in T} X_t \neq \sup_{t\in S} X_t\}$ is negligible. Set $\|X\| = \sup_{t\in S} X_t$. Then, provided $\|X\| < \infty$ almost surely,

$$\mathbb{E}\big(\exp(\alpha\|X\|^2)\big) < \infty \quad \text{if and only if} \quad \alpha < \frac{1}{2\sigma^2}$$

with $\sigma^2 = \sup_{t\in S} \mathbb{E}(X_t^2)$ $(= \sup_{t\in T} \mathbb{E}(X_t^2))$.

As still another remark, notice that the proof of Theorem 4.1 also shows that whenever $X = (X_1, \ldots, X_n)$ is a centered Gaussian random vector in $\mathbb{R}^n$, then

$$\mathrm{var}\big(\max_{1\leq i\leq n} X_i\big) \leq \max_{1\leq i\leq n} \mathrm{var}(X_i).$$

(Use again the Lipschitz map $f(x) = \max_{1\leq i\leq n} M(x)_i$ where $\Gamma = M^{\,t}M$ is the covariance matrix of X with however (2.4) instead of (2.8).) This inequality may however be deduced directly from the Poincaré type inequality

$$\int \Big|f - \int f d\gamma_n\Big|^2 d\gamma_n \leq \int |\nabla f|^2 d\gamma_n \quad (\leq \|f\|_{\mathrm{Lip}}^2)$$

which is elementary (by an expansion in Hermite polynomials for example).

Our aim will now be to extend the isoperimetric and concentration inequalities to the setting of an infinite dimensional Gaussian measure μ as before. Let us mention however before that the fundamental inequalities are the ones in finite dimension and that the infinite dimensional extensions we will present actually follow in a rather classical and straightforward manner from the finite dimensional case. The main tool will be the concept of abstract Wiener space and reproducing kernel Hilbert space which will define the isoperimetric neighborhoods or enlargements in this framework. We follow essentially C. Borell [Bo3] in the construction below.

Let μ be a mean zero Gaussian measure on a real separable Banach space E. Consider then the abstract Wiener space factorization [Gr1], [B-C], [Ku], [Bo3] (for recent accounts, cf. [Bog], [Lif3]),

$$E^* \xrightarrow{\ j\ } \mathrm{L}^2(\mu) \xrightarrow{\ j^*\ } E.$$

First note that since E is separable and μ is a Borel probability measure on E, μ is Radon, that is, for every $\varepsilon > 0$ there is a compact set K in E such that $\mu(K) \geq 1 - \varepsilon$. Let $(K_n)_{n\in\mathbb{N}}$ be a sequence of compact sets such that $\mu(K_n) \to 1$. If φ is an element of $\mathrm{L}^2(\mu)$, $j^*(\varphi I_{K_n})$ belongs to E since it may be identified with the expectation, in the strong sense, $\int_{K_n} x\varphi(x)d\mu(x)$. Now, the sequence $\left(\int_{K_n} x\varphi(x)d\mu(x)\right)_{n\in\mathbb{N}}$ is Cauchy in E since,

$$\sup_{\xi\in E^*,\|\xi\|\leq 1} \langle \xi, \textstyle\int_{K_n} x\varphi(x)d\mu(x) - \int_{K_m} x\varphi(x)d\mu(x)\rangle \leq \sigma\left(\int \varphi^2 |I_{K_n} - I_{K_m}| d\mu\right)^{1/2} \to 0.$$

It therefore converges in E to the weak integral $\int x\varphi(x)d\mu(x) = j^*(\varphi) \in E$.

Define now the reproducing kernel Hilbert space $\mathcal{H}$ of μ as the subspace $j^*(\mathrm{L}^2(\mu))$ of E. Since $j(E^*)^\perp = \mathrm{Ker}(j^*)$, j^* restricted to the closure E_2^* of E^* in $\mathrm{L}^2(\mu)$ is linear and bijective onto $\mathcal{H}$. For simplicity in the notation, we set below $\widetilde{h} = ({j^*}_{|E_2^*})^{-1}(h)$. Under μ, $\widetilde{h}$ is Gaussian with variance $|h|^2$. Note that σ of (4.1) is then also $\sup_{x\in\mathcal{K}} \|x\|$ where $\mathcal{K}$ is the closed unit ball of $\mathcal{H}$ for its Hilbert space scalar product given by

$$\langle j^*(\varphi), j^*(\psi)\rangle_{\mathcal{H}} = \langle \varphi, \psi\rangle_{\mathrm{L}^2(\mu)}, \quad \varphi, \psi \in \mathrm{L}^2(\mu).$$

In particular, for every x in $\mathcal{H}$, $\|x\| \leq \sigma|x|$ where $|x| = |x|_{\mathcal{H}} = \langle x, x\rangle_{\mathcal{H}}^{1/2}$. Moreover, $\mathcal{K}$ is a compact subset of E. Indeed, if $(\xi_n)_{n\in\mathbb{N}}$ is a sequence in the unit ball of E^*, there is a subsequence $(\xi_{n'})_{n'\in\mathbb{N}}$ which converges weakly to some ξ in E^*. Now, since the ξ_n are Gaussian under μ, $\xi_{n'} \to \xi$ in $\mathrm{L}^2(\mu)$ so that j is a compact operator. Hence j^* is also a compact operator which is the claim.

For γ_n the canonical Gaussian measure on $\mathbb{R}^n$ (equipped with some arbitrary norm), it is plain that $\mathcal{H} = \mathbb{R}^n$ with its Euclidean structure, that is $\mathcal{K}$ is the Euclidean unit ball $B(0,1)$. If $X = (X_1, \ldots, X_n)$ is a centered Gaussian measure on $\mathbb{R}^n$ with nondegenerate covariance matrix $\Gamma = M\,{}^tM$, it is easily seen that the unit ball $\mathcal{K}$ of the reproducing kernel Hilbert space associated to the distribution of X is the ellipsoid $M(B(0,1))$. As another example, let us mention the classical Wiener space associated with Brownian motion, say on [0,1] and with real values for simplicity. Let thus E be the Banach space $C_0([0,1])$ of all real continuous functions x on [0,1] vanishing at the origin equipped with the supnorm (the Wiener space) and let μ be the distribution of a standard Brownian motion, or Wiener process, $W = (W(t))_{t\in[0,1]}$ starting at the origin (the Wiener measure). If m is a finitely supported measure on [0,1], $m = \sum_i c_i\delta_{t_i}$, $c_i \in \mathbb{R}$, $t_i \in [0,1]$, clearly $h = j^*j(m)$ is the element of E given by

$$h(t) = \sum_i c_i(t_i \wedge t), \quad t \in [0,1];$$

it satisfies

$$\int_0^1 h'(t)^2\,dt = \int \langle m, x\rangle^2 d\mu(x) = |h|_{\mathcal{H}}^2.$$

By a standard extension, the reproducing kernel Hilbert space $\mathcal{H}$ associated to the Wiener measure μ on E may then be identified with the Cameron-Martin Hilbert space of the absolutely continuous elements h of $C_0([0,1])$ such that

$$\int_0^1 h'(t)^2\,dt < \infty.$$

Moreover, if $h \in \mathcal{H}$, $\widetilde{h} = (j^*{}_{|E_2^*})^{-1}(h) = \int_0^1 h'(t)dW(t)$. While we equipped the Wiener space $C_0([0,1])$ with the uniform topology, other choices are possible. Let F be a separable Banach space such that the Wiener process W belongs almost surely to F. Using probabilistic notation, we know from the previous abstract Wiener space theory that if φ is a real valued random variable with $\mathbb{E}(\varphi^2) < \infty$, then $h = \mathbb{E}(W\varphi) \in F$. Since $\mathbb{P}\{W \in F \cap C_0([0,1])\} = 1$, it immediately follows that the Cameron-Martin Hilbert space may be identified with a subset of F and is also the reproducing kernel Hilbert space of Wiener measure on F. For h in the Cameron-Martin space, $\widetilde{h} = (j^*{}_{|F_2^*})^{-1}(h)$ may be identified with $\int_0^1 h'(t)dW(t)$ as soon as there is a sequence $(\xi_n)_{n\in\mathbb{N}}$ in F^* such that

$$\mathbb{E}\left(\left|\int_0^1 h'(t)dW(t) - \langle \xi_n, W\rangle\right|^2\right) \to 0.$$

This is the case if, for every $t \in [0,1]$, there is $(\xi_n)_{n\in\mathbb{N}}$ in F^* with

$$\mathbb{E}\left(\left|W(t) - \langle \xi_n, W\rangle\right|^2\right) \to 0.$$

Examples include the Lebesgue spaces $\mathrm{L}^p([0,1])$, $1 \le p < \infty$, or the Hölder spaces (see below). Actually, since the preceding holds for the L^1-norm, this will be the case for a norm $\|\cdot\|$ on $C_0([0,1])$ as soon as, for some constant $C > 0$, $\|x\| \ge C\int_0^1 |x(t)|dt$ for every x in $C_0([0,1])$.

The next proposition is a useful series representation of Gaussian measures and random vectors which can be used efficiently in proofs by finite dimensional approximation. This proposition puts forward the fundamental Gaussian measurable structure consisting of the canonical Gaussian product measure on $\mathbb{R}^{\mathbb{N}}$ with reproducing kernel Hilbert space ℓ^2.

Proposition 4.2. *Let μ be as before. Let $(g_i)_{i\ge1}$ denote an orthonormal basis of the closure E_2^* of E^* in $\mathrm{L}^2(\mu)$ and set $e_i = j^*(g_i)$, $i \ge 1$. Then $(e_i)_{i\ge1}$ defines an orthonormal basis of $\mathcal{H}$ and the series $X = \sum_{i=1}^\infty g_ie_i$ converges in E μ-almost everywhere and in every L^p and is distributed as μ.*

Proof. Since μ is a Radon measure, the space $\mathrm{L}^2(\mu)$ is separable and E_2^* consists of Gaussian random variables on the probability space $(E, \mathcal{B}, \mu)$. Hence, $(g_i)_{i\ge1}$ defines on this space a sequence of independent standard Gaussian random variables. The sequence $(e_i)_{i\ge1}$ is clearly a basis in $\mathcal{H}$. Recall from Theorem 4.1 that the integral $\int \|x\|d\mu(x)$ is finite. Denote then by $\mathcal{B}_n$ the σ-algebra generated by $g_1, \ldots, g_n$. It is easily seen that the conditional expectation of the identity map on (E,μ) with respect to $\mathcal{B}_n$ is equal to $X_n = \sum_{i=1}^n g_i e_i$. By the vector valued martingale convergence theorem (cf. [Ne2]), the series $\sum_{i=1}^\infty g_i e_i$ converges almost surely. Since $\int \|x\|^p d\mu(x) < \infty$ for every $p > 0$, the convergence also takes place in any L^p-space. Since moreover

$$\int \langle \xi, X\rangle^2 d\mu = \sum_{i=1}^{\infty} \langle \xi, e_i\rangle^2 = \sum_{i=1}^{\infty} \langle j(\xi), g_i\rangle^2 = \int \langle \xi, x\rangle^2 d\mu(x)$$

for every ξ in E^*, X has law μ. Proposition 4.2 is proved. □

According to Proposition 4.2, we use from time to time below more convenient probabilistic notation and consider $(g_i)_{i\geq 1}$ as a sequence of independent standard Gaussian random variables on some probability space $(\Omega, \mathcal{A}, \mathbb{P})$ and X as a random variable on $(\Omega, \mathcal{A}, \mathbb{P})$ with values in E and law μ.

As a consequence of Proposition 4.2, note that the closure $\overline{\mathcal{H}}$ of $\mathcal{H}$ in E coincides with the support of μ (for the topology given by the norm on E). Indeed, by Proposition 4.2, $\mathrm{supp}(\mu) \subset \overline{\mathcal{H}}$. Conversely, it suffices to prove that $\mu(B(h, \eta)) > 0$ for every h in $\mathcal{H}$ and every $\eta > 0$ where $B(h, \eta)$ is the ball in E with center h and radius η. By the Cameron-Martin translation formula (see below), it suffices to prove it for $h = 0$. Now, for every $a \in E$, by symmetry and independence,

$$\begin{aligned}\mu\big(B(a,\eta)\big)^2 &= \mu\big(x; \|x-a\| \leq \eta\big)\mu\big(x; \|x+a\| \leq \eta\big) \\ &\leq \mu \otimes \mu\big((x,y); \big\|(x-a)+(y+a)\big\| \leq 2\eta\big) \\ &= \mu\big(B(0, \eta\sqrt{2})\big)\end{aligned}$$

since $x+y$ under $\mu\otimes\mu$ is distributed as $\sqrt{2}x$ under μ. Now, assume that $\mu(B(h, \eta_0)) = 0$ for some $\eta_0 > 0$. Since μ is Radon, there is a sequence $(a_n)_{n\in\mathbb{N}}$ in E such that

$$\mu\big(x; \exists\, n, \|x - a_n\| \leq \eta_0/\sqrt{2}\big) = 1.$$

Then,

$$1 \leq \sum_n \mu\big(B(a_n, \eta_0/\sqrt{2})\big) \leq \sum_n \mu\big(B(0, \eta_0)\big)^{1/2} = 0$$

which is a contradiction (cf. also [D-HJ-S]).

To complete this brief description of the reproducing kernel Hilbert space of a Gaussian measure, let us mention the dual point of view more commonly used by analysts on Wiener spaces (see [Ku], [Bog] for further details). Let $\mathcal{H}$ be a real separable Hilbert space with norm $|\cdot|$ and let $e_1, e_2, \ldots$ be an orthormal basis of $\mathcal{H}$. Define a simple additive measure ν on the cylinder sets in $\mathcal{H}$ by

$$\nu\big(x \in \mathcal{H}; \big(\langle x, e_1\rangle, \ldots, \langle x, e_n\rangle\big) \in A\big) = \gamma_n(A)$$

for all Borel sets A in $\mathbb{R}^n$. Let $\|\cdot\|$ be a measurable seminorm on $\mathcal{H}$ and denote by E the completion of $\mathcal{H}$ with respect to $\|\cdot\|$. Then $(E, \|\cdot\|)$ is a real separable Banach space. If $\xi \in E^*$, we consider $\xi_{|\mathcal{H}} : \mathcal{H} \to \mathbb{R}$ that we identify with an element h in $\mathcal{H} = \mathcal{H}^*$ (in our language, $h = j^*j(\xi)$). Let then μ be the (σ-additive) extension of ν

on the Borel sets of E. In particular, the distribution of $\xi \in E^*$ under μ is Gaussian with mean zero and variance $|h|^2$. Therefore, μ is a Gaussian Radon measure on E with reproducing kernel Hilbert space $\mathcal{H}$. With respect to this approach, our construction priviledges the point of view of the measure.

We are now ready to state and prove the isoperimetric inequality in $(E, \mathcal{H}, \mu)$. As announced, the isoperimetric neighborhoods A_r, $r \geq 0$, of a set A in E will be understood in this setting as the Minkowski sum $A + r\mathcal{K} = \{x + ry; x \in A, y \in \mathcal{K}\}$ where we recall that $\mathcal{K}$ is the unit ball of the reproducing kernel Hilbert space $\mathcal{H}$ associated to the Gaussian measure μ. In this form, the result is due to C. Borell [Bo2].

Theorem 4.3. *Let A be a Borel set in E such that $\mu(A) \geq \Phi(a)$ for some real number a. Then, for every $r \geq 0$*

$$\mu_*(A + r\mathcal{K}) \geq \Phi(a + r).$$

It might be worthwhile mentioning that if the support of μ is infinite dimensional, $\mu(\mathcal{H}) = 0$ so that the infinite dimensional version of the Gaussian isoperimetric inequality might be somewhat more surprising than its finite dimensional statement. [The use of inner measure in Theorem 4.3 is not stricly necessary since at it is known from the specialists, somewhat deep arguments from measure theory may be used to show that $A + r\mathcal{K}$ is actually μ-measurable in this setting. These arguments are however completely out of the scope of this work, and anyway, Theorem 4.3 as stated is the best possible inequality one may hope for. We therefore do not push further in these measurability questions and use below inner measure. Of course, for example, if A is closed, $A + r\mathcal{K}$ is also closed (since $\mathcal{K}$ is compact) and thus measurable.]

Proof. As announced, it is based on a classical finite dimensional approximation procedure. We use the series representation $X = \sum_{i=1}^{\infty} g_i e_i$ of Proposition 4.2 and, accordingly, probabilistic notations. We may assume that $-\infty < a < +\infty$. Let $r \geq 0$ be fixed. Let also $\varepsilon > 0$. Since μ is a Radon measure, there exists a compact set $K \subset A$ such that

$$\mathbb{P}\{X \in K\} = \mu(K) > \Phi(a - \varepsilon).$$

For every $\eta > 0$, let $K^\eta = \{x \in E; \inf_{y \in K} \|x - y\| \leq \eta\}$. Recall $X_n = \sum_{i=1}^{n} g_i e_i$. Since $\mathbb{P}\{\|X - X_n\| > \eta\} \to 0$, for some n_0 and every $n \geq n_0$, $\mathbb{P}\{X_n \in K^\eta\} \geq \Phi(a - 2\varepsilon)$ and

$$\mathbb{P}\{X \in K^{3\eta} + r\mathcal{K}\} \geq \mathbb{P}\{X_n \in K^{2\eta} + r\mathcal{K}\} - \varepsilon.$$

Now, let $\mathcal{K}_n$ be the unit ball of the reproducing kernel Hilbert space of the (finite dimensional) Gaussian random vector X_n, or rather of its distribution on E. $\mathcal{K}_n$ consists of those elements in E of the form $\mathbb{E}(X_n \varphi)$ with $\|\varphi\|_2 \leq 1$. Clearly,

$$\|\mathbb{E}(X\varphi) - \mathbb{E}(X_n\varphi)\| \leq \left(\mathbb{E}\|X - X_n\|^2\right)^{1/2} \to 0$$

independently of φ, $\|\varphi\|_2 \le 1$. Hence, for some $n_1 \ge n_0$, and every $n \ge n_1$,

$$\mathbb{P}\{X \in K^{3\eta} + r\mathcal{K}\} \ge \mathbb{P}\{X_n \in K^\eta + r\mathcal{K}_n\} - \varepsilon.$$

Let Q be the map from $\mathbb{R}^n$ into E defined by $Q(z) = \sum_{i=1}^n z_i e_i$, $z = (z_1, \dots, z_n)$. Therefore

$$\gamma_n\big(Q^{-1}(K^\eta)\big) = \mathbb{P}\{X_n \in K^\eta\} \ge \Phi(a - 2\varepsilon).$$

Since the distribution of X_n is the image by Q of γ_n and since similarly $\mathcal{K}_n$ is the image by Q of the Euclidean unit ball, it follows from Theorem 1.3 that

$$\mathbb{P}\{X_n \in K^\eta + r\mathcal{K}_n\} = \gamma_n\big(\big(Q^{-1}(K^\eta)\big)_r\big) \ge \Phi(a - 2\varepsilon + r).$$

Summarizing, for every $\eta > 0$,

$$\mu(K^{3\eta} + r\mathcal{K}) \ge \Phi(a - 2\varepsilon + r) - \varepsilon.$$

Since K and $\mathcal{K}$ are compact in E, letting η decrease to zero yields

$$\mu_*(A + r\mathcal{K}) \ge \mu(K + r\mathcal{K}) \ge \Phi(a - 2\varepsilon + r) - \varepsilon.$$

Since $\varepsilon > 0$ is arbitrary, the theorem is proved. □

The approximation procedure developed in the proof of Theorem 4.3 may be used exactly in the same way on the basis of inequality (2.15) to show that, for every $r \ge 0$,

$$\mu_*(A + r\mathcal{K}) \ge 1 - \exp\left(-\frac{r^2}{2} + r\delta\big(\mu(A)\big)\right) \tag{4.6}$$

where we recall that

$$\delta(v) = \int_0^\infty \min\big(1 - v, \mathrm{e}^{-t^2v^2/2}\big)\,dt, \quad 0 \le v \le 1.$$

The point here is that inequality (2.15) (and thus also inequality (4.6)) was obtained at the very cheap price of Proposition 2.1. In what follows, inequality (4.6) will be good enough for almost all the applications we have in mind.

Theorem 4.3, or inequality (4.6), of course allows us to recover the integrability properties described in Theorem 4.1. For example, if $f : E \to \mathbb{R}$ is measurable and Lipschitz in the direction of $\mathcal{H}$, that is

$$\big|f(x + h) - f(x)\big| \le |h| \quad \text{for all } x \in E,\ h \in \mathcal{H}, \tag{4.7}$$

and if m is median of f for μ, exactly as in the finite dimensional case (2.4),

$$\mu(f \ge m + r) \le 1 - \Phi(r) \le \mathrm{e}^{-r^2/2} \tag{4.8}$$

for every $r \geq 0$. In the same way, a finite dimensional argument on (2.8) shows that $\int f d\mu$ exists and that, for all $r \geq 0$,

$$\mu\Big(f \geq \int f d\mu + r\Big) \leq e^{-r^2/2}. \tag{4.9}$$

Indeed, assume first that f is bounded. We follow Proposition 4.2 and its notation. Let f_n, $n \geq 1$, be the conditional expectation of f with respect to $\mathcal{B}_n$. Define $\widetilde{f}_n : \mathbb{R}^n \to \mathbb{R}$ by

$$\widetilde{f}_n(z) = \int f\bigg(\sum_{i=1}^n z_i e_i + y\bigg) d\mu^n(y), \quad z = (z_1, \dots, z_n) \in \mathbb{R}^n,$$

where μ^n is the distribution of $\sum_{i=n+1}^{\infty} g_i e_i$. Then f_n under μ has the same distribution as $\widetilde{f}_n$ under γ_n. Moreover, it is clear by (4.7) that $\widetilde{f}_n$ is Lipschitz in the usual sense on $\mathbb{R}^n$ with $\|\widetilde{f}_n\|_{\mathrm{Lip}} \leq 1$. Therefore, by (2.8) applied to $\widetilde{f}_n$, for every $r \geq 0$,

$$\mu\Big(f_n \geq \int f_n d\mu + r\Big) \leq e^{-r^2/2}.$$

Letting n tend to infinity, we see that (4.9) is satisfied for bounded functionals f on E satisfying (4.7). When f is not bounded, set, for every integer N,

$$f^N = \min\big(\max(f, -N), N\big).$$

Then f^N still satisfies (4.7) for each N so that

$$\mu\Big(f^N \geq \int f^N d\mu + r\Big) \leq e^{-r^2/2}$$

for every $r \geq 0$. Of course, the same result holds for $|f^N|$. Let then m be such that $\mu(|f| \leq m) \geq \frac{3}{4}$. There exists N_0 such that for every $N \geq N_0$, $\mu(|f^N| \leq m+1) \geq \frac{1}{2}$. Let $r_0 \geq 0$ be such that $e^{-r_0^2/2} < \frac{1}{2}$. Together with the preceding inequality for $|f^N|$, we thus get that for every $N \geq N_0$,

$$\int |f^N| d\mu \leq m + 1 + r_0.$$

Moreover, $\mu(|f^N| \geq m + 1 + r_0 + r) \leq e^{-r^2/2}$, $r \geq 0$. Hence, in particular, the supremum $\sup_N \int |f^N|^2 d\mu$ is finite. The announced claim (4.9) now easily follows by uniform integrability.

Let us also mention that the preceding inequalities (4.8) and (4.9) may of course be applied to $f(x) = \|x\|$, $x \in E$, since, as we have seen,

$$\big|\|x + h\| - \|x\|\big| \leq \|h\| \leq \sigma|h|, \quad x \in E, h \in \mathcal{H}.$$

It should be noticed that the $\mathcal{H}$-Lipschitz hypothesis (4.7) has recently been shown [E-S] to be equivalent to the fact that the Malliavin derivative Df of f exists and satisfies $\||Df|_{\mathcal{H}}\|_\infty \leq 1$. (Due to the preceding simple arguments, the hypothesis that f be in $L^2(\mu)$ in the paper [E-S] is easily seen to be superfluous.) But actually, that (4.7) holds when $\| |Df|_{\mathcal{H}} \|_\infty \le 1$ is the easy part of the argument so that the preceding result is as general as possible. One could also prove (4.9) along the lines of Proposition 2.1 in infinite dimension with the Ornstein-Uhlenbeck semigroup associated to μ. One however runs into the question of differentiability in infinite dimension (Gross-Malliavin derivatives) that is not really needed here.

In the preceding spirit, it might be worthwhile to briefly describe some related inequalities due to B. Maurey and G. Pisier [Pi1]. Let f be of class C^1 on $\mathbb{R}^n$ with gradient ∇f. Let furthermore V be a convex function on $\mathbb{R}$. To avoid integrability questions, assume first that f is bounded. By Jensen's inequality,

$$\int V\Big(f - \int f d\gamma_n\Big) d\gamma_n \le \int\!\!\int V\big(f(x) - f(y)\big) d\gamma_n(x) d\gamma_n(y).$$

Now, for x, y in $\mathbb{R}^n$, and every real number θ, set

$$x(\theta) = x \sin\theta + y\cos\theta, \quad x'(\theta) = x\cos\theta - y\sin\theta.$$

We have

$$f(x) - f(y) = \int_0^{\pi/2} \frac{d}{d\theta} f\big(x(\theta)\big) d\theta = \int_0^{\pi/2} \big\langle \nabla f\big(x(\theta)\big), x'(\theta)\big\rangle d\theta.$$

Hence, using Jensen's inequality one more time but now with respect to the variable θ,

$$\int V\Big(f - \int f d\gamma_n\Big) d\gamma_n \le \frac{2}{\pi}\int_0^{\pi/2}\int\!\!\int V\Big(\frac{\pi}{2}\big\langle \nabla f\big(x(\theta)\big), x'(\theta)\big\rangle\Big) d\gamma_n(x) d\gamma_n(y) d\theta.$$

By the fundamental rotational invariance of Gaussian measures, for any θ, the couple $(x(\theta), x'(\theta))$ has the same distribution as the original independent couple (x, y). Therefore, we obtained that

$$\text{(4.10)} \qquad \int V\Big(f - \int f d\gamma_n\Big) d\gamma_n \le \int\!\!\int V\Big(\frac{\pi}{2}\langle \nabla f(x), y\rangle\Big) d\gamma_n(x) d\gamma_n(y).$$

We leave it to the interested reader to properly extend this type of inequality to unbounded functions. It also easily extends to infinite dimensional Gaussian measures μ. Indeed, let f be smooth enough, more precisely differentiable in the direction of $\mathcal{H}$ or in the sense of Gross-Malliavin (cf. e.g. [Bel], [Wa], [Nu]...). With the same notation as in the proof of (4.9),

$$\nabla \tilde{f}_n = (D_{e_1} f, \ldots, D_{e_n} f),$$

where $D_h f$ is the derivative of f in the direction of $h \in \mathcal{H}$. Therefore, for every n,

$$\int V\Big(f_n - \int f_n d\mu\Big) d\mu \le \int\!\!\int V\Big(\frac{\pi}{2}\sum_{i=1}^n y_i D_{e_i} f(x)\Big) d\mu(x) d\gamma_n(y).$$

Hence, by Fatou's lemma and Jensen's inequality, (4.10) yields in an infinite dimensional setting that

$$\int V\left(f - \int f d\mu\right) d\mu \leq \int\int V\left(\frac{\pi}{2}\sum_{i=1}^{\infty} y_i D_{e_i} f(x)\right) d\mu(x) d\gamma_\infty(y)$$

where γ_∞ is the canonical Gaussian product measure on $\mathbb{R}^{\mathbb{N}}$. If V is an exponential function $e^{\lambda x}$, we may perform partial integration in the variable y to get that

$$\int \exp\left[\lambda\left(f - \int f d\mu\right)\right] d\mu \leq \int \exp\left(\frac{\lambda^2\pi^2}{4} |Df|_{\mathcal{H}}^2\right) d\mu.$$

In particular, if f is Lipschitz, we recover in this way an inequality similar to (2.8) (or (4.9)) with however a worse constant. Inequality (4.10) is however more general and applies moreover to vector valued functions (cf. [Pi1]).

So far, we only used isoperimetry and concentration in a very mild way for the application to the integrability properties. As we have seen, there is however a strong difference between these integrability properties (Theorem 4.1) and, for example, inequalities (4.4), (4.8) or (4.9). In these inequalities indeed, two parameters, and not only one, on the Gaussian measure enter into the problem, namely the median or the mean of the $\mathcal{H}$-Lipschitz map f and its Lipschitz norm (the supremum σ of weak variances in the case of a norm). These can be very different even in simple examples.

We now present another application of the Gaussian isoperimetric inequality due to M. Talagrand [Ta1]. It is a powerful strengthening on Theorem 4.1 that makes critical use of the preceding comment. (See also [G-K1] for some refinement.) More on Theorem 4.4 may be found in Chapter 7.

Theorem 4.4. *Let μ be a Gaussian measure on E. For every $\varepsilon > 0$, there exists $r_0 = r_0(\varepsilon)$ such that for every $r \geq r_0$,*

$$\mu(x; \|x\| \geq \varepsilon + \sigma r) \leq \exp\left(-\frac{r^2}{2} + \varepsilon r\right).$$

Ehrhard's inequality (1.8) (or rather its infinite dimensional extension) indicates that the map $F(r) = \Phi^{-1}(\mu(x; \|x\| \leq r))$, $r \geq 0$, is concave. While Theorem 4.1 expresses that $\lim_{r\to\infty} F(r)/r = 0$, Theorem 4.4 yields $\lim_{r\to\infty}[F(r) - (r/\sigma)] = \frac{1}{\sigma}$. In other words, the line r/σ is an asymptote at infinity to F. Notice furthermore that Theorem 4.4 implies (is equivalent to saying) that

$$\int \exp\left(\frac{1}{2\sigma^2}\left(\|x\| - \varepsilon\right)^2\right) d\mu(x) < \infty$$

for all $\varepsilon > 0$.

Proof. Recall the series $X = \sum_{i=1}^{\infty} g_i e_i$ of Proposition 4.2 which we consider on some probability space $(\Omega, \mathcal{A}, \mathbb{P})$. Let $X_n = \sum_{i=1}^{n} g_i e_i$ and $X^n = X - X_n$, $n \geq 1$. Let

$\varepsilon > 0$ be fixed and set $A = \{x \in E; \|x\| < \varepsilon\}$. For every $r \geq 0$ and every integer $n \geq 1$, we can write

$$\begin{aligned}\mathbb{P}\{\|X\| \geq \varepsilon + \sigma r\} &\leq \mathbb{P}\{X \notin A + r\mathcal{K}\} \\ &\leq \mathbb{P}\{|X_n| > r\} + \mathbb{P}\{|X_n| \leq r, X \notin A + r\mathcal{K}\}.\end{aligned}$$

On the set $\{|X_n| \leq r\}$, $X \notin A + r\mathcal{K}$ implies that

$$X^n \notin A + \left(r^2 - |X_n|^2\right)^{1/2} \mathcal{K}^n$$

where $\mathcal{K}^n$ is the unit ball of the reproducing kernel Hilbert space associated to the distribution of X^n. Indeed, if this is not the case,

$$X^n = a + \left(r^2 - |X_n|^2\right)^{1/2} h^n$$

for some $a \in A$ and $h^n \in \mathcal{K}^n$. This would imply that

$$X = X_n + X^n = a + X_n + \left(r^2 - |X_n|^2\right)^{1/2} h^n = a + k$$

where, by orthogonality, $|k| \leq r$. Therefore,

$$\mathbb{P}\{X \notin A + r\mathcal{K}\} \leq \mathbb{P}\{|X_n| > r\} + \mathbb{P}\left\{|X_n| \leq r, X^n \notin A + \left(r^2 - |X_n|^2\right)^{1/2} \mathcal{K}^n\right\}.$$

Recall now the function δ of (2.12) or (4.6) and choose n large enough in order that $\delta(\mathbb{P}\{\|X^n\| < \varepsilon\}) \leq \varepsilon$. Now, X_n and X^n are independent and $|X_n| = \left(\sum_{i=1}^n g_i^2\right)^{1/2}$. Hence, by inequality (4.6),

$$\begin{aligned}&\mathbb{P}\left\{|X_n| \leq r, X^n \notin A + \left(r^2 - |X_n|^2\right)^{1/2} \mathcal{K}^n\right\} \\ &\qquad \leq \int_{\{|X_n| \leq r\}} \exp\left(-\frac{1}{2}\left(r^2 - |X_n|^2\right) + \varepsilon\left(r^2 - |X_n|^2\right)^{1/2}\right) d\mathbb{P} \\ &\qquad \leq C_n r^n \exp\left(-\frac{r^2}{2} + \varepsilon r\right)\end{aligned}$$

where $C_n > 0$ only depends on n. In summary,

$$\mathbb{P}\{\|X\| \geq \varepsilon + \sigma r\} \leq \mathbb{P}\left\{\sum_{i=1}^n g_i^2 > r^2\right\} + C_n r^n \exp\left(-\frac{r^2}{2} + \varepsilon r\right)$$

from which the conclusion immediately follows. Theorem 4.4 is established. □

We now present some further applications of isoperimetry and concentration to the study of large deviations of Gaussian measures. As an introduction to these ideas, we first present the elementary concentration proof, due to S. Chevet [Che], of the upper bound in the large deviation principle for Gaussian measures.

Let μ be as before a mean zero Gaussian measure on a separable Banach space E with reproducing kernel Hilbert space $\mathcal{H}$. For a subset A of E, let

$$\mathcal{I}(A) = \inf\{\tfrac{1}{2}|h|^2; h \in A \cap \mathcal{H}\}$$

be the classical large deviation rate functional in this setting. Set $\mu_\varepsilon(\cdot) = \mu(\varepsilon^{-1}(\cdot))$, $\varepsilon > 0$. Let now A be closed in E and take r such that $0 < r < \mathcal{I}(A)$. By the very definition of $\mathcal{I}(A)$,

$$A \cap \sqrt{2r}\mathcal{K} = \emptyset.$$

Since A is closed and the balls in $\mathcal{H}$ are compact in E, there exists $\eta > 0$ such that we still have

$$A \cap \left[\sqrt{2r}\mathcal{K} + B_E(0,\eta)\right] = \emptyset$$

where $B_E(0,\eta)$ is the ball with center the origin and with radius η for the norm $\|\cdot\|$ in E. Since

$$\lim_{\varepsilon\to 0} \mu\big(B_E(0,\varepsilon^{-1}\eta)\big) = \lim_{\varepsilon\to 0} \mu_\varepsilon\big(B_E(0,\eta)\big) = 1,$$

it is then an immediate consequence of (4.6) (or Theorem 4.3) that for every $\varepsilon > 0$ small enough

$$\mu_\varepsilon(A) \le \mu\big(\big[\varepsilon^{-1}\sqrt{2r}\mathcal{K} + B_E(0,\varepsilon^{-1}\eta)\big]^c\big) \le \exp\left(-\frac{r}{\varepsilon^2} + \frac{\sqrt{2r}}{\varepsilon}\right).$$

Therefore, since $r < \mathcal{I}(A)$ is arbitrary,

$$\limsup_{\varepsilon\to 0} \varepsilon^2 \log \mu_\varepsilon(A) \le -\mathcal{I}(A).$$

This simple proof may easily be modified to yield some version of the large deviation theorem with only "measurable operations" on the sets. One may indeed ask about the role of the topology in a large deviation statement. As we will see, the isoperimetric and concentration ideas in this Gaussian setting are powerful enough to state a large deviation principle without any topological operations of closure or interior.

Let, as before, $(E,\mathcal{H},\mu)$ be an abstract Wiener space. If A and B are subsets of E, and if λ is a real number, we set

$$\lambda A + B = \{\lambda x + y; x \in A, y \in B\},$$
$$A \ominus B = \{x \in A; x + B \subset A\}.$$

Crucial to the approach is the class $\mathcal{V}$ of all Borel subsets V of E such that

$$\liminf_{\varepsilon\to 0} \mu_\varepsilon(V) > 0.$$

Notice that if $V \in \mathcal{V}$, then $\lambda V \in \mathcal{V}$ for every $\lambda > 0$. Typically, the balls $B_E(0,\eta)$, $\eta > 0$, for the norm $\|\cdot\|$ on E belong to $\mathcal{V}$ while the balls in the reproducing kernel Hilbert space $\mathcal{H}$ do not (when the support of μ is infinite dimensional). A starlike subset V of E of positive measure belongs to $\mathcal{V}$.

In the example of Wiener measure on $C_0([0,1])$, the balls centered at the origin for the Hölder norm $\|\cdot\|_\alpha$ of exponent α, $0<\alpha<\frac{1}{2}$, given by

$$\|x\|_\alpha = \sup_{0\le s\ne t\le 1} \frac{|x(s)-x(t)|}{|s-t|^\alpha}, \quad x\in C_0([0,1]),$$

do belong to the class $\mathcal{V}$. Actually, the balls of any reasonable norm on Wiener space for which Wiener measure is Radon are in $\mathcal{V}$. Using the properties of the Brownian paths, many other examples of elements of $\mathcal{V}$ may be imagined (cf. [B-BA-K]).

Provided with the preceding notation, we introduce new rate functionals, on the subsets of E rather than the points. For a Borel subset A of E, set

$$r(A) = \sup\{r\ge 0; \exists V\in\mathcal{V}, (V + r\mathcal{K})\cap A = \emptyset\}$$

($r(A)=0$ if $\{\ \} = \emptyset$) and

$$s(A) = \inf\{s\ge 0; \exists V\in\mathcal{V}, (A\ominus V)\cap(s\mathcal{K})\ne\emptyset\}$$

($s(A)=\infty$ if $\{\ \}=\emptyset$). The functionals $r(\cdot)$ and $s(\cdot)$ are decreasing for the inclusion. Furthermore, it is elementary to check that $\frac{1}{2}r(A)^2 \ge \mathcal{I}(A)$ when A is closed in $(E,\|\cdot\|)$ and that $\frac{1}{2}s(A)^2\le\mathcal{I}(A)$ when A is open. These inequalities correspond to the choice of a ball $B_E(0,\eta)$ as an element of $\mathcal{V}$ in the definitions of $r(A)$ and $s(A)$ (cf. also the previous elementary proof of the classical large deviation principle). Let us briefly verify this claim. Assume first that A is closed and let r be such that $0<r<\mathcal{I}(A)$ (there is nothing to prove if $\mathcal{I}(A)=0$). Then $A\cap r\mathcal{K}=\emptyset$ and since A is closed and $\mathcal{K}$ is compact in E, there exists $\eta>0$ such that $A\cap(r\mathcal{K}+B_E(0,\eta))$ is still empty. Now $B_E(0,\eta)\in\mathcal{V}$ so that $r(A)\ge\sqrt{2r}$. Since $r<\mathcal{I}(A)$ is arbitrary, the first assertion follows. When A is open, let h be in $A\cap\mathcal{H}$ (there is nothing to prove if there is no such h). There exists $\eta>0$ such that $B_E(h,\eta)\subset A$ which means that

$$\big(A\ominus B_E(0,\eta)\big)\cap\big(|h|\mathcal{K}\big)\ne\emptyset.$$

Therefore, $s(A)\le|h|$ and since h is arbitrary in $A\cap\mathcal{H}$, $\frac{1}{2}s(A)^2\le\mathcal{I}(A)$. It should be noticed that the compactness of $\mathcal{K}$ is only used in the argument concerning the functional $r(\cdot)$. One may also note that if we restrict (without loss of generality) the class $\mathcal{V}$ to those elements V for which $0\in V$, then for any set A, $\frac{1}{2}r(A)^2\le\mathcal{I}(A)\le\frac{1}{2}s(A)^2$. In particular, $\frac{1}{2}r(A)^2$ (respectively $\frac{1}{2}s(A)^2$) coincide with $\mathcal{I}(A)$ if A is closed (respectively open).

The next theorem [BA-L1] is the main result concerning the measurable large deviation principle. The proof of the upper bound is entirely similar to the preceeding sketch of proof of the classical large deviation theorem. The lower bound amounts to the classical argument based on Cameron-Martin translates. Recall that the Cameron-Martin translation formula [C-M] (cf. [Ne1], [Ku], [Fe5], [Lif3]...) indicates that, for any h in $\mathcal{H}$, the probability measure $\mu(h+\cdot)$ is absolutely continuous with respect to μ with density given by the formula

$$\mu(h+A) = \exp\Big(-\frac{|h|^2}{2}\Big)\int_A \exp(-\tilde{h})\,d\mu \tag{4.11}$$

for every Borel set A in E (where we recall that $\widetilde{h} = (j^*_{|E_2^*})^{-1}(h))$.

Theorem 4.5. *For every Borel set A in E,*

$$\limsup_{\varepsilon\to 0} \varepsilon^2 \log \mu_\varepsilon(A) \le -\tfrac{1}{2} r(A)^2 \tag{4.12}$$

and

$$\liminf_{\varepsilon\to 0} \varepsilon^2 \log \mu_\varepsilon(A) \ge -\tfrac{1}{2} s(A)^2. \tag{4.13}$$

By the preceding comments, this result generalizes the classical large deviations theorem for the Gaussian measure μ (due to M. Schilder [Sc] for Wiener measure and to M. Donsker and S. R. S. Varadhan [D-V] in general – see e.g. [Az], [D-S], [Var]...) which expresses that

$$\limsup_{\varepsilon\to 0} \varepsilon^2 \log \mu_\varepsilon(A) \le -\mathcal{I}(\bar{A}), \tag{4.14}$$

where $\bar{A}$ is the closure of A (in $(E, \|\cdot\|)$) and

$$\liminf_{\varepsilon\to 0} \varepsilon^2 \log \mu_\varepsilon(A) \ge -\mathcal{I}(\mathring{A}) \tag{4.15}$$

where $\mathring{A}$ is the interior of A. It is rather easy to find examples of sets A such that $\frac{1}{2} r(A)^2 > \mathcal{I}(\bar{A})$ and $\frac{1}{2} s(A)^2 < \mathcal{I}(\mathring{A})$. (For example, if we fix the uniform topology on Wiener space, and if $A = \{x; \|x\|_\alpha \ge 1\}$ where $\|\cdot\|_\alpha$ is the Hölder norm of index α, then $r(A) > 0$ but $\mathcal{I}(\bar{A}) = 0$. (In this case of course, one can simply consider Wiener measure on the corresponding Hölder space.) More significant examples are described in [B-BA-K].) Therefore, Theorem 4.5 improves upon the classical large deviations for Gaussian measures.

Proof of Theorem 4.5. We start with (4.12). Let $r \ge 0$ be such that $(V + r\mathcal{K}) \cap A = \emptyset$ for some V in $\mathcal{V}$. Then

$$\mu_\varepsilon(A) = \mu(\varepsilon^{-1} A) \le 1 - \mu_*(\varepsilon^{-1} V + \varepsilon^{-1} r\mathcal{K}).$$

Since $V \in \mathcal{V}$, there exists $\alpha > 0$ such that $\mu(\varepsilon^{-1} V) \ge \alpha$ for every $\varepsilon > 0$ small enough. Hence, by (4.6) (or Theorem 4.3),

$$\mu_\varepsilon(A) \le \exp\left(-\frac{r^2}{2\varepsilon^2} + \frac{r}{\varepsilon}\,\delta(\alpha)\right)$$

from which (4.12) immediately follows in the limit.

As announced, the proof of (4.13) is classical. Let $s \ge 0$ be such that

$$(A \ominus V) \cap (s\mathcal{K}) \ne \emptyset$$

for some V in $\mathcal{V}$. Therefore, there exists h in $\mathcal{H}$ with $|h| \le s$ such that $h + V \subset A$. Hence, for every $\varepsilon > 0$,

$$\mu_\varepsilon(A) = \mu(\varepsilon^{-1} A) \ge \mu(\varepsilon^{-1}(h + V)).$$

By Cameron-Martin's formula (4.11) (one could also use (1.10) in this argument),

$$\mu(\varepsilon^{-1}(h+V)) = \exp\Big(-\frac{|h|^2}{2\varepsilon^2}\Big)\int_{\varepsilon^{-1}V}\exp\Big(-\frac{\widetilde{h}}{\varepsilon}\Big)d\mu.$$

Since $V \in \mathcal{V}$, there exists $\alpha > 0$ such that $\mu(\varepsilon^{-1}V) \geq \alpha$ for every $\varepsilon > 0$ small enough. By Jensen's inequality,

$$\int_{\varepsilon^{-1}V}\exp\Big(-\frac{\widetilde{h}}{\varepsilon}\Big)d\mu \geq \mu(\varepsilon^{-1}V)\exp\Big(-\int_{\varepsilon^{-1}V}\frac{\widetilde{h}}{\varepsilon}\cdot\frac{d\mu}{\mu(\varepsilon^{-1}V)}\Big).$$

Now,

$$\int_{\varepsilon^{-1}V}\widetilde{h}\,d\mu \leq \int|\widetilde{h}|d\mu \leq \Big(\int\widetilde{h}^2 d\mu\Big)^{1/2} = |h|.$$

We have thus obtained that, for every $\varepsilon > 0$ small enough,

$$\mu_\varepsilon(A) \geq \mu(\varepsilon^{-1}(h+V)) \geq \alpha\exp\Big(-\frac{|h|^2}{2\varepsilon^2} - \frac{|h|}{\alpha\varepsilon}\Big)$$

from which we deduce that

$$\liminf_{\varepsilon\to 0}\varepsilon^2\log\mu_\varepsilon(A) \geq -\frac{1}{2}|h|^2 \geq -\frac{1}{2}s^2.$$

The claim (4.13) follows since s may be chosen arbitrary less than $s(A)$. The proof of Theorem 4.5 is complete. □

It is a classical result in the theory of large deviations, due to S. R. S. Varadhan (cf. [Az], [D-S], [Var]...), that the statements (4.14) and (4.15) on sets may be translated essentially equivalently on functions. More precisely, if $F : E \to \mathbb{R}$ is bounded and continuous on E,

$$\lim_{\varepsilon\to 0}\varepsilon^2\log\Big(\int\exp\Big(-\frac{1}{\varepsilon^2}F(\varepsilon x)\Big)d\mu(x)\Big) = -\inf_{x\in E}\big(F(x)+\mathcal{I}(x)\big).$$

One consequence of measurable large deviations is that it allows us to weaken the continuity hypothesis into a continuity "in probability".

Corollary 4.6. *Let $F : E \to \mathbb{R}$ be measurable and bounded on E and such that, for every $r > 0$ and every $\eta > 0$,*

$$\limsup_{\varepsilon\to 0}\mu\big(x;\ \sup_{|h|\leq r}|F(h+\varepsilon x) - F(h)| > \eta\big) < 1.$$

Then,

$$\lim_{\varepsilon\to 0}\varepsilon^2\log\Big(\int\exp\Big(-\frac{1}{\varepsilon^2}F(\varepsilon x)\Big)d\mu(x)\Big) = -\inf_{x\in E}\big(F(x)+\mathcal{I}(x)\big).$$

It has to be mentioned that the continuity assumption in Corollary 4.6 is not of the Malliavin calculus type since limits are taken along the elements of E and not the elements of $\mathcal{H}$.

Proof. Set

$$L(\varepsilon) = \int \exp\Big(-\frac{1}{\varepsilon^2}\,F(\varepsilon x)\Big)\,d\mu(x), \quad \varepsilon > 0.$$

By a simple translation, we may assume that $F \geq 0$. For simplicity in the notation, let us assume moreover that $0 \leq F \leq 1$. For every integer $n \geq 1$, set

$$A_k^n = \big\{\tfrac{k-1}{n} < F \leq \tfrac{k}{n}\big\}, \quad k = 2, \ldots, n, \quad A_1^n = \big\{F \leq \tfrac{1}{n}\big\}.$$

Since

$$L(\varepsilon) \leq \sum_{k=1}^{n} \exp\Big(-\frac{k-1}{\varepsilon^2 n}\Big)\mu^\varepsilon(A_k^n),$$

we get that

$$\limsup_{\varepsilon\to 0} \varepsilon^2 \log L(\varepsilon) \leq -\min_k\Big(\frac{k-1}{n} + \frac{1}{2}\,r(A_k^n)^2\Big).$$

Since $r(A_k^n) \geq r(\{F \leq \frac{k}{n}\})$ and since n is arbitrary, it follows that

$$\limsup_{\varepsilon\to 0} \varepsilon^2 \log L(\varepsilon) \leq -\inf_{t\in\mathbb{R}}\Big(t + \frac{1}{2}\,r(\{F \leq t\})^2\Big). \tag{4.16}$$

Now, we show that the right hand side of (4.16) is less than or equal to

$$-\inf_{x\in E}(F(x) + \mathcal{I}(x)).$$

Let $t \in \mathbb{R}$ be fixed. Let $\eta > 0$ and $r > r(\{F \leq t\})$ (assumed to be finite). Set

$$V = \big\{x;\ \sup_{|h|\leq r}\big|F(h+x) - F(h)\big| \leq \eta\big\}.$$

By the hypothesis, $V \in \mathcal{V}$. By the definition of r,

$$(V + r\mathcal{K}) \cap \{F \leq t\} \neq \emptyset.$$

Therefore, there exist v in V and $|h| \leq r$ such that $F(h+v) \leq t$. By definition of V, $F(h) \leq t + \eta$. Hence

$$\inf_{x\in E}\big(F(x) + \mathcal{I}(x)\big) \leq t + \eta + \frac{r^2}{2}.$$

Since $\eta > 0$ and $r > r(\{F \leq t\})$ are arbitrary, the claim follows and thus, together with (4.16),

$$\limsup_{\varepsilon\to 0} \varepsilon^2 \log L(\varepsilon) \leq -\inf_{x\in E}\big(F(x) + \mathcal{I}(x)\big). \tag{4.17}$$

The proof of the lower bound is similar. We have, for every $n \geq 1$,

$$\begin{aligned} L(\varepsilon) &\geq \sum_{k=1}^{n} \exp\Big(-\frac{k}{\varepsilon^2 n}\Big)\mu^\varepsilon(A_k^n) \\ &\geq \sum_{k=1}^{n-1}\Big[\exp\Big(-\frac{k}{\varepsilon^2 n}\Big) - \exp\Big(-\frac{k+1}{\varepsilon^2 n}\Big)\Big]\mu^\varepsilon(\{F \leq \tfrac{k}{n}\}) \\ &\geq \frac{1}{2}\sum_{k=1}^{n-1} \exp\Big(-\frac{k}{\varepsilon^2 n}\Big)\mu^\varepsilon(\{F \leq \tfrac{k}{n}\}) \end{aligned}$$

at least for all $\varepsilon > 0$ small enough. Therefore,

$$\liminf_{\varepsilon \to 0} L(\varepsilon) \geq -\inf_{t \in \mathbb{R}}\Big(t + \frac{1}{2}s(\{F \leq t\})^2\Big). \tag{4.18}$$

Now, let h be in $\mathcal{H}$ and set $t = F(h)$. Let $\eta > 0$ and $0 < s < s(\{F \leq t + \eta\})$. We will show that $|h| > s$. Let

$$V = \{x; F(h+x) \leq F(h) + \eta\}.$$

By the hypothesis, $V \in \mathcal{V}$ and by the definition of s,

$$(\{F \leq t + \eta\} \ominus V) \cap (s\mathcal{K}) = \emptyset.$$

It is clear that $h \in \{F \leq t + \eta\} \ominus V$. Hence $|h| > s$, and since s is arbitrary, we have $|h| \geq s(\{F \leq t + \eta\})$. Now, if $t > -\infty$,

$$t + \frac{1}{2}s(\{F \leq t + \eta\})^2 \leq F(h) + \mathcal{I}(h).$$

If $t = -\infty$, $0 \leq s(\{F = -\infty\}) \leq |h| < \infty$, and the preceding also holds. In any case,

$$-\inf_{t \in \mathbb{R}}\Big(t + \frac{1}{2}s(\{F \leq t\})^2\Big) \leq \inf_{x \in E}(F(x) + \mathcal{I}(x)).$$

Together with (4.18) and (4.17), the proof of Corollary 4.6 is complete. □

In the last part of this chapter, we prove a large deviation principle for the Ornstein-Uhlenbeck process due to S. Kusuoka [Kus]. If μ is a Gaussian measure on E, define, for every say bounded measurable function f on E, and every $x \in E$ and $t \geq 0$,

$$P_t f(x) = \int_E f\big(e^{-t}x + (1 - e^{-2t})^{1/2}y\big)\,d\mu(y).$$

If A and B are Borel subsets of E, set then, as at the end of Chapter 2,

$$K_t(A, B) = \int_A P_t(I_B)\,d\mu, \quad t \geq 0.$$

We will be interested in the large deviation behavior of $K_t(A,B)$ in terms of the $\mathcal{H}$-distance between A and B. Set indeed

$$d_{\mathcal{H}}(A,B) = \inf\{|h-k|; h \in A, k \in B, h-k \in \mathcal{H}\}.$$

One defines in the same way $d_{\mathcal{H}}(x,A)$, $x \in E$, and notices that $d_{\mathcal{H}}(x,A) < \infty$ μ-almost everywhere if and only if $\mu(A+\mathcal{H}) = 1$. By the isoperimetric inequality, this is immediately the case as soon as $\mu(A) > 0$.

The main result is the following. S. Kusuoka's proof uses the wave equation. We folllow here the approach by S. Fang [Fa] (who actually establishes a somewhat stronger statement by using a slightly different distance on the subsets of E).

Theorem 4.7. *Let A and B be Borel subsets in E such that $\mu(A) > 0$ and $\mu(B) > 0$. Then*

$$\limsup_{t\to 0} 4t \log K_t(A,B) \le -d_{\mathcal{H}}(A,B)^2.$$

If moreover A and B are open, then

$$\liminf_{t\to 0} 4t \log K_t(A,B) \ge -d_{\mathcal{H}}(A,B)^2.$$

Proof. We start with the upper bound which is thus based on Proposition 2.3. We use the same approximation procedure as the one described in the proof of Theorem 4.3 and, accordingly the probabilistic notation put forward in Proposition 4.2. Denote in particular by Y an independent copy of X (with thus distribution μ). Assume that $d_{\mathcal{H}}(A,B) > r > 0$. Choose $K \subset A$ and $L \subset B$ compact subsets of positive measure. Let $t \ge 0$ be fixed and let $\varepsilon > 0$. We can write that, for every $n \ge n_0$ large enough,

$$\begin{aligned}
&\mathbb{P}\{X \in K, \mathrm{e}^{-t}X + (1-\mathrm{e}^{-2t})^{1/2}Y \notin K^{3\varepsilon} + r\mathcal{K}\} \\
&\qquad \le \mathbb{P}\{X_n \in K^{\varepsilon}, \mathrm{e}^{-t}X_n + (1-\mathrm{e}^{-2t})^{1/2}Y_n \notin K^{2\varepsilon} + r\mathcal{K}\} + \varepsilon \\
&\qquad \le \mathbb{P}\{X_n \in K^{\varepsilon}, \mathrm{e}^{-t}X_n + (1-\mathrm{e}^{-2t})^{1/2}Y_n \notin K^{\varepsilon} + r\mathcal{K}_n\} + \varepsilon.
\end{aligned}$$

Hence, according to (2.16) of Chapter 2,

$$\mathbb{P}\{X \in K, \mathrm{e}^{-t}X + (1-\mathrm{e}^{-2t})^{1/2}Y \notin K^{3\varepsilon} + r\mathcal{K}\} \le \exp\left(-\frac{r^2}{4(1-\mathrm{e}^{-t})}\right) + \varepsilon.$$

Letting ε decrease to zero, by compactness,

$$\mathbb{P}\{X \in K, \mathrm{e}^{-t}X + (1-\mathrm{e}^{-2t})^{1/2}Y \notin K + r\mathcal{K}\} \le \exp\left(-\frac{r^2}{4(1-\mathrm{e}^{-t})}\right)$$

and thus, by definition of r,

$$\mathbb{P}\{X \in K, \mathrm{e}^{-t}X + (1-\mathrm{e}^{-2t})^{1/2}Y \in L\} \le \exp\left(-\frac{r^2}{4(1-\mathrm{e}^{-t})}\right).$$

The first claim of Theorem 4.7 is proved.

The lower bound relies on Cameron-Martin translates. Let $r = d_{\mathcal{H}}(A,B) \geq 0$ (assumed to be finite). Let also $h \in A \cap \mathcal{H}$ and $k \in B \cap \mathcal{H}$. Since A and B are open, there exists $\eta > 0$ such that $B_E(h, 2\eta) \subset A$ and $B_E(k, 2\eta) \subset B$. Therefore, for every $t \geq 0$,

$$K_t(A,B) \geq K_t\big(B_E(h,2\eta), B_E(k,2\eta)\big).$$

By the Cameron-Martin translation formula (4.11),

$$\begin{aligned} K_t&\big(B_E(h,2\eta), B_E(k,2\eta)\big) \\ &= \exp\bigg(-\frac{|h-k|^2}{2(1-\mathrm{e}^{-2t})}\bigg) \int_{(x,y)\in C} \exp\bigg(\frac{\widetilde{h}(y)-\widetilde{k}(y)}{(1-\mathrm{e}^{-2t})^{1/2}}\bigg) d\mu(x)d\mu(y) \end{aligned}$$

where $C = \{(x,y) \in E \times E; x \in B_E(h,2\eta), \mathrm{e}^{-t}x + (1-\mathrm{e}^{-2t})^{1/2}y \in B_E(h,2\eta)\}$. Now, for $t \leq t_0(\eta, h)$ small enough,

$$B_E(h,\eta) \times B_E(0,1) \subset C$$

so that

$$\begin{aligned} K_t(A,B) &\geq \exp\bigg(-\frac{|h-k|^2}{2(1-\mathrm{e}^{-2t})}\bigg)\mu\big(B_E(h,\eta)\big) \int_{B_E(0,1)} \exp\bigg(\frac{\widetilde{h}-\widetilde{k}}{(1-\mathrm{e}^{-2t})^{1/2}}\bigg) d\mu \\ &\geq \exp\bigg(-\frac{|h-k|^2}{2(1-\mathrm{e}^{-2t})}\bigg)\mu\big(B_E(h,\eta)\big)\mu\big(B_E(0,1)\big) \end{aligned}$$

by Jensen's inequality. Therefore,

$$\liminf_{t\to 0} 4t \log K_t(A,B) \geq -|h-k|^2$$

and the result follows since h and k are arbitrary in A and B respectively. The proof of Theorem 4.7 is complete. □

Notes for further reading. There is an extensive literature on precise estimates on the tail behavior of norms of Gaussian random vectors (involving in particular the tool of entropy – cf. Chapter 6). We refer in particular the interested reader to the works [Ta4], [Ta13], [Lif2] and the references therein (see also [Lif3]). In the paper [Lif2], a Laplace method is developed to yield some unexpected irregular behaviors. Large deviations without topology may be applied to Strassen's law of the iterated logarithm for Brownian motion [B-BA-K], [BA-L1], [D-L]. In [D-L], a complete description of norms on Wiener space for which the law of the iterated logarithm holds is provided.

5. LARGE DEVIATIONS OF WIENER CHAOS

The purpose of this chapter is to further demonstrate the usefulness and interest of isoperimetric and concentration methods in large deviation theorems in the context of Wiener chaos. This chapter intends actually to present some aspects of the remarkable work of C. Borell on homogeneous chaos whose early ideas strongly influenced the subsequent developments. We present here, closely following the material in [Bo5], [Bo9], a simple isoperimetric proof of the large deviations properties of homogeneous Gaussian chaos (even vector valued). We take again the exposition of [Led2].

Let, as in the preceding chapter, $(E, \mathcal{H}, \mu)$ be an abstract Wiener space. According to Proposition 4.2, for any orthonormal basis $(g_i)_{i\in\mathbb{N}}$ of the closure E_2^* of E^* in $\mathrm{L}^2(\mu)$, μ has the same distribution as the series $\sum_i g_i j^*(g_i)$. It will be convenient here (although this is not strictly necessary) to consider this basis in E^*. Let thus $(\xi_i)_{i\in\mathbb{N}} \subset E^*$ be any fixed orthonormal basis of E_2^* (take any weak-star dense sequence of the unit ball of E^* and orthonormalize it with respect to μ using the Gram-Schmidt procedure). Denote by $(h_k)_{k\in\mathbb{N}}$ the sequence of the Hermite polynomials defined from the generating series

$$e^{\lambda x - \lambda^2/2} = \sum_{k=0}^{\infty} \lambda^k h_k(x), \quad \lambda, x \in \mathbb{R}.$$

$(\sqrt{k!}\, h_k)_{k\in\mathbb{N}}$ is an orthonormal basis of $\mathrm{L}^2(\gamma_1)$ where γ_1 is the canonical Gaussian measure on $\mathbb{R}$. If $\alpha = (\alpha_0, \alpha_1, \ldots) \in \mathbb{N}^{(\mathbb{N})}$, i.e. $|\alpha| = \alpha_0 + \alpha_1 + \cdots < \infty$, set

$$H_\alpha = \sqrt{\alpha!} \prod_i h_{\alpha_i} \circ \xi_i$$

(where $\alpha! = \alpha_0!\alpha_1!\cdots$). Then the family (H_α) constitutes an orthonormal basis of $L^2(\mu)$.

Let now B be a real separable Banach space with norm $\|\cdot\|$ (we denote in the same way the norm on E and the norm on B). $\mathrm{L}^p((E, \mathcal{B}, \mu); B) = \mathrm{L}^p(\mu; B)$

$(0 \leq p < \infty)$ is the space of all Bochner measurable functions F on (E, μ) with values in B $(p = 0)$ such that $\int \|F\|^p d\mu < \infty$ $(0 < p < \infty)$. For each integer $d \geq 1$, set

$$\mathcal{W}^{(d)}(\mu; B) = \{F \in \mathrm{L}^2(\mu; B); \langle F, H_\alpha \rangle = \int F H_\alpha d\mu = 0 \text{ for all } \alpha \text{ such that } |\alpha| \neq d\}.$$

$\mathcal{W}^{(d)}(\mu; B)$ defines the B-valued homogeneous Wiener chaos of degree d [Wi]. An element Ψ of $\mathcal{W}^{(d)}(\mu; B)$ can be written as

$$\Psi = \sum_{|\alpha|=d} \langle \Psi, H_\alpha \rangle \, H_\alpha$$

where the multiple sum is convergent (for any finite filtering) μ-almost everywhere and in $\mathrm{L}^2(\mu; B)$. (Actually, as a consequence of [Bo5], [Bo9] (see also [L-T2], or the subsequent main result), this convergence also takes place in $\mathrm{L}^p(\mu; B)$ for any p.) To see it, we simply follow the proof of Proposition 4.2. Let, for each n, $\mathcal{B}_n$ be the sub-σ-algebra of $\mathcal{B}$ generated by the functions $\xi_0, \ldots, \xi_n$ on E and let Ψ_n be the conditional expectation of Ψ with respect to $\mathcal{B}_n$. Recall that $\mathcal{B}$ may be assumed to be generated by $(\xi_i)_{i \in \mathbf{N}}$. Then

$$\Psi_n = \sum_{\substack{|\alpha|=d \\ \alpha_k = 0, k > n}} \langle \Psi, H_\alpha \rangle \, H_\alpha \tag{5.1}$$

as can be checked on linear functionals, and therefore, by the vector valued martingale convergence theorem (cf. [Ne2]), the claim follows. One could actually take this series representation as the definition of a homogeneous chaos, which would avoid the assumption $F \in \mathrm{L}^2(\mu; B)$ in $\mathcal{W}^{(d)}(\mu; B)$. By the preceding comment, both definitions actually agree (cf. [Bo5], [Bo9]).

As a consequence of the Cameron-Martin formula, we may define for every F in $L^0(\mu; B)$ and every h in $\mathcal{H}$, a new element $F(\cdot + h)$ of $\mathrm{L}^0(\mu; B)$. Furthermore, if F is in $\mathrm{L}^2(\mu; B)$, for any $h \in \mathcal{H}$,

$$\int \|F(x+h)\| \, d\mu(x) \leq \exp\left(\frac{|h|^2}{2}\right) \left(\int \|F(x)\|^2 d\mu(x)\right)^{1/2}. \tag{5.2}$$

Indeed,

$$\int \|F(x+h)\| \, d\mu(x) = \int \exp\left(-\widetilde{h}(x) - \frac{|h|^2}{2}\right) \|F(x)\| d\mu(x)$$

from which (5.2) follows by Cauchy-Schwarz inequality and the fact that $\widetilde{h} = (j^*_{|E_2^*})^{-1}(h)$ is Gaussian with variance $|h|^2$.

Let F be in $\mathrm{L}^2(\mu; B)$. By (5.2), for any h in $\mathcal{H}$, we can define an element $F^{(d)}(h)$ of B by setting

$$F^{(d)}(h) = \int F(x+h) d\mu(x).$$

If $\Psi \in \mathcal{W}^{(d)}(\mu;B)$, $\Psi^{(d)}(h)$ is homogeneous of degree d. To see it, we can work by approximation on the Ψ_n's and use then the easy fact (checked on the generating series for example) that, for any real number λ and any integer k,

$$\int h_k(x+\lambda)d\gamma_1(x) = \frac{1}{k!}\lambda^k.$$

Actually, $\Psi^{(d)}(h)$ can be written as the convergent multiple sum

$$\Psi^{(d)}(h) = \sum_{|\alpha|=d} \frac{1}{\alpha!}\langle \Psi, H_\alpha\rangle h^\alpha$$

where h^α is meant as $\langle \xi_0, h\rangle^{\alpha_0}\langle \xi_1, h\rangle^{\alpha_1}\cdots$.

Given thus Ψ in $\mathcal{W}^{(d)}(\mu;B)$, for any s in B, set $\mathcal{I}_\Psi(s) = \inf\{\frac{1}{2}|h|^2; s = \Psi^{(d)}(h)\}$ if there exists h in $\mathcal{H}$ such that $s = \Psi^{(d)}(h)$, $\mathcal{I}_\Psi(s) = \infty$ otherwise. For a subset A of B, set $\mathcal{I}_\Psi(A) = \inf_{s\in A}\mathcal{I}_\Psi(s)$.

We can now state the large deviation properties for the elements Ψ of $\mathcal{W}^{(d)}(\mu;B)$. The case $d=1$ of course corresponds to the classical large deviation result for Gaussian measures (cf. (4.14) and (4.15) for $B=E$ and Ψ the identity map on E). From the point of view of isoperimetry and concentration, the proof for higher order chaos is actually only the appropriate extension of the case $d=1$.

Theorem 5.1. *Let $\mu_\varepsilon(\cdot) = \mu(\varepsilon^{-1}(\cdot))$, $\varepsilon > 0$. Let d be an integer and let Ψ be an element of $\mathcal{W}^{(d)}(\mu;B)$. Then, if A is a closed subset of B,*

$$\limsup_{\varepsilon\to 0} \varepsilon^2 \log \mu_\varepsilon(x; \Psi(x)\in A) \le -\mathcal{I}_\Psi(A). \tag{5.3}$$

If A is an open subset of B,

$$\liminf_{\varepsilon\to 0} \varepsilon^2 \log \mu_\varepsilon(x; \Psi(x)\in A) \ge -\mathcal{I}_\Psi(A). \tag{5.4}$$

The proof of (5.4) follows rather easily from the Cameron-Martin translation formula. (5.3) is rather easy too, but our approach thus rests on the tool of isoperimetric and concentration inequalities. The proof of (5.3) also sheds some light on the structure of Gaussian polynomials as developed by C. Borell, and in particular the homogeneous structures. As it is clear indeed from [Bo5] (and the proof below), the theorem may be shown to hold for all Gaussian polynomials, i.e. elements of the closure in $L^0(\mu;B)$ of all continuous polynomials from E into B of degree less than or equal to d. As we will see, $\mathcal{W}^{(d)}(\mu;B)$ may be considered as a subspace of the closure of all homogeneous Gaussian polynomials of degree d (at least if the support of μ is infinite dimensional), and hence, the elements of $\mathcal{W}^{(d)}(\mu;B)$ are μ-almost everywhere d-homogeneous. In particular, (5.3) and (5.4) of the theorem are equivalent to saying that (changing moreover ε into t^{-1})

$$\limsup_{t\to\infty} \frac{1}{t^2}\log \mu(x; \Psi(x)\in t^d A) \le -\mathcal{I}_\Psi(A) \tag{5.5}$$

(A closed) and

(5.6) $$\liminf_{t\to\infty} \frac{1}{t^2} \log \mu(x; \Psi(x) \in t^d A) \geq -\mathcal{I}_\Psi(A),$$

(A open) and these are the properties we will actually establish.

Before turning to the proof of Theorem 5.1, let us mention some application. If we take A in the theorem to be the complement U^c of the (open or closed) unit ball U of B, one immediately checks that

$$\mathcal{I}_\Psi(U^c) = \frac{1}{2}\big(\sup_{h\in\mathcal{K}} \|\Psi^{(d)}(h)\|\big)^{-2/d}.$$

We may therefore state the following corollary of Theorem 5.1 which was actually established directly from the isoperimetric inequality by C. Borell [Bo5] (see also [Bo8], [L-T2]). It is the analogue for chaos of Theorem 4.1.

Corollary 5.2. *Let Ψ be an element of $\mathcal{W}^{(d)}(\mu; B)$. Then*

$$\lim_{t\to\infty} \frac{1}{t^{2/d}} \log \mu(x; \|\Psi(x)\| \geq t) = -\frac{1}{2}\big(\sup_{h\in\mathcal{K}} \|\Psi^{(d)}(h)\|\big)^{-2/d}.$$

As in Theorem 4.1, we have that

$$\int \exp\big(\alpha \|\Psi\|^{2/d}\big) d\mu < \infty \quad \text{if and only if} \quad \alpha < \frac{1}{2}\big(\sup_{h\in\mathcal{K}} \|\Psi^{(d)}(h)\|\big)^{-2/d}.$$

Furthermore, the proof of the theorem will show that all moments of Ψ are equivalent (see also Chapter 8, (8.23)).

In the setting of the classical Wiener space $E = C_0([0,1])$ equipped with the Wiener measure μ, and when $B = E$, K. Itô [It] (see also [Ne1] and the recent approach [Str]) identified the elements Ψ of $\mathcal{W}^{(d)}(\mu; E)$ with the multiple stochastic integrals

$$\Psi = \left(\int_0^t \int_0^{t_1} \cdots \int_0^{t_{d-1}} k(t_1, \ldots, t_d)\, dW(t_1) \cdots dW(t_d)\right)_{t\in[0,1]}$$

where k deterministic is such that

$$\int_0^1 \int_0^{t_1} \cdots \int_0^{t_{d-1}} k(t_1, \ldots, t_d)^2 dt_1 \cdots dt_d < \infty.$$

If h belongs to the reproducing kernel Hilbert space of the Wiener measure, then

$$\Psi^{(d)}(h) = \left(\int_0^t \int_0^{t_1} \cdots \int_0^{t_{d-1}} k(t_1, \ldots, t_d)\, h'(t_1) \cdots h'(t_d)\, dt_1 \cdots dt_d\right)_{t\in[0,1]}.$$

Proof of Theorem 5.1. Let us start with the simpler property (5.4). Recall Ψ_n from (5.1). We can write (explicitly on the Hermite polynomials), for all x in E, h in $\mathcal{H}$ and t real number,

$$\Psi_n(x+th) = \sum_{k=0}^{d} t^k \Psi_n^{(k)}(x,h).$$

If $P(t) = a_0 + a_1 t + \cdots + a_d t^d$ is a polynomial of degree d in $t \in \mathbb{R}$ with vector coefficients $a_0, a_1, \ldots, a_d$, there exist real constants $c(i,k,d)$, $0 \le i,k \le d$, independent of P, such that, for every $k = 0, \ldots, d$,

$$a_k = c(0,k,d)P(0) + \sum_{i=1}^{d} c(i,k,d)P(2^{i-1}).$$

Hence, for every $h \in \mathcal{H}$,

$$\Psi_n^{(k)}(\cdot,h) = c(0,k,d)\Psi_n(\cdot) + \sum_{i=1}^{d} c(i,k,d)\Psi_n(\cdot + 2^{i-1}h)$$

from which we deduce together with (5.2) that, for every $k = 0, \ldots, d$,

$$\int \|\Psi_n^{(k)}(x,h)\| \, d\mu(x) \le C(k,d;h) \left(\int \|\Psi_n(x)\|^2 d\mu(x) \right)^{1/2}$$

for some constants $C(k,d;h)$ thus only depending on k, d and $h \in \mathcal{H}$. In the limit, we conclude that there exist, for every h in $\mathcal{H}$ and $k = 0, \ldots, d$, elements $\Psi^{(k)}(\cdot,h)$ of $\mathrm{L}^1(\mu; B)$ such that

$$\Psi(\cdot + th) = \sum_{k=0}^{d} t^k \Psi^{(k)}(\cdot,h)$$

for every $t \in \mathbb{R}$, with

$$\int \|\Psi^{(k)}(x,h)\| \, d\mu(x) \le C(k,d;h) \left(\int \|\Psi(x)\|^2 d\mu(x) \right)^{1/2}$$

and $\Psi^{(0)}(\cdot,h) = f(\cdot)$, $\Psi^{(d)}(\cdot,h) = \Psi^{(d)}(h)$ (since $\int f(x+th)\, d\mu(x) = t^d f^{(d)}(h)$). As a main consequence, we get that, for every h in $\mathcal{H}$,

$$\lim_{t\to\infty} \frac{1}{t^d} \int \|\Psi(x+th) - t^d \Psi^{(d)}(h)\| d\mu(x) = 0. \tag{5.7}$$

This limit can be made uniform in $h \in \mathcal{K}$ but we will not use this observation in this form later (that is in the proof of (5.3); we use instead a stronger property, (5.9) below).

To establish (5.4), let A be open in B and let $s = \Psi^{(d)}(h)$, $h \in \mathcal{H}$, belong to A (if no such s exists, then $\mathcal{I}_\Psi(A) = \infty$ and (5.4) then holds trivially). Since A is open, there is $\eta > 0$ such that the ball $B(s,\eta)$ in B with center s and radius η is contained in A. Therefore, if $V = V(t) = \{x \in E; \Psi(x) \in t^d B(s,\eta)\}$, by the Cameron-Martin translation formula (4.11),

$$\mu\big(x; \Psi(x) \in t^d A\big) \geq \mu(V) = \int_{V-th} \exp\left(t\widetilde{h} - \frac{t^2|h|^2}{2}\right) d\mu.$$

Furthermore, by Jensen's inequality,

$$\mu(V) \geq \exp\left(-\frac{t^2|h|^2}{2}\right)\mu(V-th)\exp\left(\frac{t}{\mu(V-th)}\int_{V-th}\widetilde{h}\,d\mu\right).$$

By (5.7),

$$\mu(V-th) = \mu\big(x; \big\|\Psi(x+th) - t^d\Psi^{(d)}(h)\big\| \leq \eta t^d\big) \geq \frac{1}{2}$$

for all $t \geq t_0$ large enough. We have

$$\int_{V-th}\widetilde{h}\,d\mu \geq -\int |\widetilde{h}|\,d\mu \geq -\left(\int \widetilde{h}^2 d\mu\right)^{1/2} = -|h|.$$

Thus, for all $t \geq t_0$,

$$\frac{t}{\mu(V-th)}\int_{V-th}\widetilde{h}\,d\mu \geq -2t|h|,$$

and hence, summarizing,

$$\mu\big(x; \Psi(x) \in t^d A\big) \geq \frac{1}{2}\exp\left(-\frac{t^2|h|^2}{2} - 2t|h|\right).$$

It follows that

$$\liminf_{t\to\infty}\frac{1}{t^2}\log\mu\big(x; \Psi(x) \in t^d A\big) \geq -\frac{1}{2}|h|^2 = -\mathcal{I}_\Psi(s)$$

and since s is arbritrary in A, property (5.6) is satisfied. As a consequence of what we will develop now, (5.4) will be satisfied as well.

Now, we turn to (5.3) and in the first part of this investigation, we closely follow C. Borell [Bo5], [Bo9]. We start by showing that every element Ψ of $\mathcal{W}^{(d)}(\mu; B)$ is limit (at least if the dimension of the support of μ is infinite), μ-almost everywhere and in $L^2(\mu; B)$, of a sequence of d-homogeneous polynomials. In particular, Ψ is μ-almost everywhere d-homogeneous justifying therefore the equivalences between (5.3) and (5.4) and respectively (5.5) and (5.6). Assume thus in the following that μ is infinite dimensional. We can actually always reduce to this case by appropriately tensorizing μ, for example with the canonical Gaussian measure on $\mathbb{R}^{\mathbb{N}}$. Recall that Ψ is limit almost surely and in $L^2(\mu; B)$ of the Ψ_n's of (5.1). The finite sums Ψ_n can be decomposed into their homogeneous components as

$$\Psi_n = \Psi_n^{(d)} + \Psi_n^{(d-2)} + \cdots,$$

where, for any x in E,

$$\Psi_n^{(k)}(x) = \sum_{i_1,\ldots,i_k=0}^{\infty} b_{i_1,\ldots,i_k} \langle \xi_{i_1}, x\rangle \langle \xi_{i_2}, x\rangle \cdots \langle \xi_{i_k}, x\rangle \tag{5.8}$$

with only finitely many $b_{i_1,\ldots,i_k}$ in B nonzero. The main observation is that the constant 1 is limit of homogeneous polynomials of degree 2: indeed, simply take by the law of large numbers

$$p_n(x) = \frac{1}{n+1} \sum_{k=0}^{n} \langle \xi_k, x\rangle^2 .$$

Since p_n and $\Psi_n^{(k)}$ belong to $\mathrm{L}^p(\mu)$ and $\mathrm{L}^p(\mu; B)$ respectively for every p, and since $p_n - 1$ tends there to 0, it is easily seen that there exists a subsequence m_n of the integers such that $(p_{m_n} - 1)(\Psi_n^{(d-2)} + \Psi_n^{(d-4)} + \cdots)$ converges to 0 in $\mathrm{L}^2(\mu; B)$. This means that Ψ is the limit in $\mathrm{L}^2(\mu; B)$ of $\Psi_n^{(d)} + p_{m_n}(\Psi_n^{(d-2)} + \Psi_n^{(d-4)} + \cdots)$, that is limit of a sequence of polynomials Ψ'_n whose decomposition in homogeneous polynomials

$$\Psi'_n = \Psi'^{\,(d)}_n + \Psi'^{\,(d-2)}_n + \cdots$$

is such that $\Psi'^{\,(1)}_n$, or $\Psi'^{\,(0)}_n$ and $\Psi'^{\,(2)}_n$, according as d is odd or even, can be taken to be 0. Repeating this procedure, Ψ is indeed seen to be the limit in $\mathrm{L}^2(\mu; B)$ of a sequence (Ψ'_n) of d-homogeneous polynomials (i.e. polynomials of the type (5.8)).

The important property in order to establish (5.5) is the following. It improves upon (5.7) and claims that, in the preceding notations, i.e. if Ψ is limit of the sequence (Ψ'_n) of d-homogeneous polynomials,

$$\lim_{t\to\infty} \frac{1}{t^{2d}} \sup_n \int \sup_{h\in\mathcal{K}} \left\| \Psi'_n(x + th) - t^d \Psi'_n(h) \right\|^2 d\mu(x) = 0 \tag{5.9}$$

where we recall that $\mathcal{K}$ is the unit ball of the reproducing kernel Hilbert space $\mathcal{H}$ of μ. To establish this property, given

$$\Psi'_n(x) = \sum_{i_1,\ldots,i_d=0}^{\infty} b^n_{i_1,\ldots,i_d} \langle \xi_{i_1}, x\rangle \langle \xi_{i_2}, x\rangle \cdots \langle \xi_{i_d}, x\rangle$$

(with only finitely many $b^n_{i_1,\ldots,i_d}$ nonzero), let us consider the (unique) multilinear symmetric polynomial $\widehat{\Psi}'_n$ on E^d such that $\widehat{\Psi}'_n(x,\ldots,x) = \Psi'_n(x)$; $\widehat{\Psi}'_n$ is given by

$$\widehat{\Psi}'_n(x_1,\ldots,x_d) = \sum_{i_1,\ldots,i_d=0}^{\infty} \widehat{b}^n_{i_1,\ldots,i_d} \langle \xi_{i_1}, x_1\rangle \cdots \langle \xi_{i_d}, x_d\rangle, \quad x_1,\ldots,x_d \in E,$$

where

$$\widehat{b}^n_{i_1,\ldots,i_d} = \frac{1}{d!} \sum_{\sigma} b^n_{\sigma(i_1),\ldots,\sigma(i_d)},$$

the sum running over all permutations σ of $\{1,\dots,d\}$. We use the following polarization formula: letting $\varepsilon_1,\dots,\varepsilon_d$ be independent random variables taking values ±1 with probability $\frac{1}{2}$ and denoting by $\mathbb{E}$ expectation with respect to them,

$$\widehat{\Psi}'_n(x_1,\dots,x_d) = \frac{1}{d!}\,\mathbb{E}\big(\Psi'_n(\varepsilon_1 x_1 + \cdots + \varepsilon_d x_d)\,\varepsilon_1\cdots\varepsilon_d\big). \tag{5.10}$$

We adopt the notation $x^{d-k}y^k$ for the element $(x,\dots,x,y,\dots,y)$ in E^d where x is repeated $(d-k)$-times and y k-times. Then, for any x,y in E, we have

$$\Psi'_n(x+y) = \sum_{k=0}^{d}\binom{d}{k}\widehat{\Psi}'_n(x^{d-k}y^k). \tag{5.11}$$

To establish (5.9), we see from (5.11) that it suffices to show that for all $k = 1,\dots,d-1$,

$$\sup_n \int \sup_{h\in\mathcal{K}}\big\|\widehat{\Psi}'_n(x^{d-k}h^k)\big\|^2 d\mu(x) < \infty. \tag{5.12}$$

Let k be fixed. By orthogonality,

$$\begin{aligned}
&\sup_{h\in\mathcal{K}}\big\|\widehat{\Psi}'_n(x^{d-k}h^k)\big\|^2 \\
&\qquad\le \sup_{\|\zeta\|\le1}\ \sup_{h_1,\dots,h_k\in\mathcal{K}} \langle \zeta, \widehat{\Psi}'_n(x,\dots,x,h_1,\dots,h_k)\rangle^2 \\
&\qquad\le \sup_{\|\zeta\|\le1}\ \sum_{i_{d-k+1},\dots,i_d=0}^{\infty}\Big|\sum_{i_1,\dots,i_{d-k}=0}^{\infty}\langle\zeta,\widehat{b}^n_{i_1,\dots,i_d}\rangle\langle\xi_{i_1},x\rangle\cdots\langle\xi_{i_{d-k}},x\rangle\Big|^2 \\
&\qquad= \sup_{\|\zeta\|\le1}\int\cdots\int\langle\zeta,\widehat{\Psi}'_n(x,\dots,x,y_1,\dots,y_k)\rangle^2\,d\mu(y_1)\cdots d\mu(y_k) \\
&\qquad\le \int\cdots\int\big\|\widehat{\Psi}'_n(x,\dots,x,y_1,\dots,y_k)\big\|^2\,d\mu(y_1)\cdots d\mu(y_k).
\end{aligned}$$

By the polarization formula (5.10),

$$\begin{aligned}
&\widehat{\Psi}'_n(x,\dots,x,y_1,\dots,y_k) \\
&\qquad= \frac{1}{d!}\,\mathbb{E}\big(\Psi'_n((\varepsilon_{k+1}+\cdots+\varepsilon_d)x + \varepsilon_1 y_1 + \cdots + \varepsilon_k y_k)\,\varepsilon_1\cdots\varepsilon_d\big).
\end{aligned}$$

Therefore, we obtain from the rotational invariance of Gaussian distributions and homogeneity that

$$\begin{aligned}
&(d!)^2\int\sup_{h\in\mathcal{K}}\big\|\widehat{\Psi}'_n(x^{d-k}h^k)\big\|^2\,d\mu(x) \\
&\le\mathbb{E}\iint\cdots\int\big\|\Psi'_n((\varepsilon_{k+1}+\cdots+\varepsilon_d)x+\varepsilon_1y_1+\cdots+\varepsilon_ky_k)\big\|^2 d\mu(x)d\mu(y_1)\cdots d\mu(y_k) \\
&=\mathbb{E}\int\big\|\Psi'_n\big(((\varepsilon_{k+1}+\cdots+\varepsilon_d)^2+k)^{1/2}x\big)\big\|^2\,d\mu(x) \\
&=\mathbb{E}\big(((\varepsilon_{k+1}+\cdots+\varepsilon_d)^2+k)^d\big)\int\big\|\Psi'_n(x)\big\|^2 d\mu(x).
\end{aligned}$$

Hence (5.12) and therefore (5.9) are established.

We can now conclude the proof of (5.5) and thus of the theorem. It is intuitively clear that

$$\lim_{n\to\infty} \sup_{h\in\mathcal{K}} \big\| \Psi'_n(h) - \Psi^{(d)}(h) \big\| = 0. \tag{5.13}$$

This property is an easy consequence of (5.9). Indeed, for all n and $t > 0$,

$$\begin{aligned}
\sup_{h\in\mathcal{K}} \big\| \Psi'_n(h) - \Psi^{(d)}(h) \big\| & \\
\leq \sup_m \sup_{h\in\mathcal{K}} & \Big\| \Psi'_m(h) - t^{-d} \int \Psi'_m(x+th)\,d\mu(x) \Big\| \\
& + \sup_{h\in\mathcal{K}} t^{-d} \Big\| \int \Psi'_n(x+th) - \Psi(x+th) \Big\|\, d\mu(x) \\
\leq \sup_m \int & \sup_{h\in\mathcal{K}} \big\| \Psi'_m(h) - t^{-d}\Psi'_m(x+th) \big\| d\mu(x) \\
& + \sup_{h\in\mathcal{K}} t^{-d} \int \big\| \Psi'_n(x+th) - \Psi(x+th) \big\|\, d\mu(x)
\end{aligned}$$

and, using (5.2) and (5.9), the limit in n and then in t yields (5.13). Let now A be closed in B and take $0 < r < \mathcal{I}_\Psi(A)$. The definition of $\mathcal{I}_\Psi(A)$ indicates that $(2r)^{d/2}\Psi^{(d)}(\mathcal{K}) \cap A = \emptyset$ where we recall that the unit ball $\mathcal{K}$ of $\mathcal{H}$ is a compact subset of E. Therefore, since $\Psi^{(d)}(\mathcal{K})$ is clearly seen to be compact in B by (5.13), and since A is closed, one can find $\eta > 0$ such that

$$\big((2r)^{d/2}\Psi^{(d)}(\mathcal{K}) + B(0,2\eta)\big) \cap \big(A + B(0,\eta)\big) = \emptyset. \tag{5.14}$$

By (5.13), there exists $n_0 = n_0(\eta)$ large enough such that for every $n \geq n_0$,

$$(2r)^{d/2}\Psi'_n(\mathcal{K}) \subset (2r)^{d/2}\Psi^{(d)}(\mathcal{K}) + B(0,\eta). \tag{5.15}$$

Let thus $n \geq n_0$. For any $t > 0$, we can write

$$\begin{aligned}
&\mu(x; \Psi(x) \in t^d A) \\
&\quad \leq \mu\big(x; \|\Psi(x) - \Psi'_n(x)\| > \eta t^d\big) + \mu\big(x; \Psi'_n(x) \in t^d\big(A + B(0,\eta)\big)\big) \\
&\quad \leq \mu\big(x; \|\Psi(x) - \Psi'_n(x)\| > \eta t^d\big) + \mu^*\big(x; x \notin V + t\sqrt{2r}\mathcal{K}\big)
\end{aligned} \tag{5.16}$$

where

$$V = V(t,n) = \Big\{v; \sup_{h\in\mathcal{K}} t^{-d} \big\| \Psi'_n(v + t\sqrt{2r}h) - t^d(2r)^{d/2}\Psi'_n(h) \big\| \leq \eta \Big\}.$$

To justify the second inequality in (5.16), observe that if $x = v + t\sqrt{2r}h$ with $v \in V$ and $h \in \mathcal{K}$, then

$$t^{-d}\Psi'_n(x) = t^{-d}\big[\Psi'_n(v + t\sqrt{2r}h) - t^d(2r)^{d/2}\Psi'_n(h)\big] + (2r)^{d/2}\Psi'_n(h),$$

so that the claim follows by (5.14), (5.15) and the definition of V. By (5.9), let now $t_0 = t_0(\eta)$ be large enough so that, for all $t \geq t_0$,

$$\sup_n \frac{1}{t^d} \int \sup_{h \in \mathcal{K}} \left\| \Psi'_n(x + t\sqrt{2r}h) - t^d(2r)^{d/2} \Psi'_n(h) \right\|^2 d\mu(x) \leq \frac{\eta^2}{2}.$$

That is, for every n and every $t \geq t_0$, $\mu(V(t,n)) \geq \frac{1}{2}$. By Theorem 4.3 (one could use equivalently (4.6)), it follows that

$$\mu^*(x; x \notin V + t\sqrt{2r}\mathcal{K}) \leq e^{-rt^2}. \tag{5.17}$$

Fix now $t \geq t_0 = t_0(\eta)$. Choose $n = n(t) \geq n_0 = n_0(\eta)$ large enough in order that

$$\mu\left(x; \left\|\Psi(x) - \Psi'_n(x)\right\| > \eta t^d\right) \leq e^{-rt^2}.$$

Together with (5.16) and (5.17), it follows that for every $t \geq t_0$,

$$\mu\left(x; \Psi(x) \in t^d A\right) \leq 2e^{-rt^2}.$$

Since $r < \mathcal{I}_\Psi(A)$ is arbitrary, the proof of (5.5) and therefore of Theorem 5.1 is complete. □

Note that it would of course have been possible to work directly on Ψ rather than on the approximating sequence (Ψ'_n) in the preceding proof. This approach however avoids several measurability questions and makes everything more explicit.

It is probably possible to develop, as in Chapter 4, a nontopological approach to large deviations of Wiener chaos.

Notes for further reading. The reader may consult the recent paper [MW-N-PA] for a different approach to the results presented in this chapter, however also based on Borell's main contribution (Corollary 5.2). Borell's articles [Bo8], [Bo9]... contain further interesting results on chaos. See also [A-G], [G-K2], [L-T2]...

6. REGULARITY OF GAUSSIAN PROCESSES

In this chapter, we provide a complete treatment of boundedness and continuity of Gaussian processes via the tool of majorizing measures. After the work of R. M. Dudley, V. Strassen, V. N. Sudakov and X. Fernique on entropy, M. Talagrand [Ta2] gave, in 1987, necessary and sufficient conditions on the covariance structure of a Gaussian process in order that it is almost surely bounded or continuous. These necessary and sufficient conditions are based on the concept of majorizing measure introduced in the early seventies by X. Fernique and C. Preston, and inspired in particular by the "real variable lemma" of A. M. Garsia, E. Rodemich and H. Rumsey Jr. [G-R-R]. Recently, M. Talagrand [Ta7] gave a simple proof of his theorem on necessity of majorizing measures based on the concentration phenomenon for Gaussian measures. We follow this approach here. The aim of this chapter is in fact to demonstrate the actual simplicity of majorizing measures that are usually considered as difficult and obscure.

Let T be a set. A Gaussian random process (or better, random function) $X = (X_t)_{t\in T}$ is a family, indexed by T, of random variables on some probability space $(\Omega, \mathcal{A}, \mathbb{P})$ such that the law of each finite family $(X_{t_1}, \dots, X_{t_n})$, $t_1, \dots, t_n \in T$, is centered Gaussian on $\mathbb{R}^n$. Throughout this work, Gaussian will always mean centered Gaussian. In particular, the law (the distributions of the finite dimensional marginals) of the process X is uniquely determined by the covariance structure $\mathbb{E}(X_s X_t)$, $s, t \in T$. Our aim will be to characterize almost sure boundedness and continuity (whenever T is a topological space) of the Gaussian process X in terms of an as simple as possible criterion on this covariance structure. Actually, the main point in this study will be the question of boundedness. As we will see indeed, once the appropriate bounds for the supremum of X are obtained, the characterization of continuity easily follows. Due to the integrability properties of norms of Gaussian random vectors or supremum of Gaussian processes (Theorem 4.1), we will avoid, at a first stage, various cumbersome and unessential measurability questions, by considering the supremum functional

$$F(T) = \sup\{\mathbb{E}(\sup_{t\in U} X_t); U \text{ finite in } T\}.$$

(If $S \subset T$, we define in the same way $F(S)$.) Thus, $F(T) < \infty$ if and only if X is almost surely bounded in any reasonable sense. In particular, we already see that the main question will reduce to a uniform control of $F(U)$ over the finite subsets U of T.

After various preliminary results [Fe1], [De]..., the first main idea in the study of regularity of Gaussian processes is the introduction (in the probabilistic area), by R. M. Dudley, V. Strassen and V. N. Sudakov (cf. [Du1], [Du2], [Su1-4]), of the notion of ε-entropy. The idea consists in connecting the regularity of the Gaussian process $X = (X_t)_{t\in T}$ to the size of the parameter set T for the L^2-metric induced by the process itself and given by

$$d(s,t) = \left(\mathbb{E}|X_s - X_t|^2\right)^{1/2}, \quad s,t \in T.$$

Note that this metric is entirely characterized by the covariance structure of the process. It does not necessarily separate points in T but this is of no importance. The size of T is more precisely estimated by the entropy numbers: for every $\varepsilon > 0$, let $N(T,d;\varepsilon)$ denote the minimal number of (open to fix the idea) balls of radius ε for the metric d that are necessary to cover T. The two main results concerning regularity of Gaussian processes under entropy conditions, due to R. M. Dudley [Du1] for the upper bound and V. N. Sudakov [Su3] for the lower bound (cf. [Du2], [Fe4]), are summarized in the following statement.

Theorem 6.1. *There are numerical constants $C_1 > 0$ and $C_2 > 0$ such that for all Gaussian processes $X = (X_t)_{t\in T}$,*

$$C_1^{-1} \sup_{\varepsilon>0} \varepsilon\left(\log N(T,d;\varepsilon)\right)^{1/2} \le F(T) \le C_2 \int_0^\infty \left(\log N(T,d;\varepsilon)\right)^{1/2} d\varepsilon. \tag{6.1}$$

As possible numerical values for C_1 and C_2, one may take $C_1 = 6$ and $C_2 = 42$ (see below). The convergence of the entropy integral is understood for the small values of ε since it stops at the diameter $D(T) = \sup\{d(s,t); s,t \in T\}$. Actually, if any of the three terms of (6.1) is finite, then (T,d) is totally bounded and in particular $D(T) < \infty$. We will show in more generality below that the process $X = (X_t)_{t\in T}$ actually admits an almost surely continuous version when the entropy integral is finite. Conversely, if $X = (X_t)_{t\in T}$ is continuous, one can show that $\lim_{\varepsilon\to 0} \varepsilon(\log N(T,d;\varepsilon))^{1/2} = 0$ (cf. [Fe4]).

For the matter of comparison with the more refined tool of majorizing measures we will study next, we present a sketch of the proof of Theorem 6.1.

Proof. We start with the upper bound. We may and do assume that T is finite (although this is not strictly necessary). Let $q > 1$ (usually an integer). (We will consider q as a power of discretization; a posteriori, its value is completely arbitrary.) Let n_0 be the largest integer n in $\mathbb{Z}$ such that $N(T,d;q^{-n}) = 1$. For every $n \ge n_0$, we consider a family of cardinality $N(T,d;q^{-n}) = N(n)$ of balls of radius q^{-n} covering T. One may therefore construct a partition $\mathcal{A}_n$ of T of cardinality $N(n)$ on the basis of this covering with sets of diameter less than $2q^{-n}$. In each A of $\mathcal{A}_n$, fix a point of T and denote by T_n the collection of these points. For each t in T, denote by $A_n(t)$

the element of $\mathcal{A}_n$ that contains t. For every t and every n, let then $s_n(t)$ be the element of T_n such that $t \in A_n(s_n(t))$. Note that $d(t, s_n(t)) \leq 2q^{-n}$ for every t and $n \geq n_0$.

The main argument of the proof is the so-called chaining argument (which goes back to A. N. Kolmogorov in his proof of continuity of paths of processes under L^p-control of their increments): for every t,

$$X_t = X_{s_0} + \sum_{n>n_0} \left(X_{s_n(t)} - X_{s_{n-1}(t)}\right) \tag{6.2}$$

where $s_0 = s_{n_0}(t)$ may be chosen independent of $t \in T$. Note that

$$d\big(s_n(t), s_{n-1}(t)\big) \leq 2q^{-n} + 2q^{-n+1} = 2(q+1)q^{-n}.$$

Let $c_n = 4(q+1)q^{-n}(\log N(n))^{1/2}$, $n > n_0$. It follows from (6.2) that

$$\begin{aligned} F(T) &= \mathbb{E}\big(\sup_{t\in T} X_t\big) \\ &\leq \sum_{n>n_0} c_n + \mathbb{E}\Big(\sup_{t\in T} \sum_{n>n_0} \left|X_{s_n(t)} - X_{s_{n-1}(t)}\right| I_{\{|X_{s_n(t)} - X_{s_{n-1}(t)}| > c_n\}}\Big) \\ &\leq \sum_{n>n_0} c_n + \mathbb{E}\Big(\sum_{n>n_0} \sum_{u,v\in H_n} |X_u - X_v| I_{\{|X_u - X_v| > c_n\}}\Big) \end{aligned}$$

where $H_n = \{(u,v) \in T_n \times T_{n-1}; d(u,v) \leq 2(q+1)q^{-n}\}$. If G is a real centered Gaussian variable with variance less than or equal to σ^2, for every $c > 0$

$$\mathbb{E}\big(|G| I_{\{|G|>c\}}\big) \leq \sigma e^{-c^2/2\sigma^2}.$$

Hence,

$$\begin{aligned} F(T) &\leq \sum_{n>n_0} c_n + \sum_{n>n_0} \mathrm{Card}(H_n) 2(q+1)q^{-n} \exp\big(-c_n^2/8(q+1)^2 q^{-2n}\big) \\ &\leq \sum_{n>n_0} 4(q+1)q^{-n}\big(\log N(n)\big)^{1/2} + \sum_{n>n_0} 2(q+1)q^{-n} \\ &\leq 7(q+1)\sum_{n>n_0} q^{-n}\big(\log N(n)\big)^{1/2} \end{aligned}$$

where we used that $\mathrm{Card}(H_n) \leq N(n)^2$. Since

$$\begin{aligned} \int_0^\infty \big(\log N(T,d;\varepsilon)\big)^{1/2} d\varepsilon &\geq \sum_{n>n_0} \int_{q^{-n-1}}^{q^{-n}} \big(\log N(T,d;\varepsilon)\big)^{1/2} d\varepsilon \\ &\geq (1-q^{-1}) \sum_{n>n_0} q^{-n}\big(\log N(n)\big)^{1/2}, \end{aligned}$$

the conclusion follows. If $q = 2$, we may take $C_2 = 42$.

The proof of the lower bound relies on a comparison principle known as Slepian's lemma [Sl]. We use it in the following modified form due to V. N. Sudakov, S. Chevet and X. Fernique (cf. [Su1], [Su2], [Fe4], [L-T2]): if $Y = (Y_1, \ldots, Y_n)$ and $Z = (Z_1, \ldots, Z_n)$ are two Gaussian random vectors in $\mathbb{R}^n$ such that $\mathbb{E}|Y_i - Y_j|^2 \leq \mathbb{E}|Z_i - Z_j|^2$ for all i, j, then

$$\mathbb{E}\big(\max_{1\leq i\leq n} Y_i\big) \leq \mathbb{E}\big(\max_{1\leq i\leq n} Z_i\big). \tag{6.3}$$

Fix $\varepsilon > 0$ and let $n \leq N(T, d; \varepsilon)$. There exist therefore $t_1, \ldots, t_n$ in T such that $d(t_i, t_j) \geq \varepsilon$. Let then $g_1, \ldots, g_n$ be independent standard normal random variables. We have, for every $i, j = 1, \ldots, n$,

$$\mathbb{E}\left|\frac{\varepsilon}{\sqrt{2}} g_i - \frac{\varepsilon}{\sqrt{2}} g_j\right|^2 = \varepsilon^2 \leq d(t_i, t_j) = \mathbb{E}|X_{t_i} - X_{t_j}|^2.$$

Therefore, by (6.3),

$$F(T) \geq \mathbb{E}\big(\max_{1\leq i\leq n} X_{t_i}\big) \geq \frac{\varepsilon}{\sqrt{2}}\, \mathbb{E}\big(\max_{1\leq i\leq n} g_i\big).$$

Now, it is classical and easily seen that

$$\mathbb{E}\big(\max_{1\leq i\leq n} g_i\big) \geq c\,(\log n)^{1/2}$$

for some numerical $c > 0$ (one may choose c such that $\sqrt{2}/c \leq 6$). Since n is arbitrary less than or equal to $N(T, d; \varepsilon)$, the conclusion trivially follows. Theorem 6.1 is established. □

As an important remark for further purposes, note that simple proofs of Sudakov's minoration avoiding the rather rigid Slepian's lemma are now available. These are based on a dual Sudakov inequality [L-T2], p. 82-83, and duality of entropy numbers [TJ].They allow the investigation of minoration inequalities outside the Gaussian setting (cf. [Ta10], [Ta12]). Note furthermore that we will only use the Sudakov inequality in the proof of the majorizing measure minoration principle (cf. Lemma 6.4).

A simple example of application of Theorem 6.1 is Brownian motion $(W(t))_{0\leq t\leq 1}$ on $T = [0,1]$. Since $d(s,t) = \sqrt{|s-t|}$, the entropy numbers $N(T, d; \varepsilon)$ are of the order of ε^{-2} as ε goes to zero and the entropy integral is trivially convergent. Together with the proof of continuity presented below in the framework of majorizing measures, Theorem 6.1 is certainly the shortest way to prove boundedness and continuity of the Brownian paths.

In Theorem 6.1, the difference between the upper and lower bounds is rather tight. It however exists. The examples of a standard orthogaussian sequence or of the canonical Gaussian process indexed by an ellipsoid in a Hilbert space (see [Du1], [Du2], [L-T2], [Ta13]) are already instructive. We will see later on that the convergence of Dudley's entropy integral however characterizes $F(T)$ when T has a group structure and the metric d is translation invariant, an important result of X. Fernique [Fe4].

If one tries to imagine what can be used instead of the entropy numbers in order to sharpen the conclusions of Theorem 6.1, one realizes that one feature of entropy is that is attributes an equal weight to each piece of the parameter set T. One is then naturally led to the possible following definition. Let, as in the proof of Theorem 6.1, q be (an integer) larger than 1. Let $\mathcal{A} = (\mathcal{A}_n)_{n\in\mathbb{Z}}$ be an increasing sequence (i.e. each $A \in \mathcal{A}_{n+1}$ is contained in some $B \in \mathcal{A}_n$) of finite partitions of T such that the diameter $D(A)$ of each element A of $\mathcal{A}_n$ is less than or equal to $2q^{-n}$. If $t \in T$, denote by $A_n(t)$ the element of $\mathcal{A}_n$ that contains t. Now, for each partition $\mathcal{A}_n$, one may consider nonnegative weights $\alpha_n(A)$, $A \in \mathcal{A}_n$, such that $\sum_{A\in\mathcal{A}_n} \alpha_n(A) \le 1$. Set then

$$\Theta_{\mathcal{A},\alpha} = \sup_{t\in T} \sum_n q^{-n} \left(\log \frac{1}{\alpha_n(A_n(t))}\right)^{1/2}. \tag{6.4}$$

It is worthwhile mentioning that for $2q^{-n} \ge D(T)$, one can take $\mathcal{A}_n = \{T\}$ and $\alpha_n(T) = 1$. Denote by $\Theta(T)$ the infimum of the functional $\Theta_{\mathcal{A},\alpha}$ over all possible choices of partitions $(\mathcal{A}_n)_{n\in\mathbb{Z}}$ and weights $\alpha_n(A)$. In this definition, we may take equivalently

$$\Theta_{\mathcal{A},m} = \sup_{t\in T} \sum_n q^{-n} \left(\log \frac{1}{m(A_n(t))}\right)^{1/2}$$

where m is a probability measure on (T, d). Indeed, if $\Theta_{\mathcal{A},\alpha} < \infty$, it is easily seen that $D(T) < \infty$. Let then n_0 be the largest integer n in $\mathbb{Z}$ such that $2q^{-n} \le D(T)$. Fix a point in each element of $\mathcal{A}_n$ and denote by T_n, $n \ge n_0$, the collection of these points. It is then clear that if m is a (discrete) probability measure such that

$$m \ge (1 - q^{-1}) \sum_{n\ge n_0} q^{-n+n_0} \sum_{t\in T_n} \alpha_n(A_n(t))\delta_t,$$

where δ_t is point mass at t, the functional $\Theta_{\mathcal{A},m}$ is of the same order as $\Theta_{\mathcal{A},\alpha}$ (see also below). We need not actually be concerned with these technical details and consider for simplicity the functionals $\Theta_{\mathcal{A},\alpha}$. Furthermore, the number $q > 1$ should be thought as a universal constant.

The condition $\Theta(T) < \infty$ is called a majorizing measure condition and the main result of this section is that $C^{-1}\Theta(T) \le F(T) \le C\Theta(T)$ for some constant $C > 0$ only depending on q. In order to fully appreciate this definition, it is worthwhile comparing it to the entropy integral. As we used it in the proof of Theorem 6.1, the entropy integral is equivalent (for any q) to the series

$$\sum_{n>n_0} q^{-n} \left(\log N(T, d; q^{-n})\right)^{1/2}.$$

We then construct an associated sequence $(\mathcal{A}_n)_{n\in\mathbb{Z}}$ of increasing partitions of T and weights $\alpha_n(A)$ in the following way. Let $\mathcal{A}_n = \{T\}$ and $\alpha_n(T) = 1$ for every

$n \le n_0$. Once $\mathcal{A}_n$ $(n > n_0)$ has been constructed, partition each element A of $\mathcal{A}_n$ with a covering of A of cardinality at most $N(A, d; q^{-n-1}) \le N(T, d; q^{-n-1})$ and let $\mathcal{A}_{n+1}$ be the collection of all the subsets of T obtained in this way. To each A in $\mathcal{A}_n$, $n > n_0$, we give the weight

$$\alpha_n(A) = \left(\prod_{i=n_0+1}^{n} N(T, d; q^{-i}) \right)^{-1}$$

$(\alpha(T) = 1)$. Clearly $\sum_{A \in \mathcal{A}_n} \alpha_n(A) \le 1$. Moreover, for each t in T,

$$\begin{aligned} \sum_{n > n_0} q^{-n} \left(\log \frac{1}{\alpha(A_n(t))} \right)^{1/2} &\le \sum_{n > n_0} \sum_{i=n_0+1}^{n} q^{-n} \left(\log N(T, d; q^{-i}) \right)^{1/2} \\ &\le (q-1)^{-1} \sum_{i > n_0} q^{-i} \left(\log N(T, d; q^{-i}) \right)^{1/2}. \end{aligned}$$

In other words,

$$\Theta(T) \le C \int_0^\infty \left(\log N(T, d; \varepsilon) \right)^{1/2} d\varepsilon$$

where $C > 0$ only depends on $q > 1$.

It is clear from this construction how entropy numbers give a uniform weight to each subset of T and how the possible refined tool of majorizing measures can allow a better understanding of the metric properties of T. (Actually, one has rather to think about entropy numbers as the equal weight that is put on each piece of a partition of the parameter set T.) This is what we will investigate now. First however, we would like to briefly comment on the name "majorizing measure" as well as the dependence on $q > 1$ in the definition of the functional $\Theta(T)$. Classically, a majorizing measure m on T is a probability measure on the Borel sets of T such that

$$\sup_{t \in T} \int_0^\infty \left(\log \frac{1}{m(B(t, \varepsilon))} \right)^{1/2} d\varepsilon < \infty \tag{6.5}$$

where $B(t, \varepsilon)$ is the ball in T with center t and radius $\varepsilon > 0$. As the definition of the entropy integral, a majorizing measure condition only relies on the metric structure of T and the convergence of the integral is for the small values of ε. In order to connect this definition with the preceding one (6.4), let $q > 1$ and let $(\mathcal{A}_n)_{n \in \mathbb{Z}}$ be an increasing sequence of finite partitions of T such that the diameter $D(A)$ of each element A of $\mathcal{A}_n$ is less than or equal to $2q^{-n}$. Let furthermore m be a probability measure on T. Note that $A_n(t) \subset B(t, 2q^{-n})$ for every t. Therefore

$$\begin{aligned} \int_0^\infty \left(\log \frac{1}{m(B(t, \varepsilon))} \right)^{1/2} d\varepsilon &\le C \sum_n q^{-n} \left(\log \frac{1}{m(B(t, 2q^{-n}))} \right)^{1/2} \\ &\le C \sum_n q^{-n} \left(\log \frac{1}{m(A_n(t))} \right)^{1/2} \end{aligned}$$

where $C > 0$ only depends on q. Since m is a probability measure, we can set $\alpha_n(A) = m(A)$ for every A in $\mathcal{A}_n$ and every n. It immediately follows that, for every $q > 1$,

$$\inf_m \sup_{t\in T} \int_0^\infty \left(\log \frac{1}{m(B(t,\varepsilon))}\right)^{1/2} d\varepsilon \le C\Theta(T)$$

where C only depends on q. One can prove the reverse inequality in the same spirit with the help however of a somewhat technical and actually nontrivial discretization lemma (cf. [L-T2], Proposition 11.10). In particular, the various functionals $\Theta(T)$ when q varies are all equivalent. We actually need not really be concerned with these technical details since our aim is to show that $F(T)$ and $\Theta(T)$ are of the same order (for some $q > 1$). (It will actually follow from the proofs presented below that the functionals $\Theta(T)$ are equivalent up to constants depending only on $q \ge q_0$ for some universal q_0 large enough.)

Now, we start our investigation of the regularity properties of a Gaussian process $X = (X_t)_{t\in T}$ under majorizing measure conditions. The first part of our study concerns upper bounds and sufficient conditions for boundedness and continuity of X. The following theorem is due, in this form and with this proof, to X. Fernique [Fe3], [Fe4]. It follows independently from the work of C. Preston [Pr1], [Pr2].

Theorem 6.2. *Let $X = (X_t)_{t\in T}$ be a Gaussian process indexed by a set T. Then, for every $q > 1$,*

$$F(T) \le C\Theta(T)$$

where $C > 0$ only depends on q. If, in addition to $\Theta_{\mathcal{A},\alpha} < \infty$ for some partition $\mathcal{A}$ and weights α, one has

$$\lim_{k\to\infty} \sup_{t\in T} \sum_{n\ge k} q^{-n} \left(\log \frac{1}{\alpha(A_n(t))}\right)^{1/2} = 0, \tag{6.6}$$

then X admits a version with almost all sample paths bounded and uniformly continuous on (T,d).

Proof. It is very similar to the proof of the upper bound in Theorem 6.1. We first establish the inequality $F(T) \le C\Theta_{\mathcal{A},\alpha}(T)$ for any partition $\mathcal{A}$ and any family of weights α. We may asssume that T is finite. Let n_0 be the largest integer n in $\mathbb{Z}$ such that the diameter $D(T)$ of T is less than or equal to $2q^{-n}$. For every $n \ge n_0$, fix a point in each element of the partition $\mathcal{A}_n$ and denote by T_n the (finite) collection of these points. We may take $T_{n_0} = \{s_0\}$ for some fixed s_0 in T. For every t in T, denote by $s_n(t)$ the element of T_n which belongs to $A_n(t)$. As in (6.2), for every t,

$$X_t = X_{s_0} + \sum_{n>n_0} \left(X_{s_n(t)} - X_{s_{n-1}(t)}\right).$$

Since the partitions $\mathcal{A}_n$ are increasing,

$$s_n(t) \in A_{n-1}\big(s_n(t)\big) = A_{n-1}(t), \quad n > n_0.$$

In particular, $d(s_n(t), s_{n-1}(t)) \le 2q^{-n+1}$. Now, for every t in T and every $n > n_0$, let

$$c_n(t) = 2\sqrt{2}q^{-n+1}\left(\log \frac{1}{\alpha(A_n(t))}\right)^{1/2}.$$

With respect to the entropic proof, note here the dependence of c_n on t which is the main feature of the majorizing measure technique. Actually, the partitions $\mathcal{A}$ and weights α are used to bound, in the chaining argument, the "heaviest" portions of the process. We can now write, almost as in the proof of Theorem 6.1,

$$\begin{aligned} &F(T) \\ &\le \sup_{t\in T} \sum_{n>n_0} c_n(t) + \mathbb{E}\Big(\sup_{t\in T} \sum_{n>n_0} |X_{s_n(t)} - X_{s_{n-1}(t)}| I_{\{|X_{s_n(t)}-X_{s_{n-1}(t)}|>c_n(t)\}}\Big) \\ &\le \sup_{t\in T} \sum_{n>n_0} c_n(t) + \mathbb{E}\Big(\sum_{n>n_0} \sum_{u\in T_n} |X_u - X_{s_{n-1}(u)}| I_{\{|X_u-X_{s_{n-1}(u)}|>c_n(u)\}}\Big) \\ &\le \sup_{t\in T} \sum_{n>n_0} c_n(t) + \sum_{n>n_0} \sum_{u\in T_n} 2q^{-n+1} \exp\big(-c_n^2(u)/8q^{-2n+2}\big). \end{aligned}$$

Therefore

$$\begin{aligned} F(T) &\le \sup_{t\in T} \sum_{n>n_0} c_n(t) + \sum_{n>n_0} 2q^{-n+1} \sum_{u\in T_n} \alpha\big(A_n(u)\big) \\ &\le \sup_{t\in T} \sum_{n>n_0} c_n(t) + 2(q-1)^{-1} q^{-n_0+1}. \end{aligned}$$

Since

$$\Theta_{\mathcal{A},\alpha} \ge (\log 2)^{1/2} q^{-n_0-1},$$

the first claim of Theorem 6.2 follows.

We turn to the sample path continuity. Let $\eta > 0$. For each k $(> n_0)$, set

$$\begin{aligned} V = V_k = \big\{(x,y) \in T_k \times T_k;\, \exists\, u,v \text{ in } T \text{ such that} \\ d(u,v) \le \eta \text{ and } s_k(u) = x,\, s_k(v) = y\big\}. \end{aligned}$$

If $(x,y) \in V$, we fix $u_{x,y}$, $v_{x,y}$ in T such that $s_k(u_{x,y}) = x$, $s_k(v_{x,y}) = y$ and $d(u_{x,y}, v_{x,y}) \le \eta$. Now, let s,t in T with $d(s,t) \le \eta$. Set $x = s_k(s)$, $y = s_k(t)$. Clearly $(x,y) \in V$. By the triangle inequality,

$$\begin{aligned} |X_s - X_t| &\le |X_s - X_{s_k(s)}| + |X_{s_k(s)} - X_{u_{x,y}}| + |X_{u_{x,y}} - X_{v_{x,y}}| \\ &\qquad + |X_{v_{x,y}} - X_{s_k(t)}| + |X_{s_k(t)} - X_t| \\ &\le \sup_{(x,y)\in V} |X_{u_{x,y}} - X_{v_{x,y}}| + 4 \sup_{r\in T} |X_r - X_{s_k(r)}|. \end{aligned}$$

Clearly,

$$\mathbb{E}\Big(\sup_{(x,y)\in V} |X_{u_{x,y}} - X_{v_{x,y}}|\Big) \le \eta\big(\mathrm{Card}(T_k)\big)^2.$$

Now, the chaining argument in the proof of boundedness similarly shows that

$$\mathbb{E}\big(\sup_{t\in T}|X_t - X_{s_k(t)}|\big) \leq C\sup_{t\in T}\sum_{n\geq k} q^{-n}\left(\log\frac{1}{\alpha(A_n(t))}\right)^{1/2}$$

for some constant $C > 0$ (independent of k). Therefore, hypothesis (6.6) and the preceding inequalities ensure that for each $\varepsilon > 0$ there exists $\eta > 0$ such that, for every finite and thus also countable subset U of T,

$$\mathbb{E}\big(\sup_{s,t\in U, d(s,t)\leq\eta}|X_s - X_t|\big) \leq \varepsilon.$$

Since (T,d) is totally bounded, there exists U countable and dense in T. Then, set $\widetilde{X}_t = X_t$ if $t \in U$ and $\widetilde{X}_t = \lim X_t$ where the limit, in probability or in L^1, is taken for $u \to t$, $u \in U$. Then $(\widetilde{X}_t)_{t\in T}$ is a version of the process X with uniformly continuous sample paths on (T,d). Indeed, let, for each integer n, $\eta_n > 0$ be such that

$$\mathbb{E}\big(\sup_{d(s,t)\leq\eta_n}|\widetilde{X}_s - \widetilde{X}_t|\big) \leq 4^{-n}.$$

Then, if $C_n = \{\sup_{d(s,t)\leq\eta_n}|\widetilde{X}_s - \widetilde{X}_t| \geq 2^{-n}\}$, $\sum_n \mathbb{P}(C_n) < \infty$ and the claim follows from the Borel-Cantelli lemma. The proof of Theorem 6.2 is complete. □

We now turn to the theorem of M. Talagrand [Ta2] on necessity of majorizing measures. This result was conjectured by X. Fernique back in 1974. As announced, we follow the simplified proof of the author [Ta7] based on concentration of Gaussian measures. This new proof moreover allows us to get some insight on the weights α of the "minorizing" measure.

Theorem 6.3. *There exists a universal value $q_0 \geq 2$ such that for every $q \geq q_0$ and every Gaussian process $X = (X_t)_{t\in T}$ indexed by T,*

$$\Theta(T) \leq CF(T)$$

where $C > 0$ is a constant only depending on q.

Proof. The key step is provided by the following minoration principle based on concentration and Sudakov's inequality. It may actually be considered as a strengthening of the latter.

Lemma 6.4. *There exists a numerical constant $0 < c < \frac{1}{2}$ with the following property. If $\varepsilon > 0$ and if $t_1, \ldots, t_N$ are points in T such that $d(t_k, t_\ell) \geq \varepsilon$, $k \neq \ell$, $N \geq 2$, and if $B_1, \ldots, B_N$ are subsets of T such that $B_k \subset B(t_k, c\varepsilon)$, $k = 1, \ldots N$, we have*

$$\mathbb{E}\big(\max_{1\leq k\leq N}\sup_{t\in B_k} X_t\big) \geq c\varepsilon(\log N)^{1/2} + \min_{1\leq k\leq N}\mathbb{E}\big(\sup_{t\in B_k} X_t\big).$$

Proof. We may assume that B_k is finite for every k. Set $Y_k = \sup_{t\in B_k}(X_t - X_{t_k})$, $k = 1, \ldots, N$. Then,

$$\sup_{t\in B_k} X_t = (Y_k - \mathbb{E}Y_k) + \mathbb{E}Y_k + X_{t_k}$$

and thus

$$\text{(6.7)}\qquad \max_{1\le k\le N} X_{t_k} \le \max_{1\le k\le N}\sup_{t\in B_k} X_t + \max_{1\le k\le N}|Y_k - \mathbb{E}Y_k| - \min_{1\le k\le N}\mathbb{E}\big(\sup_{t\in B_k} X_t\big)\,.$$

Integrate both sides of this inequality. By Sudakov's minoration (Theorem 6.1),

$$\mathbb{E}\big(\max_{1\le k\le N} X_{t_k}\big) \ge C_1^{-1}\varepsilon(\log N)^{1/2}.$$

Furthermore, the concentration inequalities, in the form for example of (2.9) or (4.2), (4.3), show that, for every $r\ge 0$, and every k,

$$\mathbb{P}\big\{|Y_k - \mathbb{E}Y_k| \ge r\big\} \le 2\mathrm{e}^{-r^2/2c^2\varepsilon^2}.$$

This estimate easily and classically implies that

$$\mathbb{E}\big(\max_{1\le k\le N}|Y_k - \mathbb{E}Y_k|\big) \le C_3 c\varepsilon(\log N)^{1/2}$$

where $C_3>0$ is numerical. Indeed, by the integration by parts formula, for every $\delta>0$,

$$\begin{aligned}\mathbb{E}\big(\max_{1\le k\le N}|Y_k - \mathbb{E}Y_k|\big) &\le \delta + \int_\delta^\infty \mathbb{P}\big\{\max_{1\le k\le N}|Y_k - \mathbb{E}Y_k| \ge r\big\}dr\\ &\le \delta + 2N\int_\delta^\infty \mathrm{e}^{-r^2/2c^2\varepsilon^2}dr\end{aligned}$$

and the conclusion follows by letting δ be of the order of $c\varepsilon(\log N)^{1/2}$. Hence, coming back to (6.7), we see that if $c>0$ is such that $\frac{1}{C_1} - cC_3 = c$, the minoration inequality of the lemma holds. The value of q_0 in Theorem 6.3 only depends on this choice. (Since we may take $C_1 = 6$ and $C_3 = 20$ (for example), we see that $c = .007$ will work.) Lemma 6.4 is proved. □

We now start the proof of Theorem 6.3 itself and the construction of a partition $\mathcal{A}$ and weights α. Assume that $F(T)<\infty$ otherwise there is nothing to prove. In particular, (T,d) is totally bounded. We further assume that $q\ge q_0$ where $q_0 = c^{-1}$ has been determined by Lemma 6.4.

For each n and each subset of T of diameter less than or equal to $2q^{-n}$, we will construct an associated partition in sets of diameter less than or equal to $2q^{-n-1}$. Let thus S be a subset of T with $D(S)\le 2q^{-n}$. We first construct by induction a (finite) sequence $(t_k)_{k\ge1}$ of points in S. t_1 is chosen so that $F(S\cap B(t_1,q^{-n-2}))$ is maximal. Assume that $t_1,\dots,t_{k-1}$ have been constructed and set

$$H_k = \bigcup_{\ell<k}\big(S\cap B(t_\ell, q^{-n-1})\big).$$

If $H_k = S$, the construction stops (and it will eventually stop since (T,d) is totally bounded). If not, choose t_k in $S\setminus H_k$ such that $F(B_k)$ is maximal where we set $B_k = (S\setminus H_k)\cap B(t_k,q^{-n-2})$. For every k, let

$$A_k = (S\setminus H_k)\cap B(t_k,q^{-n-1}).$$

Clearly $D(A_k) \leq 2q^{-n-1}$ and the A_k's define a partition of S. One important feature of this construction is that, for every t in A_k,

$$F\big(A_k \cap B(t, q^{-n-2})\big) \leq F(B_k). \tag{6.8}$$

On the other hand, the minoration lemma 6.4 applied with $\varepsilon = q^{-n-1}$ yields (since $q \geq c^{-1}$), for every k,

$$F(S) \geq cq^{-n-1}(\log k)^{1/2} + F(B_k). \tag{6.9}$$

We denote by $\mathcal{A}(S)$ this ordered finite partition $\{A_1, \ldots, A_k, \ldots\}$ of S. (6.8) and (6.9) together yield: for every $A_k \in \mathcal{A}(S)$ and every $U \in \mathcal{A}(A_k)$,

$$F(S) \geq cq^{-n-1}(\log k)^{1/2} + F(U). \tag{6.10}$$

We now complete the construction. Let n_0 be the largest in $\mathbb{Z}$ with $D(T) \leq 2q^{-n_0}$. Set $\mathcal{A}_n = \{T\}$ and $\alpha_n(T) = 1$ for every $n \leq n_0$. Suppose that $\mathcal{A}_n$ and $\alpha_n(S)$, $S \in \mathcal{A}_n$, $n > n_0$, have been constructed. We define

$$\mathcal{A}_{n+1} = \bigcup \{\mathcal{A}(S); S \in \mathcal{A}_n\}.$$

If $U \in \mathcal{A}_{n+1}$, there exists $S \in \mathcal{A}_n$ such that $U = A_k \in \mathcal{A}(S)$. We then set $\alpha_{n+1}(U) = \alpha_n(A)/2k^2$. Let t be fixed in T. With this notation, (6.10) means that for all $n \geq n_0$,

$$F\big(A_n(t)\big) \geq c\,2^{-1/2} q^{-n-1} \left(\log \frac{\alpha_n(A_n(t))}{2\alpha_{n+1}(A_{n+1}(t))}\right)^{1/2} + F\big(A_{n+2}(t)\big)$$

where we recall that we denote by $A_n(t)$ the element of $\mathcal{A}_n$ that contains t. Summing these inequalities separately on the even and odd integers, we get

$$2F(T) \geq c\,2^{-1/2} \sum_{n > n_0} q^{-n-1} \left(\log \frac{\alpha_n(A_n(t))}{2\alpha_{n+1}(A_{n+1}(t))}\right)^{1/2}$$

and thus

$$c(q-1)^{-1} q^{-n_0} + 2F(T) \geq c\,2^{-1/2}(1 - q^{-1}) \sum_{n > n_0} q^{-n} \left(\log \frac{1}{\alpha_n(A_n(t))}\right)^{1/2}.$$

Since $2q^{-n_0} \leq D(T)$, and since

$$\begin{aligned} 2F(T) &= \sup\{\mathbb{E}\big(\sup_{s,t \in U} |X_s - X_t|\big); U \text{ finite in } T\} \\ &\geq \sup_{s,t \in T} \mathbb{E}|X_s - X_t| = \left(\frac{2}{\sqrt{\pi}}\right)^{1/2} D(T), \end{aligned}$$

it follows that, for some constant $C > 0$ only depending on q,

$$CF(T) \geq c \sum_{n > n_0} q^{-n} \left(\log \frac{1}{\alpha_n(A_n(t))} \right)^{1/2}.$$

Theorem 6.3 is therefore established. □

It may be shown that if the Gaussian process X in Theorem 6.3 is almost surely continuous on (T, d), then there is a majorizing measure satisfying (6.6). We refer to [Ta2] or [L-T2] for the details.

Theorem 6.3 thus solved the question of the regularity properties of any Gaussian process. Prior to this result however, X. Fernique showed [Fe4] that the convergence of Dudley's entropy integral was necessary for a stationary Gaussian process to be almost surely bounded or continuous. One can actually easily show (cf. [L-T2]) that, in this case, the entropy integral coincides with a majorizing measure integral with respect to the Haar measure on the underlying parameter set T endowed with a group structure. One may however also provide a direct and transparent proof of the stationary case on the basis of the above minoration principle (Lemma 6.4). We would like to conclude this chapter with a brief sketch of this proof.

Let thus T be a locally compact Abelian group. Let $X = (X_t)_{t \in T}$ be a stationary centered Gaussian process indexed by T, in the sense that the L^2-metric d induced by X is translation invariant on T. As announced, we aim to prove directly that for some numerical constant $C > 0$,

$$\int_0^\infty \left(\log N(T, d; \varepsilon)\right)^{1/2} d\varepsilon \leq CF(T).$$

(cf. [Fe4], [M-P], [L-T2] for more general statements along these lines.) Since d is translation invariant,

$$\mathbb{E}\big(\sup_{s \in B(t,\varepsilon)} X_s\big) \quad \text{and} \quad N\big(B(t,\varepsilon), d; \eta\big), \quad \varepsilon, \eta > 0,$$

are independent of the point t. They will therefore be simpler denoted as

$$\mathbb{E}\big(\sup_{s \in B(\varepsilon)} X_s\big) \quad \text{and} \quad N\big(B(\varepsilon), d; \eta\big).$$

Let $n \in \mathbb{Z}$. Choose in a ball $B(q^{-n})$ a maximal family $(t_1, \ldots, t_M)$ under the relations $d(t_k, t_\ell) \geq q^{-n-1}$, $k \neq \ell$. Then the balls $B(t_k, q^{-n-1})$, $1 \leq k \leq M$, cover $B(q^{-n})$ so that $M \geq N(B(q^{-n}), d; q^{-n-1})$. Apply then Lemma 6.4 with $\varepsilon = q^{-n-1}$, $q \geq q_0 = c^{-1}$ and $B_k = B(t_k, q^{-n-2})$. We thus get

$$\mathbb{E}\big(\sup_{t \in B(q^{-n})} X_t\big) \geq c q^{-n-1} \left(\log N(B(q^{-n}), d; q^{-n-1})\right)^{1/2} + \mathbb{E}\big(\sup_{t \in B(q^{-n-2})} X_t\big).$$

Summing as before these inequalities along the even and the odd integers yields

$$F(T) \geq C^{-1} \sum_n q^{-n} (\log N(B(q^{-n}), d; q^{-n-1})^{1/2}.$$

Since

$$N(T,d;q^{-n-1}) \le N(T,d;q^{-n})N\big(B(q^{-n}),d;q^{-n-1})\big),$$

the proof is complete.

To conclude, let us mention the following challenging open problem. Let x_i, $i \in \mathbb{N}$, be real valued functions on a set T such that $\sum_i x_i(t)^2 < \infty$ for every $t \in T$. Let furthermore $(\varepsilon_i)_{i\in\mathbb{N}}$ be a sequence of independent symmetric Bernoulli random variables and set, for each $t \in T$, $X_t = \sum_i \varepsilon_i x_i(t)$ which converges almost surely. The question of characterizing those "Bernoulli" processes $(X_t)_{t\in T}$ which are almost surely bounded is almost completely open (cf. [L-T2], [Ta14]). The Gaussian study of this chapter of course corresponds to the choice for $(\varepsilon_i)_{i\in\mathbb{N}}$ of a standard Gaussian sequence.

Notes for further reading. On the history of entropy and majorizing measures, one may consult respectively [Du2], [Fe4] and [He], [Fe4], [Ta2], [Ta18]. The first proof of Theorem 6.3 by M. Talagrand [Ta2] was quite different from the proof presented here following [Ta7]. Another proof may be found in [L-T2]. These proofs are based on the fundamental principle, somewhat hidden here, that the size of a metric space with respect to the existence of a majorizing measure can be measured by the size of the well separated subsets it contains (see [Ta10], [Ta12] for more on this principle). More on majorizing measures and minoration of random processes may be found in [L-T2] and in the papers [Ta10], [Ta12], and in the recent survey [Ta18] where in particular new examples of applications are described. It is shown in [L-T2] how the upper bound techniques based on entropy or majorizing measures (Theorems 6.1 and 6.2) can yield deviation inequalities of the type (4.2), which are optimal by Theorem 6.3. Sharp bounds on the tail of the supremum of a Gaussian process can be obtained with these methods (see e.g. [Ta13], [Lif2], [Lif3] and the many references therein). On construction of majorizing measures, see [L-T2], [Ta14], [Ta18]. For the applications of the Dudley-Fernique theorem on stationary Gaussian processes to random Fourier series, see [M-P], [L-T2].

7. SMALL BALL PROBABILITIES FOR GAUSSIAN MEASURES AND RELATED INEQUALITIES AND APPLICATIONS

While, as we saw in Chapter 4, the behavior of (the complement of) large balls for Gaussian measures is relatively well described, small ball probabilities are much less known. This problem has gone recently a quick development and we intend to present in this chapter some significant recent results, although it seems that there is still a long way to the final word (if there is any). In the first part of this chapter, we describe a simple method to evaluate small Brownian balls and to establish various sharper concentration inequalities. Then, we present some more abstract and general results (due to J. Kuelbs and W. Li and M. Talagrand) which show in particular, on the basis of the isoperimetric tool, that small ball probabilities for Gaussian measures are closely related to some entropy numbers. In particular, we establish an important concentration inequality for enlarged balls due to M. Talagrand. To conclude this chapter, we briefly discuss some correlation inequalities which have been used recently to extend the support of a diffusion theorem, the large deviations of dynamical systems as well as the existence of Onsager-Machlup functionals for stronger norms or topologies on Wiener space.

We introduce the question of small ball probabilities for Gaussian measures with the example of Wiener measure. Let $W = (W(t))_{t\geq 0}$ be Brownian motion with values in $\mathbb{R}^d$. Denote by $\|x\|_\infty = \sup_{t\in[0,1]}|x(t)|$ the supnorm on the space of continuous functions $C_0([0,1];\mathbb{R}^d)$ (vanishing at the origin) where we equip, for example, $\mathbb{R}^d$ with its usual Euclidean norm $|\cdot|$. Let $\varepsilon > 0$. By the scaling property, for every $\lambda > 0$,

$$\mathbb{P}\{\|W\|_\infty \leq \varepsilon\} = \mathbb{P}\{\sup_{0\leq t\leq \lambda}|W(t)| \leq \varepsilon\sqrt{\lambda}\}.$$

Choosing $\lambda = \varepsilon^{-2}$, we see that

$$\mathbb{P}\{\|W\|_\infty \leq \varepsilon\} = \mathbb{P}\{\tau \geq \varepsilon^{-2}\}$$

where τ is the exit time of W from the unit ball B of $\mathbb{R}^d$. It is known (cf. [I-W]) that $u(t,x) = \mathbb{E}(f(W(t)+x)I_{\{\tau\geq t\}})$, $x \in B$, $t \geq 0$, is the solution of the initial value

Dirichlet problem

$$\frac{\partial u}{\partial t} = \frac{1}{2}\,\Delta u \text{ in } B, \quad u_{|\partial B} = 0, \quad u_{|t=0} = f.$$

Therefore

$$u(t,x) = \sum_{n=1}^{\infty} e^{-\lambda_n t}\phi_n(x)\int_B \phi_n(y)f(y)dy,$$

where $0 < \lambda_1 \leq \lambda_2 \leq \ldots$ are eigenvalues and $\phi_1, \phi_2, \ldots$ are corresponding eigenfunctions of the eigenvalue problem

$$\frac{1}{2}\,\Delta\phi + \lambda\phi = 0 \text{ in } B, \quad \phi_{|\partial B} = 0.$$

In particular

$$\mathbb{P}\{\tau \geq \varepsilon^{-2}\} = \sum_{n=1}^{\infty} e^{-\lambda_n/\varepsilon^2}\phi_n(0)\int_B \phi_n(y)dy$$

and thus

$$\mathbb{P}\{\|W\|_\infty \leq \varepsilon\} \sim e^{-\lambda_1/\varepsilon^2}\phi_1(0)\int_B \phi_1(y)dy. \tag{7.1}$$

In particular, it is known that $\lambda_1 = \pi^2/8$ for $d = 1$.

While the proof of (7.1) relies on some very specific properties of both the Brownian paths and the supnorm, one may wonder for the behavior of small ball probabilities for some other norms on Wiener space, such as for example L^p-norms or the classical Hölder norms of index α for every $0 < \alpha < \frac{1}{2}$. In what follows, we will describe some small ball Brownian probabilities, including the ones just mentioned, using some more abstract tools (which could eventually generalize to other Gaussian measures). We will however only work at the logarithmic scale. We use series expansions of Brownian motion in the Haar basis of $[0,1]$. We present the various results following the exposition of W. Stolz [St1], inspired by the works [B-R], [Ta9] and [Ta14] (among others). For simplicity, we work below with a one-dimensional Brownian motion and write $C_0([0,1])$ for $C_0([0,1];\mathbb{R})$.

We only concentrate here on the Brownian case. We mention at the end of the chapter references of extensions to some more general processes. Of course, a lot is known on small Hilbert balls for arbitrary Gaussian measures (cf. e.g. [Sy], [Zo], [K-L-L], [Li]...).

Let h_0, h_m, $m = 2^n + k - 1$, $n \geq 0$, $k = 1, \ldots, 2^n$ be the Haar functions on $[0,1]$. That is, $h_0 \equiv 1$,

$$h_1 = I_{[0,1/2)} - I_{[1/2,1]},$$

and, for every $m = 2^n + k - 1$, $n \geq 1$, $k = 1, \ldots, 2^n$,

$$h_m(t) = 2^{n/2}h_1(2^n t - k + 1), \quad 0 \leq t \leq 1.$$

Define then the Schauder functions φ_m, $m \in \mathbb{N}$, on $[0,1]$ by setting $\varphi_m(t) = \int_0^t h_m(s)ds$. The Schauder functions form a basis of the space of continous functions on $[0,1]$. In particular, Lévy's representation of Brownian motion may be expressed by saying that, almost surely,

$$W(t) = \sum_{m=0}^{\infty} g_m \varphi_m(t), \quad t \in [0,1], \tag{7.2}$$

where $(g_m)_{m\in\mathbf{N}}$ is a standard Gaussian sequence and where the convergence takes place uniformly on $[0,1]$ (cf. Proposition 4.2).

In what follows, $\|\cdot\|$ is a measurable norm on $C_0([0,1])$ for which the Wiener measure is a Radon measure, in other words for which the series (7.2) converges almost surely (cf. Proposition 4.2).

Theorem 7.1. *Let $0 \le \alpha < \frac{1}{2}$. If, for some constant $C > 0$,*

$$\left\| \sum_{k=1}^{2^n} a_k \varphi_{2^n+k-1} \right\| \le C 2^{-(\frac{1}{2}-\alpha)n} \max_{1\le k\le 2^n} |a_k|$$

for all real numbers $a_1, \dots, a_{2^n}$ and every $n \ge 0$, then

$$\log \mathbb{P}\{\|W\| \le \varepsilon\} \ge -C' \varepsilon^{-2/1-2\alpha}, \quad 0 < \varepsilon \le 1,$$

where $C' > 0$ only depends on α and C.

Proof. We simply replace the ball $\{\|W\| \le \varepsilon\}$ by an appropriate cube in $\mathbb{R}^{\mathbf{N}}$ through the representation (7.2). For q integer ≥ 1, define

$$b_n = b_n(q) = \begin{cases} 2^{(\frac{3}{4}-\frac{\alpha}{2})(n-q)} & \text{if } n \le q, \\ 2^{(\frac{1}{4}-\frac{\alpha}{2})(n-q)} & \text{if } n > q. \end{cases}$$

The choice of this sequence is not unique. If $|a_{2^n+k}| \le b_n$ for every $n \ge 0$ and $k = 1, \dots, 2^n$, and $|a_0| \le b_0$, by the triangle inequality and the hypothesis,

$$\left\| \sum_{m=0}^{\infty} a_m \varphi_m \right\| \le C_1 2^{-(\frac{1}{2}-\alpha)q}$$

for some constant C_1 only depending on α. Therefore,

$$\begin{aligned} \mathbb{P}\{\|W\| \le C_1 2^{-(\frac{1}{2}-\alpha)p}\} &\ge \mathbb{P}\{|g_0| \le b_0,\, |g_{2^n+k-1}| \le b_n,\, n \ge 0,\, k = 1, \dots, 2^n\} \\ &= \mathbb{P}\{|g| \le b_0\} \prod_{n=0}^{\infty} \mathbb{P}\{|g| \le b_n\}^{2^n} \end{aligned}$$

where g is a standard normal variable. Now, we simply need evaluate this infinite product. To this aim, we use that

$$\mathbb{P}\{|g| \le u\} \ge \frac{u}{3} \quad \text{if} \quad |u| \le 1 \tag{7.3}$$

and

$$\mathbb{P}\{|g| \le u\} \ge 1 - \frac{1}{2}\, 2\mathrm{e}^{-u^2/2} \ge \exp\bigl(-2\mathrm{e}^{-u^2/2}\bigr) \quad \text{if} \quad |u| \ge 1. \tag{7.4}$$

It easily follows, after some elementary computations, that

$$\mathbb{P}\{\|W\| \le C_1 2^{-(\frac{1}{2}-\alpha)q}\} \ge \exp\left(-C_2 2^q\right)$$

from which Theorem 7.1 immediately follows. □

The next theorem gives an upper bound of the small ball probabilities under hypotheses dual to those of Theorem 7.1.

Theorem 7.2. *Let $0 \le \alpha < \frac{1}{2}$. If, for some constant $C > 0$,*

$$\left\| \sum_{k=1}^{2^n} a_k \varphi_{2^n+k-1} \right\| \ge \frac{1}{C}\, 2^{-(\frac{1}{2}-\alpha)n} \left(2^{-n} \sum_{k=1}^{2^n} |a_k| \right)$$

for all real numbers $a_1, \ldots, a_{2^n}$ and every $n \ge 0$, then

$$\log \mathbb{P}\{\|W\| \le \varepsilon\} \le -\frac{1}{C''}\, \varepsilon^{-2/1-2\alpha}, \quad 0 < \varepsilon \le 1,$$

where $C'' > 0$ only depends on α and C.

Proof. First recall Anderson's inequality [An]. Let μ be a centered Gaussian measure on a Banach space E as in Chapter 4. Then, for every convex symmetric subset A of E and every x in E,

$$\mu(x + A) \le \mu(A). \tag{7.5}$$

Note that (7.5) is an easy consequence of (1.8) or the logconcavity of Gaussian measures (1.9) (cf. [Bo1], [Bo3], [D-HJ-S]...). Indeed, the set

$$Z = \bigl\{a \in E;\ \mu(a + A) \le \mu(x + A)\bigr\}$$

is symmetric and convex by (1.9). Now $x \in Z$, so that $-x \in Z$ by symmetry, and, by convexity, $0 = \frac{1}{2}x + \frac{1}{2}(-x) \in Z$ which is the result (7.5). By the series representation (7.2), independence and Fubini's theorem, it follows that

$$\mathbb{P}\{\|W\| \le \varepsilon\} \le \mathbb{P}\left\{ \left\| \sum_{k=1}^{2^n} g_k \varphi_{2^n+k-1} \right\| \le \varepsilon \right\}$$

for every $\varepsilon > 0$ and every $n \ge 0$. Therefore, by the hypothesis,

$$\mathbb{P}\{\|W\| \le \varepsilon\} \le \mathbb{P}\left\{ \sum_{k=1}^{2^n} |g_k| \le \varepsilon C 2^{(\frac{1}{2}-\alpha)n} 2^n \right\}.$$

By Chebyshev's exponential inequality, for every integer $N \geq 1$,

$$\mathbb{P}\left\{\sum_{k=1}^{N} |g_k| \leq cN\right\} \leq e^{cN}\left(\mathbb{E}(e^{-|g|})\right)^N \leq e^{-cN}$$

where $c > 0$ is such that $e^{-2c} = \mathbb{E}(e^{-|g|}) < 1$. Take then n to be the largest integer such that $\varepsilon C 2^{(\frac{1}{2}-\alpha)n} \leq c$. The conclusion easily follows. The proof of Theorem 7.2 is complete. □

The main interest of Theorems 7.1 and 7.2 lies in the examples for which the hypotheses may easily be checked. Let us consider L^p-norms $\|\cdot\|_p$, $1 \leq p < \infty$, on $[0,1]$ for which

$$\left\|\sum_{k=1}^{2^n} a_k \varphi_{2^n+k-1}\right\|_p = \frac{1}{2}(p+1)^{-1/p} 2^{-n/2}\left(2^{-n}\sum_{k=1}^{2^n}|a_k|^p\right)^{1/p}$$

for all real numbers $a_1, \ldots, a_{2^n}$. Since

$$2^{-n}\sum_{k=1}^{2^n}|a_k| \leq \left(2^{-n}\sum_{k=1}^{2^n}|a_k|^p\right)^{1/p} \leq \max_{1\leq k\leq 2^n}|a_k|,$$

we deduce from Theorem 7.1 and 7.2 that, for some constant $C > 0$ only depending on p and every $0 < \varepsilon \leq 1$,

$$-C\varepsilon^{-2} \leq \log \mathbb{P}\left\{\|W\|_p \leq \varepsilon\right\} \leq -C^{-1}\varepsilon^{-2}. \tag{7.6}$$

More precise estimates on the constant C are obtained in [B-M].

In the same way, let $\|\cdot\|_\alpha$ be the Hölder norm of index $0 < \alpha < \frac{1}{2}$ defined by

$$\|x\|_\alpha = \sup_{0\leq s\neq t\leq 1}\frac{|x(s)-x(t)|}{|s-t|^\alpha}.$$

Again, it is easily seen that for every $a_1, \ldots, a_{2^n} \in \mathbb{R}$,

$$\frac{1}{2}2^{-(\frac{1}{2}-\alpha)n}\max_{1\leq k\leq 2^n}|a_k| \leq \left\|\sum_{k=1}^{2^n} a_k\varphi_{2^n+k-1}\right\|_\alpha \leq \sqrt{2}\,2^{-(\frac{1}{2}-\alpha)n}\max_{1\leq k\leq 2^n}|a_k|.$$

Hence, for some constant $C > 0$ only depending on α, for every $0 < \varepsilon \leq 1$,

$$-C\varepsilon^{-2/1-2\alpha} \leq \log \mathbb{P}\left\{\|W\|_\alpha \leq \varepsilon\right\} \leq -C^{-1}\varepsilon^{-2/1-2\alpha}. \tag{7.7}$$

This result is due to P. Baldi and B. Roynette [B-R].

Note that the supnorm may be included in either $p = \infty$ in (7.6) or $\alpha = 0$ in (7.7) so that we recover (7.1) with these elementary arguments, with however worse constants. More examples may be treated by these methods such as Besov's norm or Sobolev norms on the Wiener space. We refer to [St1] and [Li-S] for more details along these lines.

To try to investigate some further cases with these tools, let us consider, on some probability space $(\Omega, \mathcal{A}, \mathbb{P})$, a sequence $(G_m)_{m\in\mathbb{N}}$ of independent standard one dimensional Brownian motions (on $[0,1]$). Replace then, in the series representation (7.2), the standard Gaussian sequence by this sequence of independent Brownian motions. We define in this way a Brownian motion S with values in $C_0([0,1])$ and with "reference measure" the Wiener measure itself. This is one way of defining the Wiener sheet which thus turns out to be a centered Gaussian process $S = (S(s,t))_{s,t\in[0,1]}$ with covariance

$$\mathbb{E}\big(S(s,t)S(s',t')\big) = \min(s,s')\min(t,t').$$

In this framework, we may thus ask for the behavior of $\mathbb{P}\{\|S\|_\infty \le \varepsilon\}$, $0 < \varepsilon \le 1$, for the supnorm on $[0,1]^2$. Since this norm is also the supremum norm of W considered as a one dimensional process with values in $C_0([0,1])$ (equipped with the supnorm), some of the preceding material may be used in this investigation. In particular, we can replace, in the proof of Theorem 7.1, (7.3) by the small ball behavior of Brownian motion (7.1). By Theorem 4.1, the large ball behavior (7.4) is unchanged at the exception of possibly different numerical constants. The argument of Theorem 7.3 then implies, exactly in the same way, that for some constant $C > 0$, and every $0 < \varepsilon \le 1$,

$$\log \mathbb{P}\big\{\|S\|_\infty \le \varepsilon\big\} \ge -C\varepsilon^{-2}\big(\log \varepsilon^{-1}\big)^3.$$

A similar vector valued extension of Theorem 7.2 however only yields that

$$\log \mathbb{P}\big\{\|S\|_\infty \le \varepsilon\big\} \le -C^{-1}\varepsilon^{-2}\big(\log \varepsilon^{-1}\big).$$

These estimates, which go back to M. A. Lifshits [Lif-T] (see also [Bas]), only rely on the small ball behavior (7.1) and are best possible among all Gaussian measures having this small ball behavior. For the special case of Wiener measure and the Wiener sheet, M. Talagrand [Ta15] however proved the striking following theorem. The proof is based on a new wavelet decomposition of the space $\mathrm{L}^2([0,1]^2)$ and various combinatorial arguments from Banach space theory. The method does not allow any precise information on constants. We refer to [Ta15] for the proof.

Theorem 7.3. *There is a numerical constant $C > 0$ such that, for every $0 < \varepsilon \le 1$,*

$$-C\varepsilon^{-2}\big(\log \varepsilon^{-1}\big)^3 \le \log \mathbb{P}\big\{\|S\|_\infty \le \varepsilon\big\} \le -C^{-1}\varepsilon^{-2}\big(\log \varepsilon^{-1}\big)^3.$$

In this framework of small ball probabilities for Gaussian measures, let us now come back to some of the concentration inequalities of Chapters 2 and 4. There, we studied inequalities for general sets A and their enlargements A_r. Now, we try to take advantage of some geometric structures on A, such as for example being a ball (convex and symmetric with respect to the origin). In a first step, we will notice how some of the preceding tools may be applied successfully to improve, for example, a statement such as Theorem 4.4. In particular, we aim to prove inequalities for subsets A with small measure and to be able to keep the dependence in this measure. The next statement (cf. [Ta8]) is a first example of what can be accomplished for various norms on the Wiener space. Recall the unit ball $\mathcal{K}$ of the Cameron-Martin Hilbert space $\mathcal{H}$ of absolutely continuous functions on $[0,1]$ whose almost everywhere derivative is in L^2.

Theorem 7.4. *Let $\|\cdot\|$ be a norm on $C_0([0,1])$ for which Wiener measure is a Radon measure. Denote by U the unit ball of $\|\cdot\|$. Let furthermore $0 \le \alpha < \frac{1}{2}$ and assume that, for some constant $C > 0$,*

$$\Big\|\sum_{k=1}^{2^n} a_k \varphi_{2^n+k-1}\Big\| \le C 2^{-(\frac{1}{2}-\alpha)n} \max_{1\le k\le 2^n} |a_k|$$

for all real numbers $a_1, \ldots, a_{2^n}$ and every $n \ge 0$. Then, for every $\varepsilon > 0$ and every $r \ge 0$,

$$\mathbb{P}\{W \in \varepsilon U + r\mathcal{K}\} \ge 1 - \exp\bigg(\frac{C'}{\varepsilon^{2/1-2\alpha}} - \frac{\varepsilon r}{2\sigma} - \frac{r^2}{2\sigma^2}\bigg)$$

where $C' > 0$ only depends on α and C where we recall that $\sigma = \sup_{x\in\mathcal{K}} \|x\|$.

Proof. We take again the notation of Theorem 7.1. First note that since $\mathcal{K} \subset \sigma U$,

$$\varepsilon U + r\mathcal{K} \supset \frac{\varepsilon}{2} U + \Big(r + \frac{\varepsilon}{2\sigma}\Big)\mathcal{K}.$$

(The choice of $\varepsilon/2$ is rather arbitrary here.) Set $r' = r + \frac{\varepsilon}{2\sigma}$. Recall the sequence $(b_n)_{n\in\mathbb{N}}$ of the proof of Theorem 7.1 which depends on some integer $q \ge 1$. Define a sequence of real numbers $(c_m)_{m\in\mathbb{N}}$ by setting

$$c_0 = b_0, \quad c_{2^n+k-1} = b_n \quad \text{for all} \quad k = 1, \ldots, 2^n,\ n \ge 0.$$

Consider the set $V = V_q$ of all functions φ on $[0,1]$ that can be written as $\varphi = \sum_{m=0}^{\infty} a_m \varphi_m$ where $|a_m| \le c_m$ for every m. By the hypothesis on the norm $\|\cdot\|$ and the triangle inequality, $V \subset C_1 2^{-(\frac{1}{2}-\alpha)q} U$ for some constant $C_1 > 0$. Therefore, if q is the smallest integer such that $2C_1 2^{-(\frac{1}{2}-\alpha)q} \le \varepsilon$, then $\varepsilon U + r\mathcal{K} \supset V + r'\mathcal{K}$. Hence, by the series representation (7.2),

$$\mathbb{P}\{W \in \varepsilon U + r\mathcal{K}\} \ge \mathbb{P}\{W \in V + r'\mathcal{K}\} = \gamma_\infty(Q + r'\sigma^{-1}B_2)$$

where γ_∞ is the canonical Gaussian measure on $\mathbb{R}^{\mathbb{N}}$, B_2 the unit ball of the reproducing kernel of γ_∞, that is the unit ball of ℓ^2, and Q the "cube" in $\mathbb{R}^{\mathbb{N}}$

$$Q = \big\{x = (x_m)_{m\in\mathbb{N}} \in \mathbb{R}^{\mathbb{N}};\, |x_m| \le c_m,\, m \in \mathbb{N}\big\}.$$

Consider the function on $\mathbb{R}^{\mathbb{N}}$ given by $d(x) = \inf\{u \ge 0; x \in Q + uB_2\}$. Note that $\gamma_\infty(Q + uB_2) = \gamma_\infty(d < u)$. By Chebyshev's inequality,

$$\gamma_\infty(d \ge u) \le \mathrm{e}^{-u^2/2} \int \mathrm{e}^{d^2/2} d\gamma_\infty.$$

For every $m \geq 0$, let $d_m(x) = (|x_m| - c_m)^+$. Then

$$\begin{aligned}
\int e^{d_m^2/2} d\gamma_\infty &= \frac{2}{\sqrt{2\pi}} \int_0^\infty \exp\left(\frac{1}{2}\left(\left(|t| - c_m\right)^+\right)^2 - \frac{t^2}{2}\right) dt \\
&\leq 1 + \frac{2}{\sqrt{2\pi}} \int_{c_m}^\infty \exp\left(\frac{1}{2}(|t| - c_m)^2 - \frac{t^2}{2}\right) dt \\
&\leq 1 + \int_{c_m}^\infty \exp\left(-c_m t + \frac{c_m^2}{2}\right) dt = 1 + \frac{1}{c_m} e^{-c_m^2/2}.
\end{aligned}$$

Now, the nice feature of this geometric construction is that $d^2 = \sum_{m=0}^\infty d_m^2$. Therefore, it follows from the preceding that, for every $u \geq 0$,

$$\gamma_\infty(d \geq u) \leq \prod_{m=0}^\infty \left(1 + \frac{1}{c_m} e^{-c_m^2/2}\right) e^{-u^2/2}.$$

To conclude the proof, we need simply estimate this infinite product. By the definition of the sequence $(c_m)_{m \in \mathbb{N}}$, we see that

$$\prod_{m=0}^\infty \left(1 + \frac{1}{c_m} e^{-c_m^2/2}\right) = \prod_{n=0}^\infty \left(1 + \frac{1}{b_n} e^{-b_n^2/2}\right)^{2^n}.$$

Now, the very definition of the sequence $(b_n)_{n \in \mathbb{N}}$ implies, after elementary, though somewhat tedious, computations that the preceding infinite product is bounded above by $\exp(C_2 2^p)$ for some numerical constant $C_2 > 0$. By the choice of q, this completes the proof of Theorem 7.4 . □

With respect to the classical isoperimetric and concentration inequalities usually stated for sets with measure larger than $\frac{1}{2}$, we note here that the probability $\mathbb{P}\{W \in \varepsilon U\}$ can be very small as $\varepsilon \to 0$. Moreover, according to Theorem 7.1, the first term in the exponential extimate of Theorem 7.4 is precisely the order of $\mathbb{P}\{W \in \varepsilon U\}$. Theorem 7.4 applies to the supnorm and the Hölder norms and may be used in the study of rates of convergence in Strassen's law of the iterated logarithm. Let for example

$$Z_n(t) = \left(\frac{W(nt)}{\sqrt{2n\mathrm{LL}n}}\right)_{t \in [0,1]}, \quad n \geq 1,$$

where $\mathrm{LL}n = \log\log n$ if $n \geq 3$, $\mathrm{LL}n = 1$ if $n = 1, 2$. It is shown in [Ta8] using Theorem 7.4 that, almost surely,

$$0 < \limsup_{n \to \infty} (\mathrm{LL}n)^{2/3} d(Z_n, \mathcal{K}) < \infty$$

where $d(\cdot, \mathcal{K})$ is the uniform distance to the Strassen set (Cameron-Martin unit ball) on $C_0([0,1])$. See also [Gri] for an alternate proof and [Ta9] for further results.

Recently, M. Talagrand [Ta9] proved a deep extension of Theorem 7.4 in the abstract setting of enlarged balls. We now would like to present this statement. We will state and prove the main result for the canonical Gaussian measure γ_n on

$\mathbb{R}^n$. As in the preceding chapters, this is again the main inequality and standard tools may then be used to extend it to arbitrary Gaussian measures as in Chapter 4. The isoperimetric and concentration inequalities for γ_n yield powerful bounds of the measure of an enlarged set A_r, especially when r is large. However, the extremal sets of Gaussian isoperimetry are the half-spaces and it may well be that the concentration properties could be sharpened for sets with special geometrical structures such as for example convex symmetric bodies. The next theorem [Ta10] answers this problem.

Theorem 7.5. *Let C be a closed convex symmetric subset of $\mathbb{R}^n$. Assume that the polar C° of C may be covered by N sets $(T_i)_{1\le i\le N}$ such that $\int \sup_{y\in T_i}\langle x,y\rangle d\gamma_n(x) \le \frac{1}{2}$. Then, for every $r\ge 1$,*

$$\gamma_n(C_r) \ge 1 - 4N\log(er)\mathrm{e}^{-r^2/2}.$$

Proof. Denote for simplicity by B the closed Euclidean unit ball in $\mathbb{R}^n$. Since C is closed and B is compact, $C_r = C+rB$ is closed. By the bipolar theorem, $C+rB = U^\circ$ where $U = (C+rB)^\circ$. By definition,

$$U = \big\{x\in\mathbb{R}^n; \forall\, y\in C, \forall\, z\in B, \big|\langle x,y\rangle + r\langle x,z\rangle\big| \le 1\big\}.$$

Setting $\|x\|_C = \sup_{y\in C}\langle x,y\rangle = \sup_{y\in C}|\langle x,y\rangle|$, we see that

$$U = \big\{x\in\mathbb{R}^n; \|x\|_C + r|x| \le 1\big\}. \tag{7.8}$$

Observe also by the definition of the polar that $x\in \|x\|_C C^\circ$. If T is a subset of $\mathbb{R}^n$, we set

$$E(T) = \int \sup_{y\in T}\langle x,y\rangle d\gamma_n(x).$$

Let p_0 be the largest integer p such that $2^{p-1}\le r^2$. In particular, $p_0\le 1+4\log r$. Now, set

$$U_0 = \big\{x\in U; |x| \ge r^{-1}(1-r^{-2})\big\}$$

and, for $1\le p\le p_0$,

$$U_p = \big\{x\in U; r^{-1}(1-r^{-2}2^p) \le |x| \le r^{-1}(1-r^{-2}2^{p-1})\big\}.$$

Thus we have $U\subset \bigcup_{0\le p\le p_0} U_p$. Moreover, for $x\in U_p$, by (7.8), $\|x\|_C \le r^{-2}2^p$. Therefore, $U_p\subset r^{-2}2^pC^\circ$. It thus follows from the hypothesis on C° that U_p can be covered by subsets $(T_{p,i})_{1\le i\le N}$ where $T_{p,i} = r^{-2}2^pT_i\cap U_p$. Hence $E(T_{p,i})\le r^{-2}2^{p-1}$.

The essential step of the proof is concentration. From (4.3) for example, we get that, for every subset T of $\mathbb{R}^n$ and every $t\ge E(T)$,

$$\gamma_n\big(x; \sup_{y\in T}\langle y,x\rangle \ge t\big) \le \exp\bigg(-\frac{(t-E(T))^2}{2\sigma^2}\bigg) \tag{7.9}$$

where $\sigma = \sigma(T) = \sup_{x\in T}|x|$. Note that for $p \le p_0$, $E(T_{p,i}) \le r^{-2}2^{p-1} \le 1$. Hence, using (7.9) for $t = 1$, we get, for every $0 \le p \le p_0$, $1 \le i \le N$,

$$\gamma_n\bigl(x;\ \sup_{y\in T_{p,i}}\langle y,x\rangle \ge 1\bigr) \le \exp\biggl(-\frac{(1-r^{-2}2^{p-1})^2}{2\sigma(T_{p,i})^2}\biggr). \tag{7.10}$$

We first consider the case $p = 0$. We have $\sigma(T_{0,i}) \le r^{-1}$ so that, by summation over $1 \le i \le N$,

$$\gamma_n\bigl(x;\ \sup_{y\in U_0}\langle y,x\rangle \ge 1\bigr) \le N\exp\biggl(-\frac{r^2}{2}\bigl(1-r^{-2}\bigr)^2\biggr) \le N\mathrm{e}\,\mathrm{e}^{-r^2/2}.$$

For $p \ge 1$, by definition of U_p, we have $\sigma(T_{p,i}) \le r^{-1}(1-r^{-2}2^{p-1})$, so that, by summation of (7.10) over $1 \le i \le N$, we get

$$\gamma_n\bigl(x;\ \sup_{y\in U_p}\langle y,x\rangle \ge 1\bigr) \le N\,\mathrm{e}^{-r^2/2}.$$

Now, summation over $0 \le p \le p_0$ and the fact that $p_0 \le 1 + 4\log r$ yield

$$\begin{aligned}\gamma_n\bigl(x;\sup_{y\in U}\langle y,x\rangle \ge 1\bigr) &\le N(\mathrm{e}+1+4\log r)\,\mathrm{e}^{-r^2/2}\\ &\le 4N\log(\mathrm{e}r)\,\mathrm{e}^{-r^2/2}.\end{aligned}$$

Since $C + rB = U^\circ$, the result follows. The proof of Theorem 7.5 is complete. □

Of course, Theorem 7.5 can be useful in applications only if the number N of the statement may be appropriately bounded. We will not go far in the technical details, but one of the main conclusions of the important paper [Ta9] is that N may actually be controlled by the behavior of $\gamma_n(\varepsilon A)$ for the small values of $\varepsilon > 0$. Actually, this observation is strongly related to a remarkable result of J. Kuelbs and W. Li [K-L2] connecting the small ball probabilities to some entropy numbers related to N. We now turn to this discussion. Related work of M. A. Lifshits [Lif1] deals with the geometric tool of Kolmogorov's widths.

Given two (convex) sets A and B in $\mathbb{R}^n$ (or more generaly in a linear vector space), denote by $N(A,B)$ the smallest number of translates of B which are needed to cover A. Now, let, as in Theorem 7.5, C be a closed convex symmetric set and let B be the Euclidean unit ball in $\mathbb{R}^n$. If $x \in \mathbb{R}^n$ is such that $(x+\varepsilon B)\cap C^\circ \neq \emptyset$, for $y \in (x+\varepsilon B)\cap C^\circ$ we have

$$(x+\varepsilon B)\cap C^\circ \subset (y+2\varepsilon B)\cap C^\circ \subset y + (2\varepsilon B\cap 2C^\circ).$$

Hence, if ε is such that $E(C^\circ \cap \varepsilon B) \le \frac{1}{8}$, then the number N in Theorem 7.5 satisfies $N \le N(C^\circ, \varepsilon B)$. Now, it has been observed in local theory of Banach spaces [TJ] that the growth of the entropy numbers $N(C^\circ, \varepsilon B)$ is very similar to the growth of the dual entropy numbers $N(B, \varepsilon C)$ (cf. [L-T2], p. 82-83). The observation of J. Kuelbs and W. Li is precisely that the behavior of the small ball probabilities $\gamma_n(\varepsilon A)$ is related to these dual entropy numbers $N(B,\varepsilon C)$. They established namely the following theorem. One crucial argument in the proof is the Gaussian isoperimetric inequality. A prior partial result appeared in [Go2].

Theorem 7.6. *Under the preceding notation, if C is compact, convex and symmetric, and if $t = (2\log(\gamma_n(C)^{-1}))^{-1/2} > 0$, then*

$$\frac{1}{2\gamma_n(2C)} \leq N(B, tC) \leq \frac{1}{2\gamma_n(C/2)^2}.$$

Proof. First note, as a consequence of Cameron-Martin's formula (in finite dimension) and Anderson's inequality (7.5), that for every $x \in \mathbb{R}^n$,

$$e^{-|x|^2/2}\gamma_n(C) \leq \gamma_n(x + C) \leq \gamma_n(C). \tag{7.11}$$

Consider now $u > 0$ and a finite subset F of uB such that for any two distinct points of F, the translates of C by these points are disjoints. By (7.11), for every x in F,

$$\gamma_n(x + C) \geq e^{-u^2/2}\gamma_n(C).$$

It follows that $\mathrm{Card}(F) \leq \gamma_n(C)^{-1}e^{u^2/2}$. When F is maximal, the sets $(x + 2C)_{x \in F}$ cover uB so that

$$N(uB, 2C) \leq \gamma_n(C)^{-1}e^{u^2/2}.$$

(When $\gamma_n(C) \geq \frac{1}{2}$, this is how the dual Sudakov inequality is proved in [L-T2], p. 83.) If we choose $u = t^{-1} = (2\log(\gamma_n(C)^{-1}))^{1/2}$, we have $N(uB, 2C) \leq \gamma_n(C)^{-2}$. Since $N(uB, C) = N(B, tC)$, the right hand side of Theorem 7.6 follows by replacing C by $\frac{1}{2}C$.

Conversely, by (7.11) again,

$$N(uB, C)\gamma_n(2C) \geq N(C + uB, 2C)\gamma_n(2C) \geq \gamma_n(C + uB).$$

Now, by the isoperimetric inequality (Theorem 1.3),

$$\Phi^{-1}(\gamma_n(C + uB)) \geq \Phi^{-1}(\gamma_n(C)) + u.$$

Let again $u = t^{-1} = (2\log(\gamma_n(C)^{-1}))^{1/2}$. Since $\Phi(-u) \leq e^{-u^2/2} = \gamma_n(C)$, the isoperimetric inequality implies that $\Phi^{-1}(\gamma_n(C + uB)) \geq 0$ that is, $\gamma_n(C + uB) \geq \frac{1}{2}$. The left hand side of the inequality of the theorem is thus also satisfied. The proof is complete. □

If we set

$$\varphi(\varepsilon) = \left(2\log\frac{1}{\gamma_n(\varepsilon C)}\right)^{1/2}, \quad \varepsilon > 0,$$

we see from Theorem 7.6 that

$$\frac{1}{2}\exp\left(\frac{\varphi(2\varepsilon)^2}{2}\right) \leq N\left(B, \frac{\varepsilon}{\varphi(\varepsilon)}\right) \leq \exp\left(\varphi\left(\frac{\varepsilon}{2}\right)^2\right).$$

Therefore, if φ is regularly varying, its behavior is essentially given by the behavior of the entropy numbers $N(B, \varepsilon C)$ (and conversely). Much more on the structure of C is thus involved with respect for example with the large ball behavior (cf. Chapter 4). While Theorem 7.6 and its proof are presented in finite dimension, the infinite dimensional extension yields a rather precise equivalence between small ball probabilities for a Gaussian measure μ on a Banach space E and the entropy numbers $N(\mathcal{K}, \varepsilon U)$ where $\mathcal{K}$ is the unit ball of the reproducing kernel Hilbert space $\mathcal{H}$ and U the unit ball of E. It has been shown by J. Kuelbs and W. Li [K-L2] to have striking consequences in approximation theory. For example, using the small ball behaviors for Wiener measure, we see that if $\mathcal{K}$ is the unit ball of the Cameron-Martin Hilbert space and C the L^p-ball on $([0,1], dt)$, $1 \le p \le \infty$, then

$$\log N(\mathcal{K}, \varepsilon C) \sim \frac{1}{\varepsilon} \quad \text{as } \varepsilon \to 0.$$

Theorem 7.3 similarly shows that for $\mathcal{K}$ the unit ball of the Cameron-Martin space associated to the Wiener sheet and for C the uniform unit ball on $C([0,1]^2; \mathbb{R})$,

$$\log N(\mathcal{K}, \varepsilon C) \sim \frac{1}{\varepsilon}\Big(\log \frac{1}{\varepsilon}\Big)^{3/2} \quad \text{as } \varepsilon \to 0.$$

This deep connection between entropy numbers and small ball probabilities is further investigated in [K-L2] and [Ta9]. In particular, in [Ta9], the author obtains very general rates for the variables $(2\log n)^{-1/2} X_n$ to cluster to $\mathcal{K}$, when $(X_n)_{n \in \mathbb{N}}$ is a sequence of independent identically distributed sequence with distribution μ. These rates depend only on the behavior of the small ball probabilities $\mu(\varepsilon U)$. These results have applications to rates of convergence in Strassen's law of the iterated logarithm for Brownian motion. Prior results on the convergence of $(2\log n)^{-1/2} X_n$ to $\mathcal{K}$ at the origin of this study are due to V. Goodman [Go1].

In [Ta9], M. Talagrand also established a general lower bound on supremum of Gaussian processes under entropy conditions. At the present time, it is one of the only few general results available in this subject of small ball probabilities. We briefly describe one simple statement. Let $(X_t)_{t \in T}$ be a (centered) Gaussian process as in Chapter 6. Recall also from this chapter the entropy numbers $N(T, d; \varepsilon)$, $\varepsilon > 0$, for the Dudley metric $d(s,t) = (\mathbb{E}|X_s - X_t|^2)^{1/2}$, $s, t \in T$. Assume that there is a nonnegative function ψ on $\mathbb{R}_+$ such that

$$N(T, d; \varepsilon) \le \psi(\varepsilon), \quad \varepsilon > 0, \tag{7.11}$$

and such that for some constants $1 < c_1 \le c_2 < \infty$ and all $\varepsilon > 0$

$$c_1 \psi(\varepsilon) \le \psi\Big(\frac{\varepsilon}{2}\Big) \le c_2 \psi(\varepsilon). \tag{7.12}$$

We thus have in mind a power type behavior $\psi(\varepsilon) = \varepsilon^{-a}$ of the entropy numbers. Then, for some $K > 0$ and every $\varepsilon > 0$,

$$\mathbb{P}\Big\{ \sup_{s,t \in T} |X_s - X_t| \le \varepsilon \Big\} \ge \exp\big(-K\psi(\varepsilon)\big). \tag{7.13}$$

We prove this result following the notations introduced in the previous chapter. Let n_0 be the largest n in $\mathbb{Z}$ such that $2^{-n} \geq D(T)$ where $D(T)$ is the diameter of T, assumed to be finite (we may actually start with T finite). For every $n \geq n_0$, consider a subset T_n of T of cardinality $N(n) = N(T, d; 2^{-n})$ such that each point of T is within distance 2^{-n} of T_n. We let $T_{n_0} = \{t_0\}$ where t_0 is any fixed point in T. For $n > n_0$, choose $s_{n-1}(t) \in T_{n-1}$ such that $d(t, s_{n-1}(t)) \leq 2^{-n+1}$ and set

$$\mathcal{Y} = \{X_t - X_{s_{n-1}(t)}; t \in T_n\}.$$

Note that $\|Y\|_2 \leq 2^{-n+1}$ for every Y in $\mathcal{Y}$. Clearly, each X_t can be written as

$$X_t = X_{t_0} + \sum_{n>n_0} Y_n$$

where $Y_n \in \mathcal{Y}$, $n > n_0$. Therefore, if $(b_n)_{n>n_0}$ is a sequence of positive numbers with $\sum_{n>n_0} b_n \leq \frac{u}{2}$, $u > 0$,

$$\begin{aligned} \mathbb{P}\{\sup_{s,t\in T} |X_s - X_t| \leq u\} &\geq \mathbb{P}\{\forall n > n_0, \forall Y \in \mathcal{Y}, |Y_n| \leq b_n\} \\ &\geq \prod_{n>n_0} \left(\mathbb{P}\{|g| \leq b_n 2^{n-1}\}\right)^{N(n)} \end{aligned} \tag{7.14}$$

where g is a standard normal variable. We used here the following consequence of the main inequality of [Kh], [Sc], [Si]... (see (7.16) below): if $(Z_1, \ldots, Z_n)$ is a (centered) Gaussian random vector, for every $\lambda_1, \ldots, \lambda_n \geq 0$,

$$\mathbb{P}\{|Z_1| \leq \lambda_1, \ldots, |Z_n| \leq \lambda_n\} \geq \prod_{i=1}^{n} \mathbb{P}\{|Z_i| \leq \lambda_i\}.$$

Let q be an integer with $q > n_0$ and set

$$b_n = b_n(q) = \begin{cases} 2^{-\frac{3q}{2}+\frac{n}{2}+1} & \text{if } n_0 < n \leq q, \\ 2^{-\frac{q}{2}-\frac{n}{2}+1} & \text{if } n > q. \end{cases}$$

Then $\sum_{n>n_0} b_n \leq 2^{-q+3}$. Apply then (7.14) with $u = 2^{-q+3}$. Using (7.3) and (7.4) and the hypothesis $N(n) \leq \psi(2^{-n})$, we get

$$\mathbb{P}\{\sup_{s,t\in T} |X_s - X_t| \leq u\} \geq \prod_{n_0<n\leq q} \left(3^{-1}2^{-3(n-q)/2}\right)^{\psi(2^{-n})} \prod_{n>q} \exp\left(-2\psi(2^{-n})e^{-2^{n-q}}\right).$$

Now, by (7.12),

$$\begin{aligned} \sum_{n_0<n\leq q} \psi(2^{-n}) \log\left(3^{-1}2^{3(n-q)/2}\right) &\leq \psi(2^{-q}) \sum_{n_0<n\leq q} c_1^{n-q} \log\left(3^{-1}2^{3(n-q)/2}\right) \\ &\leq K(c_1)\psi(2^{-q}) \end{aligned}$$

while

$$\sum_{n>q} \psi(2^{-n}) \mathrm{e}^{-2^{n-q}} \leq \psi(2^{-q}) \sum_{n>q} c_2^{n-q} \mathrm{e}^{-2^{n-q}}$$
$$\leq K(c_2)\psi(2^{-q})$$

where $K(c_1), K(c_2) > 0$ only depend on c_1 and c_2 respectively. It follows that

$$\mathbb{P}\Big\{ \sup_{s,t\in T} |X_s - X_t| \leq 2^{-q+3} \Big\} \geq \exp\big(-K\psi(2^{-q})\big)$$

where $q > n_0$. Let $\varepsilon \leq 8D(T)$ and let $q > n_0$ be the largest integer such that $2^{-q+4} \geq \varepsilon$ ($\varepsilon \leq 8D(T) \leq 2^{-n_0+3}$, $2^{-q+3} \leq \varepsilon$). Then

$$\mathbb{P}\Big\{ \sup_{s,t\in T} |X_s - X_t| \leq \varepsilon \Big\} \geq \exp\big(-K\psi(2^{-q})\big) \geq \exp\big(-K\psi(\varepsilon)\big).$$

When $\varepsilon \geq 8D(T)$, by concentration,

$$\mathbb{P}\Big\{ \sup_{s,t\in T} |X_s - X_t| \leq \varepsilon \Big\} \geq 1 - 2\exp\bigg(\frac{\varepsilon^2}{2D(T)^2}\bigg) \geq \frac{1}{2} \geq \exp\big(-\psi(\varepsilon)\big)$$

since $\psi(\varepsilon) \geq N(T, d; \varepsilon) \geq 1$. (7.13) thus is established.

To conclude this chapter, we mention some related correlation and conditional inequalities and their applications. These results have been used recently in various topological questions on Wiener space briefly discussed below.

The next inequality seems to mix small ball and large ball behaviors and might be of some interest in other contexts. It is related to conjecture (7.17) below. Let $W = (W(t))_{t\geq 0}$ be Brownian motion starting at the origin with values in $\mathbb{R}^d$. By Lévy's modulus of continuity of Brownian motion, one may consider some stronger topologies on the Wiener space $C_0([0,1]; \mathbb{R}^d)$, such as Hölder topologies. For every function $x : [0,1] \to \mathbb{R}^d$, recall the Hölder norm of index $0 < \alpha < 1$ defined as

$$\|x\|_\alpha = \sup_{0\leq s\neq t\leq 1} \frac{|x(s) - x(t)|}{|s-t|^\alpha}.$$

It is known, and due to Z. Ciesielski [Ci1], that these Hölder norms are equivalent to sequence norms. More precisely, for every continuous function $x : [0,1] \to \mathbb{R}^d$ such that $x(0) = 0$, let, for $m = 2^n + k - 1$, $n \geq 0$, $k = 1, \ldots, 2^n$,

$$\xi_m(x) = \xi_{2^n+k-1}(x) = 2^{n/2}\bigg[2x\Big(\frac{2k-1}{2^{n+1}}\Big) - x\Big(\frac{k}{2^n}\Big) - x\Big(\frac{k-1}{2^n}\Big)\bigg]$$

and $\xi_0(x) = x(1)$, be the evaluation of x in the Schauder basis on $C_0([0,1]; \mathbb{R}^d)$. Set

$$\|x\|'_\alpha = \sup_{m\geq 0} (m+1)^{\alpha - \frac{1}{2}} |\xi_m(x)|.$$

Then, for every $0 < \alpha < 1$, there exists $C_\alpha^d > 0$ such that, for all $x \in C_0([0,1];\mathbb{R}^d)$,

$$(C_\alpha^d)^{-1}\|x\|'_\alpha \le \|x\|_\alpha \le C_\alpha^d\|x\|'_\alpha. \tag{7.15}$$

Note also that Wiener measure is a Radon measure on the subspace of the space of Hölder functions x such that

$$\lim_{\eta\to 0} \sup_{\substack{|s-t|\le\eta \\ 0\le s\ne t\le 1}} \frac{|x(s)-x(t)|}{|s-t|^\alpha} = 0.$$

The next proposition is the main conditional Gaussian inequality we would like to emphasize. It evaluates large oscillations of the Brownian paths conditionally on the fact that these paths are bounded [BA-G-L].

Proposition 7.7. *Let $0 < \alpha < \frac{1}{2}$. There exists a constant $C > 0$ only depending on d and α such that for every $u > 0$ and $v > 0$,*

$$\mathbb{P}\big\{\|W\|_\alpha \ge u \,\big|\, \|W\|_\infty \le v\big\} \le C\max\bigg(1, \Big(\frac{u}{v}\Big)^{1/\alpha}\bigg)\exp\bigg(-\frac{u^{1/\alpha}}{Cv^{(1/\alpha)-2}}\bigg).$$

Proof. We use (7.15) to write that, for $u, v > 0$,

$$\begin{aligned}\mathbb{P}\big\{\|W\|'_\alpha \ge u \,\big|\, \|W\|_\infty \le v\big\} &\le \sum_{m\ge 0}\mathbb{P}\big\{|\xi_m(W)| \ge u(m+1)^{\frac{1}{2}-\alpha} \,\big|\, \|W\|_\infty \le v\big\} \\ &\le \sum_{m\ge m_0}\mathbb{P}\big\{|\xi_m(W)| \ge u(m+1)^{\frac{1}{2}-\alpha} \,\big|\, \|W\|_\infty \le v\big\}\end{aligned}$$

where $m_0 = \max(0, (u/4v)^{1/\alpha} - 1)$ since, on $\{\|W\|_\infty \le v\}$, $|\xi_m(W)| \le 4v\sqrt{m+1}$. Now, if $a > 0$ and if A is a convex symmetric subset of $\mathbb{R}^n$, it has been shown in [Kh], [Si], [Sco]... (see also [DG-E-...]) that

$$\gamma_n(A\cap S) \ge \gamma_n(A)\gamma_n(S) \tag{7.16}$$

where, as usual, γ_n is the canonical Gaussian measure on $\mathbb{R}^n$ and where S is the strip $\{x \in \mathbb{R}^n; |x_1| \le a\}$. Since the ξ_m are continuous linear functionals on the Wiener space, a simple finite dimensional approximation on (7.16) (in the spirit, for example, of the approximation procedures described in Chapter 4) then shows that

$$\mathbb{P}\big\{|\xi_m(W)| \ge u(m+1)^{\frac{1}{2}-\alpha} \,\big|\, \|W\|_\infty \le v\big\} \le \mathbb{P}\big\{|\xi_m(W)| \ge u(m+1)^{\frac{1}{2}-\alpha}\big\}$$

for every m. Hence,

$$\mathbb{P}\big\{\|W\|'_\alpha \ge u \,\big|\, \|W\|_\infty \le v\big\} \le \sum_{m\ge m_0}\mathbb{P}\big\{|\xi_m(W)| \ge u(m+1)^{\frac{1}{2}-\alpha}\big\}.$$

Now, the variables $\xi_m(W)$ are distributed according to γ_d on $\mathbb{R}^d$. By the classical Gaussian exponential bound,

$$\mathbb{P}\big\{\|W\|'_\alpha \geq u \,|\, \|W\|_\infty \leq v\big\} \leq \sum_{m \geq m_0} \exp\Big(-\frac{1}{C_d} u^2 (m+1)^{1-2\alpha}\Big)$$

where $C_d > 0$ only depends on d. The conclusion then easily follows after some elementary computations. Proposition 7.7 is established. □

It might be worthwhile noting that we obtain a weaker, although already useful result, by replacing in Proposition 7.7 the conditional probability by the probability of the intersection, that is

$$\mathbb{P}\big\{\|W\|_\alpha \geq u, \|W\|_\infty \leq v\big\}.$$

The proof for this quantity is in fact easier since it does not use (7.16). M. A. Lifshits recently mentioned to me that the bound of Proposition 7.7 is actually two-sided at the logarithmic scale as the ratio $u^{1/\alpha}/v^{(1/\alpha)-2}$ is large. His argument is based on a delicate partitioning and clever use of the Markov property. One may ask for a general version of Proposition 7.7 dealing with some arbitrary norms on a Gaussian space.

In the proof of Proposition 7.7, we made crucial use of the correlation inequality (7.16). More generally than (7.16), one may ask whether, given two symmetric convex bodies A and B in $\mathbb{R}^n$,

$$\gamma_n(A \cap B) \geq \gamma_n(A)\gamma_n(B). \tag{7.17}$$

This was established when $n = 2$ by L. Pitt [Pit], and thus for arbitrary n when B is a symmetric strip in [Kh], [Si], [Sco] (see also [DG-E-...], [Bo7]...). The general case is so far open.

Proposition 7.7 was used recently in [BA-G-L] to extend the Stroock-Varadhan support of a diffusion theorem (cf. [I-W]) to the stronger Hölder topology of index $0 < \alpha < \frac{1}{2}$ on Wiener space. This result was obtained independently in [A-K-S] and [M-SS] by other methods. It was further used in [BA-L2] to extend to this topology the Freidlin-Wentzell large deviation principle for small perturbations of dynamical systems. These results may appear as attempts to understand the role of the topolgy in these classical statements. In this direction, the support theorem is established in [G-N-SS] (see also [Me]) for fairly general modulus norms (related to the description of the natural functional norms on the Brownian paths given in [Ci2]). In the context of large deviations, one may wonder for example whether some analogue of Theorem 4.5 holds for diffusion processes. An even more precise result would be a concentration inequality for diffusions.

The next theorem is due to C. Borell [Bo4] in 1977 with a proof using the logconcavity (1.9) of Gaussian measures. We follow here the alternate proof of [S-Z1] based on the correlation inequality (7.16). This result may be used to establish conditional exponential inequalities that allow one to prove existence of Onsager-Machlup functionals for tubes around every element in the Cameron-Martin space.

Theorem 7.8. *Let $(E, \mathcal{H}, \mu)$ be an abstract Wiener space and let $h \in \mathcal{H}$. Then*

$$\lim_{\varepsilon \to 0} \frac{\mu(h + B(0,\varepsilon))}{\mu(B(0,\varepsilon))} = \mathrm{e}^{-|h|^2/2}$$

where $B(0,\varepsilon)$ is the (closed) ball with center the origin and radius $\varepsilon > 0$ for the norm on E. Equivalently, by Cameron-Martin's formula (4.11),

$$\lim_{\varepsilon \to 0} \frac{1}{\mu(B(0,\varepsilon))} \int_{B(0,\varepsilon)} \mathrm{e}^{\tilde{h}} d\mu = 1$$

where we recall that $\widetilde{h} = (j^*{}_{|E_2^*})^{-1}(h)$.

On the Wiener space $C_0([0,1])$ (with the supnorm), if h is an element of the Cameron-Martin Hilbert space, we know that $\widetilde{h} = \int_0^1 h'(t)dW(t)$. As we have seen in Chapter 4, this is still the case for a norm $\|\cdot\|$ on $C_0([0,1]$ such that, for example, $\|x\| \geq C \int_0^1 |x(t)|dt$ for every x in $C_0([0,1]$ and some constant $C > 0$. See [Bog] for further results and improvements in this direction.

Proof. By symmetry and Jensen's inequality, for each $\varepsilon > 0$,

$$\frac{1}{\mu(B(0,\varepsilon))} \int_{B(0,\varepsilon)} \mathrm{e}^{\tilde{h}} d\mu \geq 1.$$

Therefore, it suffices to show that

$$\limsup_{\varepsilon \to 0} \frac{1}{\mu(B(0,\varepsilon))} \int_{B(0,\varepsilon)} \mathrm{e}^{\tilde{h}} d\mu \leq 1. \tag{7.18}$$

It is plain that (7.18) holds when $h = j^*j(\xi)$ for some $\xi \in E^*$, in other words, $\widetilde{h}(\cdot) = j(\xi)(\cdot) = \langle \xi, \cdot \rangle$ (considered as an element of $\mathrm{L}^2(\mu)$). Now, since $\mathcal{H} = j^*(E_2^*)$, where we recall that E_2^* is the closure of E^* in $\mathrm{L}^2(\mu)$ (cf. Chapter 4), there is a sequence $(\xi_n)_{n \in \mathbb{N}}$ in E^* such that $\lim_{n \to \infty} \|\widetilde{h} - j(\xi_n)\|_{\mathrm{L}^2(\mu)} = 0$. By the Cauchy-Schwarz inequality, for every $\varepsilon > 0$ and every n,

$$\int_{B(0,\varepsilon)} \mathrm{e}^{\tilde{h}} d\mu \leq \left(\int_{B(0,\varepsilon)} \mathrm{e}^{2j(\xi_n)} d\mu \right)^{1/2} \left(\int_{B(0,\varepsilon)} \mathrm{e}^{2(\tilde{h} - j(\xi_n))} d\mu \right)^{1/2}.$$

The result will therefore be established if we show that, for every $\varepsilon > 0$ and every k in $\mathcal{H}$,

$$\frac{1}{\mu(B(0,\varepsilon))} \int_{B(0,\varepsilon)} \mathrm{e}^{\tilde{k}} d\mu \leq \int \mathrm{e}^{|\tilde{k}|} d\mu. \tag{7.19}$$

Indeed, if this is the case, let $k = k_n = 2(h - j^* j(\xi_n))$. Then $\widetilde{k}_n$ is a Gaussian random variable on the probability space $(E, \mathcal{B}, \mu)$ with variance $4\|\widetilde{h} - j(\xi_n)\|^2_{L^2(\mu)} \to 0$. Therefore, by the dominated convergence theorem,

$$\lim_{n \to \infty} \int e^{|\widetilde{k}_n|} d\mu = 1.$$

Hence, letting ε tend to zero and then n tend to infinity yields the result. We are left with the proof of (7.19). We will actually establish that, for every $t \geq 0$,

$$\mu\big(|\widetilde{k}| \geq t \,|\, B(0,\varepsilon)\big) \leq \mu\big(|\widetilde{k}| \geq t\big) \tag{7.20}$$

from which (7.19) immediately follows by integration by parts. To this purpose, assume that $|k| = 1$. We may choose an orthonormal basis $(e_i)_{i \geq 1}$ of $\mathcal{H}$ such that $e_1 = k$. Recall from Proposition 4.2 that the $g_i = (j^*{}_{|E_2^*})^{-1}(e_i) = \widetilde{e}_i$, $i \geq 1$, are independent standard Gaussian random variables. By (7.16), for every convex symmetric set B in $\mathbb{R}^n$, and every $t \geq 0$,

$$\mathbb{P}\big\{|g_1| < t, (g_1, \ldots, g_n) \in B\big\} \geq \mathbb{P}\big\{|g_1| < t\big\}\mathbb{P}\big\{(g_1, \ldots, g_n) \in B\big\}.$$

If we let then $B = \{x \in \mathbb{R}^n; \|\sum_{i=1}^n x_i e_i\| \leq \varepsilon\}$, (7.20) immediately follows from this inequality by Proposition 4.2. The proof of Theorem 7.8 is complete. □

Note that the proof of Theorem 7.8 also applies to $|\widetilde{h}|$ and $c\widetilde{h}^2$ (with $c < 1/2|h|^2$) instead of $\widetilde{h}$. With this tool, L. A. Shepp and O. Zeitouni initiated in [S-Z2] the study of Onsager-Machlup functionals for some completely symmetric norms on Wiener space. In [Ca], a general result in this direction is proved for rotational invariant norms with a known small ball behavior (including in particular Hölder norms and various Sobolev type norms).

Notes for further reading. More on small ball probabilities for Gaussian measures may be found in [D-HJ-S] and in the more recent papers [Gri], [K-L2], [K-L-L], [K-L-S], [K-L-T], [Li], [Lif3], [M-R], [Sh], [S-W], [St2]... In particular, in the latter papers, the small ball behaviors are used in the study of rates of convergence in both Strassen's and Chung's law of the iterated logarithm. Some general statements towards this goal are stated in [Ta9]. Recall also the paper [D-L] on Strassen's law of the iterated logarithm for Brownian motion for arbitrary seminorms. See also the recent reference [Lif3]. More on the support of a diffusion theorem, small perturbations of dynamical systems and Onsager-Machlup functionals in stronger topologies on Wiener space can be found in the afore mentioned references [A-K-S], [B-R], [BA-G-L], [BA-L2], [Bog], [Ca], [Ci2], [G-N-SS], [Me], [M-SS], [S-Z1], [S-Z2]...

8. ISOPERIMETRY AND HYPERCONTRACTIVITY

In this last chapter, we further investigate the tight relationships between isoperimetry and semigroup techniques as started in Chapter 2. More precisely, we present some of the semigroup tools which may be used to investigate the isoperimetric inequality in Euclidean and Gauss space. In particular, we will concentrate on the isoperimetric and concentration inequalities for Gaussian measures and show how these relate to hypercontractivity of the Ornstein-Uhlenbeck semigroup. The overwhole approach is inspired by the work of N. Varopoulos in his functional approach to isoperimetric inequalities on groups and manifolds. To better illustrate the scheme of proofs, we start with the classical isoperimetry in $\mathbb{R}^n$ and observe, in particular, that the isoperimetric inequality in $\mathbb{R}^n$ is equivalent to saying that the L^2-norm of the heat semigroup acting on characteristic functions of sets increases under isoperimetric rearrangement. Then, we investigate the analogous situation with respect to the canonical Gaussian measure γ_n. As for the concentration of measure phenomenon, we will discover how the various properties of the Ornstein-Uhlenbeck semigroup such as the commutation property or hypercontractivity can yield in a simple way (a form of) the isoperimetric inequality for Gaussian measures.

Recall from Chapter 1 that the classical isoperimetric inequality in $\mathbb{R}^n$ states that among all compact subsets A with fixed volume $\mathrm{vol}_n(A)$ and smooth boundary ∂A, Euclidean balls minimize the surface measure of the boundary. In other words, whenever $\mathrm{vol}_n(A) = \mathrm{vol}_n(B)$ where B is a ball with some radius r (and $n > 1$),

$$\mathrm{vol}_{n-1}(\partial A) \geq \mathrm{vol}_{n-1}(\partial B). \tag{8.1}$$

Now, $\mathrm{vol}_{n-1}(\partial B) = nr^{n-1}\omega_n$ where ω_n is the volume of the ball of radius 1 so that (8.1) is equivalent to saying that

$$\mathrm{vol}_{n-1}(\partial A) \geq n\omega_n^{1/n}\mathrm{vol}_n(A)^{(n-1)/n}. \tag{8.2}$$

The function $n\omega_n^{1/n}x^{(n-1)/n}$ on $\mathbb{R}^+$ is the isoperimetric function of the classical isoperimetric problem on $\mathbb{R}^n$. Euclidean balls are the extremal sets and achieve equality in (8.2).

It is well-known that (8.2) may be expressed equivalently on functions by means of the coarea formula [Fed], [Maz2], [Os]. After integration by parts (see e.g. [Maz2], p. ...), it yields

$$n\omega^{1/n}\|f\|_{n/n-1} \le \big\||\nabla f|\big\|_1 \tag{8.3}$$

for every C^∞ compactly supported function f on $\mathbb{R}^n$. This inequality is equivalent to (8.2) by letting f approximate the characterisitic function I_A of a set A whose boundary ∂A is smooth enough so that $\int |\nabla f|\,dx$ approaches $\mathrm{vol}_{n-1}(\partial A)$. For simplicity, smoothness properties will be understood in this way here. Inequality (8.3) is due independently to E. Gagliardo [Ga] and L. Nirenberg [Ni] with a nice inductive proof on the dimension. This proof, however, does not seem to yield the optimal constant, and therefore the extremal character of balls. The connection between (8.2) and (8.3) through the coarea formula seems to be due to H. Federer and W. H. Fleming [F-F] and V. G. Maz'ja [Maz1] (cf. [Os]).

Inequality (8.3) of course belongs to the family of Sobolev inequalities. Replacing f (positive) by f^α for some appropriate α easily yields after an application of Hölder's inequality that, for every C^∞ compactly supported function f on $\mathbb{R}^n$,

$$\|f\|_q \le C(n,p,q)\big\||\nabla f|\big\|_p \tag{8.4}$$

with $\frac{1}{q} = \frac{1}{p} - \frac{1}{n}$ and $C(n,p,q) > 0$ a constant only depending on n, p, q, $1 \le p < n$. The family of inequalities (8.4) with $1 < p < n$ goes back to S. Sobolev [So], the inequality for $p = 1$ (which implies the others) having thus been established later on. Of particular interest is the value $p = 2$ which may be expressed equivalently by integration by parts as $(n > 2)$

$$\|f\|_{2n/n-2}^2 \le C\int |\nabla f|^2 dx = C\int f(-\Delta f)dx \tag{8.5}$$

where Δ is the usual Laplacian on $\mathbb{R}^n$. As developed in an abstract setting by N. Varopoulos [Va2] (cf. [Va5], [V-SC-C]), this Dirichlet type inequality (8.5) is closely related to the behavior of the heat semigroup $T_t = \mathrm{e}^{t\Delta}$, $t \ge 0$, as $\|T_t f\|_\infty \le Ct^{-n/2}\|f\|_1$, $t > 0$. We will come back to this below.

Our first task will be to describe, in this concrete setting, some aspects of the semigroup techniques of [Va2], [Va3], and to show how these can yield, in a very simple way, (a form of) the isoperimetric inequality. We will work with the integral representation of the heat semigroup $T_t = \mathrm{e}^{t\Delta}$, $t \ge 0$, as

$$T_t f(x) = \int_{\mathbb{R}^n} f(x + \sqrt{2t}\,y)\,d\gamma_n(y), \quad x \in \mathbb{R}^n, \quad f \in L^1(dx),$$

where γ_n is the canonical Gaussian measure on $\mathbb{R}^n$.

The following proposition is crucial for the understanding of the general principle. Set, for Borel subsets A, B in $\mathbb{R}^n$, and $t \ge 0$,

$$K_t^T(A,B) = \int_B T_t(I_A)dx.$$

A^c denotes below the complement of A.

Proposition 8.1. *For every compact set A in $\mathbb{R}^n$ with smooth boundary ∂A and every $t \geq 0$,*

$$K_t^T(A, A^c) \leq \left(\frac{t}{\pi}\right)^{1/2} \mathrm{vol}_{n-1}(\partial A).$$

Proof. Let f, g be smooth functions on $\mathbb{R}^n$. For every $t \geq 0$, we can write

$$\begin{aligned} \int g\,(T_t f - f)dx &= \int_0^t \left(\int g\,\Delta T_s f\, dx \right) ds \\ &= - \int_0^t \left(\int \langle \nabla T_s g, \nabla f \rangle\, dx \right) ds. \end{aligned}$$

Now, by integration by parts,

$$\nabla T_s g = \frac{1}{\sqrt{2s}} \int_{\mathbb{R}^n} y\, g\big(x + \sqrt{2s}\, y\big) d\gamma_n(y)\,.$$

Hence

$$\int g(T_t f - f)dx = - \int_0^t \frac{1}{\sqrt{2s}} \iint \langle \nabla f(x), y \rangle\, g\big(x + \sqrt{2s}\, y\big) dx d\gamma_n(y) ds.$$

This inequality of course extends to $g = I_{A^c}$. Since

$$\iint \langle \nabla f(x), y \rangle dx d\gamma_n(y) = 0,$$

we see that, for every s,

$$\begin{aligned} - \iint \langle \nabla f(x), y \rangle\, I_{A^c}\big(x + \sqrt{2s}\, y\big) dx d\gamma_n(y) &\leq \iint \big(\langle \nabla f(x), y \rangle\big)^{-} dx d\gamma_n(y) \\ &= \frac{1}{2} \iint \big|\langle \nabla f(x), y \rangle\big| dx d\gamma_n(y) \\ &= \frac{1}{\sqrt{2\pi}} \int |\nabla f| dx \end{aligned}$$

by partial integration with respect with respect to $d\gamma_n(y)$. The conclusion follows since, by letting f approximate I_A, $\int |\nabla f|\, dx$ approaches $\mathrm{vol}_{n-1}(\partial A)$. The proof of Proposition 8.1 is complete. □

Proposition 8.1 is sharp since it may be tested on balls. Namely, if B is an Euclidean ball, one may check that

$$\lim_{t \to 0} \left(\frac{\pi}{t}\right)^{1/2} K_t^T(B, B^c) = \mathrm{vol}_{n-1}(\partial B). \tag{8.6}$$

By translation invariance and homogeneity, one may assume that B is the unit ball with center the origin and radius 1. Then, for $t > 0$,

$$K_t^T(B, B^c) = \int_{\{|x|>1\}} \gamma_n\big(y \in \mathbb{R}^n; \big|x + \sqrt{2t}\, y\big| \leq 1\big)\, dx.$$

Using polar coordinates and the rotational invariance of γ_n,

$$\begin{aligned} K_t^T(B, B^c) &= \int_1^\infty \int_{\omega \in \partial B} \rho^{n-1}\gamma_n\big(y; \big|\rho\omega + \sqrt{2t}\, y\big| \leq 1\big)\, d\rho\, d\omega \\ &= \mathrm{vol}_{n-1}(\partial B) \int_1^\infty \rho^{n-1}\gamma_1 \otimes \gamma_{n-1}\big((y_1, \widetilde{y}); \big|\rho + \sqrt{2t} y_1\big|^2 + 2t|\widetilde{y}|^2 \leq 1\big)\, d\rho \end{aligned}$$

where $y = (y_1, \widetilde{y})$, $y_1 \in \mathbb{R}$, $\widetilde{y} \in \mathbb{R}^{n-1}$. We then use Fubini's theorem to write

$$K_t^T(B, B^c) = \mathrm{vol}_{n-1}(\partial B) \int J_t(y_1, \widetilde{y})\, d\gamma_1(y_1)\, d\gamma_{n-1}(\widetilde{y})$$

where

$$J_t(y_1, \widetilde{y}) = I_{\{2t|\widetilde{y}|^2 \leq 1; \sqrt{2t} y_1 \leq \sqrt{1 - 2t|\widetilde{y}|^2} - 1\}} \int_1^\infty \rho^{n-1} I_{\{|\rho + \sqrt{2t} y_1|^2 \leq 1 - 2t|\widetilde{y}|^2\}}\, d\rho.$$

By a simple integration of the preceding, it is easily seen that

$$\lim_{t \to 0} \frac{1}{\sqrt{t}}\, J_t(y_1, \widetilde{y}) = -\sqrt{2}\, y_1 I_{\{y_1 \leq 0\}}$$

so that, by dominated convergence,

$$\lim_{t \to 0} \frac{1}{\sqrt{t}}\, K_t^T(B, B^c) = -\,\mathrm{vol}_{n-1}(\partial B) \int_{-\infty}^0 \sqrt{2}\, y_1\, d\gamma_1(y_1) = \frac{1}{\sqrt{\pi}}\, \mathrm{vol}_{n-1}(\partial B)$$

which is the claim (8.6).

As a consequence of (8.6), the isoperimetric inequality (8.2) is equivalent to saying that, for every $t \geq 0$ and every compact subset A with smooth boundary, $K_t^T(A, A) \leq K_t^T(B, B)$ whenever B is a ball with the same volume as A. In other words, since $K_t^T(A, A) = \|T_{t/2}(I_A)\|_2^2$,

$$\|T_t(I_A)\|_2 \leq \|T_t(I_D)\|_2, \quad t \geq 0. \tag{8.7}$$

Indeed, under such a property, by Proposition 8.1, for every $t > 0$,

$$\mathrm{vol}_{n-1}(\partial A) \geq \left(\frac{\pi}{t}\right)^{1/2} K_t^T(A, A^c) \geq \left(\frac{\pi}{t}\right)^{1/2} K_t^T(B, B^c)$$

and, when $t \to 0$, $\mathrm{vol}_{n-1}(\partial A) \geq \mathrm{vol}_{n-1}(\partial B)$ by (8.6).

Inequality (8.7) is part of the Riesz-Sobolev rearrangement inequalities (cf. e.g. [B-L-L]). While we noticed its equivalence with isoperimetry, one may wonder for an independent analytic proof of (8.7).

If one does not mind bad constants, one can actually deduce (a form of) isoperimetry from Proposition 8.1 in an elementary way. We will use below this simpler argument in the context of Riemannian manifolds. Note from the uniform estimate $\|T_t f\|_\infty \leq Ct^{-n/2}\|f\|_1$, $t > 0$, that, by interpolation, $\|T_t f\|_2 \leq \sqrt{C}t^{-n/4}\|f\|_1$, $t > 0$, for every f in $L^1(dx)$. Hence, by Proposition 8.1, for every compact subset A in $\mathbb{R}^n$ with smooth boundary ∂A, and every $t > 0$,

$$\begin{aligned}\mathrm{vol}_{n-1}(\partial A) &\geq \left(\frac{\pi}{t}\right)^{1/2} K_t^T(A, A^c)\\ &\geq \left(\frac{\pi}{t}\right)^{1/2}\Big[\mathrm{vol}_n(A) - \|T_{t/2}(I_A)\|_2^2\Big]\\ &\geq \left(\frac{\pi}{t}\right)^{1/2}\Big[\mathrm{vol}_n(A) - C\left(\frac{t}{2}\right)^{-n/2}\mathrm{vol}_n(A)^2\Big].\end{aligned}$$

Optimizing over $t > 0$ then yields

$$\mathrm{vol}_{n-1}(\partial A) \geq C'\mathrm{vol}_n(A)^{(n-1)/n}$$

hence (8.2), with however a worse constant. This easy proof could appear even simpler than the one by E. Gagliardo and L. Nirenberg.

These elementary arguments may be used in the same way in greater generality, for example in Riemannian manifolds. Following [Va2], [Va3], we briefly describe how the arguments should be developed in this case.

It is known ([C-L-Y], [Va1]) that an isoperimetric inequality on a Riemannian manifold M, for example, always forces some control on the heat kernel of M. More precisely, let M be a complete connected Riemannian manifold of dimension N, and, say, noncompact and of infinite volume. Let furthermore Δ be the Laplace-Beltrami operator on M and denote by $(P_t)_{t\geq 0}$ the heat semigroup with kernel $p_t(x,y)$.

Theorem 8.2. *Assume that there exist $n > 1$ and $C > 0$ such that for all compacts subsets A of M with smooth boundary ∂A,*

$$\mathrm{vol}(A)^{(n-1)/n} \leq C\,\mathrm{vol}(\partial A). \tag{8.8}$$

Then, for some constant $C' > 0$,

$$p_t(x,y) \leq \frac{C'}{t^{n/2}} \tag{8.9}$$

for every $t > 0$ and every $x, y \in M$. Furthermore, for each $\delta > 0$, there exists $C_\delta > 0$ such that

$$p_t(x,y) \leq \frac{C_\delta}{t^{n/2}} \exp\left(-\frac{d(x,y)^2}{4(1+\delta)t^2}\right) \tag{8.10}$$

for every $t > 0$ and every $x, y \in M$.

The proof of this theorem is entirely similar to the Euclidean case. We compare both (8.8) and (8.9) on the scale of Sobolev inequalities. One of the main points is the formal equivalence, due to N. Varopoulos [Va2] (cf. [Va5], [V-SC-C] and the references therein), of the L^2-Sobolev inequality (8.5) and the uniform control of the heat semigroup or kernel

$$\|P_t f\|_\infty \le \frac{C}{t^{n/2}} \|f\|_1, \quad t > 0, \quad f \in C_o^\infty(M). \tag{8.11}$$

This result, inspired from the work of J. Nash [Na] and J. Moser [Mo] on the regularity of solutions of parabolic differential equations, is the main link between analysis and geometry. Various techniques then allow one to deduce from the uniform control (8.11) of the kernel the Gaussian off-diagonal estimates (8.10) (cf. [Da], [L-Y], [Va4]...). Theorem 8.2 may be localized in small time (from an isoperimetric inequality on sets of small volume), or in large time [C-F] (sets of large volume).

As we have seen in the classical case, it is sometimes possible to reverse the preceding procedure and to deduce some isoperimetric property from a (uniform) control of the heat kernel. To emphasize the methods rather than the result itself, let us consider only, for simplicity, Riemannian manifolds with nonnegative Ricci curvature. Owing to the Euclidean example, we need to understand how we should complement a Sobolev inequality at the level L^2 (8.5) in order to reach the level L^1 (8.3) and therefore isoperimetry. In this Riemannian setting, this step may be performed with a fundamental inequality due to P. Li et S.-T. Yau [L-Y] in their study of parabolic Harnack inequalities. This inequality is a functional translation of curvature and its proof (see e.g. [Da]) is only based, as in Chapter 2, on Bochner formula and the related curvature-dimension inequalities (cf. Proposition 2.2). We only state it in manifolds with nonnegative Ricci curvature.

Proposition 8.3. *Let M be a Riemannian manifold of dimension N and nonnegative Ricci curvature and let $(P_t)_{t\ge 0}$ be the heat semigroup on M. For every strictly positive function f in $C_o^\infty(V)$ and every $t > 0$,*

$$\frac{|\nabla P_t f|^2}{(P_t f)^2} - \frac{\Delta P_t f}{P_t f} \le \frac{N}{2t}. \tag{8.12}$$

As shown by N. Varopoulos [Va4], one easily deduces from the pointwise inequality (8.12) that, for every f smooth enough and every $t > 0$,

$$\big\| |\nabla P_t f| \big\|_\infty \le \frac{C}{\sqrt{t}} \|f\|_\infty \tag{8.13}$$

for some C only depending on the dimension N, that is a control of the spatial derivatives of the heat kernel. Indeed, according to (8.12), $(\Delta P_t f)^- \le N(2t)^{-1} P_t f$ so that $\|\Delta P_t f\|_1 \le N t^{-1} \|f\|_1$. By duality, $\|\Delta P_t f\|_\infty \le N t^{-1} \|f\|_\infty$. This estimate, used in (8.12) again, then immediately yields (8.13).

The control (8.13) of the gradient of the semigroup in $\sqrt{t}$ (similar to Proposition 8.1) is then the crucial information which, together with (8.11), allows us to reach isoperimetry. Note that the dimension only comes into (8.11) and that (8.13) is in a sense independent of this dimension (besides the constant). We will come back to this comment in the Gaussian setting next. Inequality (8.13) shows that, by duality, for every f in $C_0^\infty(M)$ and every $t > 0$,

$$\|f - P_t f\|_1 \le C\sqrt{t}\, \| |\nabla f| \|_1. \tag{8.14}$$

Indeed, for every smooth function g such that $\|g\|_\infty \le 1$,

$$\begin{aligned}\int g(f - P_t f)\,dx &= -\int_0^t \left(\int g\,\Delta P_s f\,dx\right) ds \\ &= -\int_0^t \left(\int \Delta P_s g\, f\,dx\right) ds \\ &= \int_0^t \left(\int \langle \nabla P_s g, \nabla f\rangle dx\right) ds \le \int_0^t \|\nabla P_s g\|_\infty \| |\nabla f| \|_1\, ds.\end{aligned}$$

Now (8.14) together with (8.11) imply, exactly as in the Euclidean setting, that for some constant $C > 0$ and every compact subset A of M with smooth boundary ∂A,

$$\mathrm{vol}(A)^{(n-1)/n} \le C\,\mathrm{vol}(\partial A),$$

that is the announced isoperimetry. We thus established the following theorem [Va4].

Theorem 8.4. *Let M be a Riemannian manifold with nonnegative Ricci curvature. If for some $n > 1$ and some $C > 0$,*

$$p_t(x, y) \le \frac{C}{t^{n/2}}$$

uniformly in $t > 0$ and $x, y \in M$, then, for some constant $C' > 0$ and every compact subset A of M with smooth boundary ∂A

$$\mathrm{vol}(A)^{(n-1)/n} \le C'\,\mathrm{vol}(\partial A).$$

When the Ricci curvature is only bounded below, the preceding result can only hold locally. In general, the geometry at infinity of the manifold is such that a heat kernel estimate of the type (8.11) (for large t's) only yields isoperimetry for half of the dimension (cf. [C-L] for further details).

A third most important part of the theory concerns the relation of the preceding isoperimetric and Sobolev inequalities with minorations of volumes of balls. We refer to the works of P. Li and S.-T. Yau [L-Y] and N. Varopoulos [Va4], [Va5] and to the monographs [Da], [V-C-SC].

Now, we turn to the Gaussian isoperimetric inequality and the Ornstein-Uhlenbeck semigroup. We already saw in Chapter 2 how this semigroup may be used in order to describe the concentration properties of Gaussian measures. We use it here to try to reach the full isoperimetric statement and base our approach on hypercontractivity. As we will see indeed, hypercontractivity and logarithmic Sobolev inequalities may indeed be considered as analogues of heat kernel bounds and L^2-Sobolev inequalities in this context.

Recall the canonical Gaussian measure γ_n on $\mathbb{R}^n$ with density with respect to Lebesgue measure $\varphi_n(x) = (2\pi)^{-n/2}\exp(-|x|^2/2)$. Recall also that the isoperimetric property for γ_n indicates that if A is a Borel set in $\mathbb{R}^n$ and H is a half-space

$$H = \{x \in \mathbb{R}^n; \langle x, u\rangle \le a\}, \quad |u| = 1, \quad a \in \mathbb{R},$$

such that $\gamma_n(A) = \gamma_n(H) = \Phi(a)$, then, for any real number $r \ge 0$,

$$\gamma_n(A_r) \ge \gamma_n(H_r) = \Phi(a + r).$$

In the applications to hypercontractivity and logarithmic Sobolev inequalities, we will use the Gaussian isoperimetric inequality in its infinitesimal formulation connecting the "Gaussian volume" of a set to the "Gaussian length" of its boundary (which is really isoperimetry). More precisely, given a Borel subset A of $\mathbb{R}^n$, define ([Eh3], [Fed]) the Gaussian Minkowski content of its boundary ∂A as

$$\mathcal{O}_{n-1}(\partial A) = \liminf_{r\to 0} \frac{1}{r}\big[\gamma_n(A_r) - \gamma_n(A)\big].$$

If ∂A is smooth, $\mathcal{O}_{n-1}(\partial A)$ may be obtained as the integral of the Gaussian density along ∂A (see below). In this langague, the isoperimetric inequality then expresses that if H is a half-space with the same measure as A, then

$$\mathcal{O}_{n-1}(\partial A) \ge \mathcal{O}_{n-1}(\partial H).$$

Now, one may easily compute (in dimension one) the Minkowski content of a half-space as

$$\mathcal{O}_{n-1}(\partial H) = \liminf_{r\to 0} \frac{1}{r}\big[\Phi(a+r) - \Phi(a)\big] = \varphi_1(a)$$

where $\Phi(a) = \gamma_n(H) = \gamma_n(A)$ and where $\varphi_1(x) = (2\pi)^{-1/2}\exp(-x^2/2)$, $x \in \mathbb{R}$. Hence, denoting by Φ^{-1} the inverse function of Φ, we get that for every Borel set A in $\mathbb{R}^n$,

$$\mathcal{O}_{n-1}(\partial A) \ge \varphi_1 \circ \Phi^{-1}\big(\gamma_n(A)\big). \tag{8.15}$$

The function $\varphi_1 \circ \Phi^{-1}$ is the isoperimetric function of the Gauss space $(\mathbb{R}^n, \gamma_n)$. It may be compared to the function $n\omega^{1/n}x^{(n-1)/n}$ of the classical isoperimetric inequality in $\mathbb{R}^n$. The function $\varphi_1 \circ \Phi^{-1}$ is still concave; it is defined on $[0,1]$, is symmetric with respect to the vertical line going through $\frac{1}{2}$ with a maximum there equal to $(2\pi)^{-1/2}$, and its behavior at the origin (or at 1 by symmetry) is governed by the equivalence

$$\lim_{x\to 0} \frac{\varphi_1 \circ \Phi^{-1}(x)}{x(2\log(1/x))^{1/2}} = 1. \tag{8.16}$$

This can easily be established by noting that the derivative of $\varphi_1 \circ \Phi^{-1}$ is $-\Phi^{-1}$ and by comparing $\Phi^{-1}(x)$ to $(2\log(1/x))^{1/2}$.

As in the classical case, (8.15) may be expressed equivalently on functions by means, again, of the coarea formula (see [Fed], [Maz2], [Eh3]). Writing for a smooth function f on $\mathbb{R}^n$ with gradient ∇f that

$$\int |\nabla f| d\gamma_n = \int_0^\infty \left(\int_{C_s} \varphi_n(x)\, d\mathcal{H}_{n-1}(x) \right) ds$$

where $C_s = \{x \in \mathbb{R}^n; |f(x)| = s\}$ and where $d\mathcal{H}_{n-1}$ is the Hausdorff measure of dimension $n-1$ on C_s, we deduce from (8.15) that

$$\int |\nabla f| d\gamma_n \geq \int_0^\infty \varphi_1 \circ \Phi^{-1}\big(\gamma_n(|f| \geq s)\big) ds. \tag{8.17}$$

When f is a smooth function approximating the indicator function of a set A, we of course recover (8.15) from (8.17), at least for subsets A with smooth boundary. Due to the equivalence (8.16), one sees in particular on (8.17) that a smooth function f satisfying $\int |\nabla f| d\gamma_n < \infty$ is such that $\int |f| (\log(1 + |f|))^{1/2} d\gamma_n < \infty$. Indeed, we first see from (8.17) and (8.16) that for every s_0 large enough

$$\int |\nabla f| d\gamma_n \geq \int_{s_0}^\infty \gamma_n\big(|f| \geq s\big) ds$$

from which $\int |f| d\gamma_n \leq C < \infty$ by the classical integration by parts formula. For every $s \geq 0$, $\gamma_n(|f| \geq s) \leq C/s$ so that, by (8.17) and (8.16) again, for every large s_0,

$$\int |\nabla f| d\gamma_n \geq \int_{s_0}^\infty \gamma_n\big(|f| \geq s\big) \big(\log(s/C)\big)^{1/2} ds$$

from which the claim immediately follows. In analogy with (8.3), such an inequality belongs to the family of Sobolev inequalities, but here of logarithmic type.

It is plain that inequalities (8.15) and (8.17) have analogues in infinite dimension for the appropriate notions of surface measure and gradient (as we did for example with concentration in Chapter 4). Again, the crucial inequalities are the ones in finite dimension.

We showed in Chapter 2 how the Ornstein-Uhlenbeck semigroup $(P_t)_{t \geq 0}$ (and for the large values of the time t) may be used to investigate the concentration phenomenon of Gaussian measures. Our purpose here will be to show, in the same spirit as what we presented in the classical case, that the behavior of $(P_t)_{t \geq 0}$ for the small values of t together with its hypercontractivity property may properly be combined to yield (a version of) the infinitesimal version (8.15) of the isoperimetric inequality. More precisely, we will show, with these tools, that there exists a small enough numerical constant $0 < c < 1$ such that for every A with smooth boundary,

$$\mathcal{O}_{n-1}(\partial A) \geq c\, \varphi_1 \circ \Phi^{-1}\big(\gamma_n(A)\big).$$

We doubt that this approach can lead to the exact constant $c = 1$. The line of reasoning will follow the one of the classical case, simply replacing actually the classical heat semigroup estimates and Sobolev inequalities on $\mathbb{R}^n$ by the hypercontractivity property and logarithmic Sobolev inequalities of the Ornstein-Uhlenbeck semigroup. We follow [Led5] and now turn to hypercontractivity and logarithmic Sobolev inequalities.

Let $(W(t))_{t\geq 0}$ be a standard Brownian motion starting at the origin with values in $\mathbb{R}^n$. Consider the stochastic differential equation

$$dX(t) = \sqrt{2}\,dW(t) - X(t)dt$$

with initial condition $X(0) = x$, whose solution simply is

$$X(t) = \mathrm{e}^{-t}\Big(x + \sqrt{2}\int_0^t \mathrm{e}^s dW(s)\Big), \quad t \geq 0.$$

Since $\sqrt{2}\int_0^t \mathrm{e}^s dW(s)$ has the same distribution as $W(\mathrm{e}^{2t}-1)$, the Markov semigroup $(P_t)_{t\geq 0}$ of $(X(t))_{t\geq 0}$ is given by

$$(8.18)\quad P_t f(x) = \mathbb{E}\big(f(\mathrm{e}^{-t}x + \mathrm{e}^{-t}W(\mathrm{e}^{2t}-1))\big) = \int_{\mathbb{R}^n} f\big(\mathrm{e}^{-t}x + (1-\mathrm{e}^{-2t})^{1/2}y\big)\,d\gamma_n(y)$$

for any f in $\mathrm{L}^1(\gamma_n)$ (for example), thus defining the Ornstein-Uhlenbeck or Hermite semigroup with respect to the Gaussian measure γ_n. As we have seen in Chapter 2, $(P_t)_{t\geq 0}$ is a Markovian semigroup of contractions on all $\mathrm{L}^p(\gamma_n)$-spaces, $1 \leq p \leq \infty$, symmetric and invariant with respect to γ_n, and with generator L which acts on each smooth function f on $\mathbb{R}^n$ as $Lf(x) = \Delta f(x) - \langle x, \nabla f(x)\rangle$. The generator L satisfies the integration by parts formula with respect to γ_n

$$\int f(-Lg)\,d\gamma_n = \int \langle \nabla f, \nabla g\rangle\,d\gamma_n$$

for every smooth functions f, g on $\mathbb{R}^n$.

One of the remarkable properties of the Ornstein-Uhlenbeck semigroup is the hypercontractivity property discovered by E. Nelson [Nel]: whenever $1 < p < q < \infty$ and $t > 0$ satisfy $\mathrm{e}^t \geq [(q-1)/(p-1)]^{1/2}$, then, for all functions f in $\mathrm{L}^p(\gamma_n)$,

$$(8.19)\qquad \|P_t f\|_q \leq \|f\|_p$$

where (now) $\|\cdot\|_p$ is the norm in $\mathrm{L}^p(\gamma_n)$. In other words, P_t maps $L^p(\gamma_n)$ in $L^q(\gamma_n)$ $(q > p)$ with norm one. Many simple proofs of (8.19) have been given in the literature (see [Gr4]), mainly based on its equivalent formulation as logarithmic Sobolev inequalities due to L. Gross [Gr3]. Fix $p = 2$ and take $q(t) = 1 + \mathrm{e}^{2t}$, $t \geq 0$. Given a smooth function f, set $\Psi(t) = \|P_t f\|_{q(t)}$ where $q(t) = 1 + \mathrm{e}^{2t}$. Under the hypercontractivity property (2.2), $\Psi(t) \leq \Psi(0)$ for every $t \geq 0$ and thus $\Psi'(0) \leq 0$. Performing this differentiation, we see that

$$(8.20)\qquad \begin{aligned} &\int f^2 \log|f|\,d\gamma_n - \int f^2 d\gamma_n \log\Big(\int f^2 d\gamma_n\Big)^{1/2} \\ &\qquad\qquad \leq \int |\nabla f|^2\,d\gamma_n \quad \Big(= \int f(-\mathrm{L}f)d\gamma_n\Big) \end{aligned}$$

which in turn implies (8.19) by applying it to $(P_t f)^p$ instead of $f \geq 0$ for every t and every $p \geq 1$ (cf. [B-É]). The inequality (8.20) is called a logarithmic Sobolev inequality. One may note, with respect to the classical Sobolev inequalities on $\mathbb{R}^n$, that it is only of logarithmic type, with however constants independent of the dimension, a characteristic feature of Gaussian measures.

Simple proofs of (8.20) may be found in e.g. [Ne3], [A-C], [B-É], [Bak]... (cf. [Gr4]). The one which we present now for completeness already appeared in [Led3] and only relies (see also [B-É]) on the commutation property (2.6)

$$\nabla P_t f = \mathrm{e}^{-t} P_t(\nabla f).$$

That is, the proof we will give of hypercontractivity relies on exactly the same argument which allowed us to describe the concentration of γ_n in the form of (2.7) through Proposition 2.1 and is actually very similar. We will come back to this important point. In order to establish (8.20), replacing f (positive, or better such that $0 < a \leq f \leq b$ for constants a, b) by $\sqrt{f}$, it is enough to show that

$$\int f \log f \, d\gamma_n - \int f \, d\gamma_n \log \left(\int f \, d\gamma_n \right) \leq \frac{1}{2} \int \frac{1}{f} |\nabla f|^2 \, d\gamma_n. \tag{8.21}$$

To this aim, we can write by the semigroup properties and integration by parts that

$$\begin{aligned}
\int f \log f \, d\gamma_n - \int f \, d\gamma_n \log \left(\int f \, d\gamma_n \right) &= - \int_0^\infty \left(\frac{d}{dt} \int P_t f \log P_t f \, d\gamma_n \right) dt \\
&= - \int_0^\infty \left(\int L P_t f \log P_t f \, d\gamma_n \right) dt \\
&= \int_0^\infty \left(\int \langle \nabla P_t f, \nabla (\log P_t f) \rangle \, d\gamma_n \right) dt \\
&= \int_0^\infty \left(\int \frac{1}{P_t f} |\nabla P_t f|^2 \, d\gamma_n \right) dt.
\end{aligned}$$

Now, setting

$$F(t) = \int \frac{1}{P_t f} |\nabla P_t f|^2 \, d\gamma_n \quad t \geq 0,$$

the commutation property $\nabla P_t f = \mathrm{e}^{-t} P_t(\nabla f)$ and Cauchy-Schwarz inequality on the integral representation of P_t show that, for every $t \geq 0$,

$$\begin{aligned}
F(t) &= \mathrm{e}^{-2t} \sum_{i=1}^n \int \frac{1}{P_t f} \left(P_t \frac{\partial f}{\partial x_i} \right)^2 d\gamma_n \\
&\leq \mathrm{e}^{-2t} \sum_{i=1}^n \int P_t \left(\frac{1}{f} \left(\frac{\partial f}{\partial x_i} \right)^2 \right) d\gamma_n = \mathrm{e}^{-2t} \int \frac{1}{f} |\nabla f|^2 \, d\gamma_n
\end{aligned}$$

which immediately yields (8.21). Therefore, hypercontractivity is established in this way.

While our aim is to investigate isoperimetric inequalities via semigroup techniques, it is of interest however to notice that the Gaussian isoperimetric inequality (8.15) or (8.17) may be used to establish the logarithmic Sobolev inequality (8.20) and therefore hypercontractivity. This was observed in [Led1] in analogy with the classical case discussed in the first part of this chapter. Let f be a smooth positive function on $\mathbb{R}^n$ with $\|f\|_2 = 1$. Apply then (8.17) to $g = f^2(\log(1+f^2))^{1/2}$. Using (8.16), one obtains after some elementary, although cumbersome, computations that for every $\varepsilon > 0$ there exists $C(\varepsilon) > 0$ only depending on ε such that

$$\int f^2 \log(1+f^2)\, d\gamma_n \le (1+\varepsilon)\Big(2\int |\nabla f|^2 d\gamma_n\Big)^{1/2}\Big(\int f^2 \log(1+f^2)\, d\gamma_n + 2\Big)^{1/2} + C(\varepsilon).$$

It follows that

$$\begin{aligned} 2\int f^2 \log f\, d\gamma_n &\le \int f^2 \log(1+f^2)\, d\gamma_n \\ &\le 2(1+\varepsilon)^4 \int |\nabla f|^2 d\gamma_n + 2(1+\varepsilon)^2\Big(\int |\nabla f|^2 d\gamma_n\Big)^{1/2} + C'(\varepsilon) \end{aligned}$$

where $C'(\varepsilon) = (1+\varepsilon)C(\varepsilon)/\varepsilon$. To get rid of the extra terms on the left of this inequality, we use a tensorization argument of A. Ehrhard [Eh4]: this inequality namely holds with constants independent of the dimension n; therefore, applying it to $f^{\otimes k}$ in $(\mathbb{R}^n)^k = \mathbb{R}^{nk}$ yields

$$k\int f^2 \log f\, d\gamma_n \le k(1+\varepsilon)^4 \int |\nabla f|^2 d\gamma_n + \sqrt{k}\,(1+\varepsilon)^2\Big(\int |\nabla f|^2 d\gamma_n\Big)^{1/2} + C'(\varepsilon).$$

Divide then by k, let k tend to infinity and then ε to zero and we obtain (8.20).

Now, we would like to try to understand how hypercontractivity and logarithmic Sobolev inequalities may be used in order to reach isoperimetry in this Gaussian setting. Of course, our approach to known results and theorems is only formal, but it could be of some help in more abstract frameworks.

Before turning to the main argument, let us briefly discuss, on two specific questions, why hypercontractiviy should be of potential interest to isoperimetry and concentration. The following comments are not presented in the greatest generality.

Recall the Hermite polynomials $\{\sqrt{k!}\,h_k; k \in \mathbb{N}\}$ which forms an orthonormal basis of $\mathrm{L}^2(\gamma_1)$ (cf. the introduction of Chapter 5). In the same way, for any fixed $n \ge 1$, set, for every $\underline{k} = (k_1, \ldots, k_n) \in \mathbb{N}^n$ and every $x = (x_1, \ldots, x_n) \in \mathbb{R}^n$,

$$H_{\underline{k}}(x) = \prod_{i=1}^{n} \sqrt{k_i!}\, h_{k_i}(x_i).$$

Then, $\{H_{\underline{k}}; \underline{k} \in \mathbb{N}^n\}$ is an orthonormal basis of $\mathrm{L}^2(\gamma_n)$. Therefore, as in Chapter 5 in greater generality, a function f in $\mathrm{L}^2(\gamma_n)$ can be written as

$$f = \sum_{\underline{k} \in \mathbb{N}^n} f_{\underline{k}} H_{\underline{k}}$$

where $f_{\underline{k}} = \int f H_{\underline{k}} d\gamma_n$. This sum may also be written as

$$f = \sum_{d=0}^{\infty} \Big(\sum_{|\underline{k}|=d} f_{\underline{k}} H_{\underline{k}} \Big) = \sum_{d=0}^{\infty} \Psi^{(d)}(f)$$

where $|\underline{k}| = k_1 + \cdots + k_n$. $\Psi^{(d)}(f)$ is known as the chaos of degree d of f. Since $h_0 \equiv 1$, $\Psi^{(0)}(f)$ is simply the mean of f; $h_1(x) = x$, so chaos of order or degree 1 are (in probabilistic notation) Gaussian sums $\sum_{i=1}^{n} g_i a_i$ (where $(g_1, \ldots, g_n)$ are independent standard Gaussian random variables and a_i real numbers) etc (cf. Chapter 5).

Now, it is easily seen that, for every $t \geq 0$,

$$P_t \Psi^{(d)}(f) = \mathrm{e}^{-dt} \Psi^{(d)}(f). \tag{8.22}$$

But we can then apply the hypercontractivity property of $(P_t)_{t \geq 0}$. Fix for example $p = 2$ and let $q = q(t) = 1 + \mathrm{e}^{2t}$. Then, combining (8.22) and (8.19) we get that, for every $q \geq 2$ or $t \geq 0$,

$$(q-1)^{-d/2} \big\| \Psi^{(d)}(f) \big\|_q = \mathrm{e}^{-dt} \big\| \Psi^{(d)}(f) \big\|_q = \big\| P_t \Psi^{(d)}(f) \big\|_q \leq \big\| \Psi^{(d)}(f) \big\|_2 \tag{8.23}$$

The next step in this development is that (8.23) applies in the same way to vector valued functions. Let E be a Banach space with norm $\|\cdot\|$. Given f on $\mathbb{R}^n$ with values in E, the previous chaotic decomposition is entirely similar. We need then simply apply hypercontractivity to the real valued function $\|f\|$ and Jensen's inequality immediately shows that (8.19) also holds for E-valued functions, with the $\mathrm{L}^p(\gamma_n)$-norms replaced by $\mathrm{L}^p(\gamma_n; E)$-norms. In particular, if $e_1, \ldots, e_n$ are elements of E, the vector valued version of (8.23) for $d = 1$ for example implies that, for every $q \geq 2$,

$$\Big\| \sum_{i=1}^{n} g_i e_i \Big\|_q \leq (q-1)^{1/2} \Big\| \sum_{i=1}^{n} g_i e_i \Big\|_2 . \tag{8.24}$$

These inequalities are exactly the moment equivalences (4.5) which we obtain next to Theorem 4.1, with the same behavior of the constant as q increases to infinity (and with even a better numerical value). Since this constant is independent of n, it is not difficult to see (although we will not go into these details) that (8.24) essentially allows us to recover the integrability properties and tail behaviors of norms of Gaussian random vectors (Theorem 4.1) as well as of Wiener chaos (cf. Chapter 5). This very interesting and powerful line of reasoning was extensively developed by C. Borell to which we refer the interested reader ([Bo8], [Bo9]). Note that these hypercontractivity ideas may also be used in the context of the two point space to recover, for example, inequalities (3.6) [Bon], [Bo6].

Recently, a parallel approach was developed by S. Aida, T. Masuda and I. Shigekawa [A-M-S], but on the basis of logarithmic Sobolev inequalities rather than hypercontractivity. As we already noticed it, we established both the concentration of measure phenomenon for γ_n (Proposition 2.1) and the logarithmic Sobolev inequality (8.20) on the basis of the same commutation property $\nabla P_t f = \mathrm{e}^{-t} P_t(\nabla f)$ of the Ornstein-Uhlenbeck semigroup. In [A-M-S], it is actually shown that concentration follows from a logarithmic Sobolev inequality and hypercontractivity. Although the paper [A-M-S] is concerned with logarithmic Sobolev inequalities in an abstract Dirichlet space setting, let us restrict again to the Gaussian case to sketch the idea and show how (2.7) may be deduced from the logarithmic Sobolev inequality. (The implication is thus only formal, as this whole chapter actually.) Let thus f be a Lipschitz map on $\mathbb{R}^n$ with $\|f\|_{\mathrm{Lip}} \leq 1$ and mean zero. Let us apply the logarithmic Sobolev inequality (8.20) to $e^{\lambda f/2}$ for every $\lambda \in \mathbb{R}$. Setting

$$\varphi(\lambda) = \int e^{\lambda f} d\gamma_n, \quad \lambda \in \mathbb{R},$$

we see that

$$\lambda\varphi'(\lambda) - \varphi(\lambda)\log\varphi(\lambda) \leq \tfrac{1}{2}\,\lambda^2\varphi(\lambda), \quad \lambda \in \mathbb{R}.$$

We need then simply integrate this differential inequality (this was first done in [Da-S], originally by I. Herbst). Set $\psi(\lambda) = \frac{1}{\lambda}\log\varphi(\lambda)$, $\lambda > 0$. Hence, for every $\lambda > 0$, $\psi'(\lambda) \leq \frac{1}{2}$. Since $\psi(0) = \varphi'(0)/\varphi(0) = \int f d\gamma_n = 0$, it follows that

$$\psi(\lambda) \leq \frac{\lambda}{2}$$

for every $\lambda \geq 0$. Therefore, we have obtained (2.7), that is

$$\int e^{\lambda f} d\gamma_n \leq e^{\lambda^2/2}$$

for every $\lambda \geq 0$ and, replacing f by $-f$, also for all $\lambda \in \mathbb{R}$.

As we discussed it in Chapter 2, there is however a long way from concentration to true isoperimetry. To complete this chapter, we turn to the isoperimetric inequality (8.15) itself which we would like to analyze with the Ornstein-Uhlenbeck semigroup as we did in the classical case in the first part of this chapter. The next proposition, implicit in [Pil, p. 180], is the first step towards our goal and is the analogue of Proposition 8.1. Given Borel sets A, B in $\mathbb{R}^n$ and $t \geq 0$, we set

$$K_t(A, B) = \int_A P_t(I_B)\, d\gamma_n.$$

Note that $K_t(A, A) = \|P_{t/2}(I_A)\|_2^2$. The notation K_t is used in analogy with that of a kernel. Large deviation estimates of the kernel $K_t(A, B)$ for the Wiener measure when $d(A, B) > 0$ are developed at the end of Chapter 4.

Proposition 8.5. *For every Borel set A in $\mathbb{R}^n$ with smooth boundary ∂A and every $t \geq 0$,*

$$K_t(A, A^c) \leq (2\pi)^{-1/2} \arccos(e^{-t})\, \mathcal{O}_{n-1}(\partial A).$$

Proof. It is similar to the proof of Proposition 8.1. Let f, g be smooth functions on $\mathbb{R}^n$. For every $t \geq 0$, we can write

$$\int g(P_t f - f)\, d\gamma_n = \int_0^t \left(\int g\, LP_s f\, d\gamma_n \right) ds$$
$$= - \int_0^t \left(\int \langle \nabla P_s g, \nabla f \rangle d\gamma_n \right) ds.$$

Now, by integration by parts on the representation of P_s using the Gaussian density,

$$\nabla P_s f = \frac{e^{-s}}{(1 - e^{-2s})^{1/2}} \int_{\mathbb{R}^n} y\, g\big(e^{-s} x + (1 - e^{-2s})^{1/2} y\big)\, d\gamma_n(y).$$

Hence

$$\int g(P_t f - f)\, d\gamma_n$$
$$= - \int_0^t \frac{e^{-s}}{(1 - e^{-2s})^{1/2}} \iint \langle \nabla f(x), y \rangle\, g\big(e^{-s} x + (1 - e^{-2s})^{1/2} y\big)\, d\gamma_n(x) d\gamma_n(y) ds.$$

This identity of course extends to $g = I_{A^c}$. Since

$$\iint \langle \nabla f(x), y \rangle d\gamma_n(x) d\gamma_n(y) = 0,$$

we see that, for every s,

$$- \iint \langle \nabla f(x), y \rangle I_{A^c}\big(e^{-s} x + (1 - e^{-2s})^{1/2} y\big) d\gamma_n(x) d\gamma_n(y)$$
$$\leq \iint \big(\langle \nabla f(x), y \rangle\big)^- d\gamma_n(x) d\gamma_n(y)$$
$$= \frac{1}{2} \iint \big|\langle \nabla f(x), y \rangle\big| d\gamma_n(x) d\gamma_n(y)$$
$$= \frac{1}{\sqrt{2\pi}} \int |\nabla f| d\gamma_n .$$

The conclusion follows by letting f approximate I_A since then $\int |\nabla f| d\gamma_n$ will approach $\mathcal{O}_{n-1}(\partial A)$ when ∂A is smooth enough. Proposition 8.5 is established. □

The inequality of the proposition is sharp in many respects. When $t \to \infty$, it reads

$$\mathcal{O}_{n-1}(\partial A) \geq 2 \left(\frac{2}{\pi} \right)^{1/2} \gamma_n(A)\big(1 - \gamma_n(A)\big), \tag{8.25}$$

that is, when $\gamma_n(A) = \frac{1}{2}$, the maximum of the function $\varphi_1 \circ \Phi^{-1}(x)$ at $x = \frac{1}{2}$. Inequality (8.25) may actually be interpreted as Cheeger's isoperimetric constant [Ch] of the Gauss space $(\mathbb{R}^n, \gamma_n)$. It is responsible for the optimal factor $\pi/2$ which appears in the vector valued inequalities (4.10). Indeed, one may integrate (8.25) by the coarea formula (see [Ya]) to get that for every smooth function f with mean zero,

$$\int |f|\, d\gamma_n \le \left(\frac{\pi}{2}\right)^{1/2} \int |\nabla f|\, d\gamma_n ,$$

an inequality which is easily seen to be best possible (take $n = 1$ and f on $\mathbb{R}$ be defined by $f(x) = x/\varepsilon$ for $|x| \le \varepsilon$, $f(x) = x/|x|$ elsewhere, and let $\varepsilon \to 0$).

Proposition 8.5 may also be tested on half-spaces, as we did on balls in the classical case. Namely, if we let $H = \{x \in \mathbb{R}^n ; \langle x, u\rangle \le a\}$, $|u| = 1$, $a \in \mathbb{R}$, it is easily checked (start in dimension one and use polar coordinates) that

$$\begin{aligned} K_t(H, H^c) &= \frac{1}{2\pi} \iint_{\mathbb{R}^2} \mathrm{e}^{-x^2/2} \mathrm{e}^{-y^2/2} I_{\{x \le |a|,\, \mathrm{e}^{-t}x + (1-\mathrm{e}^{-2t})^{1/2} y > |a|\}} dx dy \\ &= \frac{1}{2\pi} \int_0^{2\pi} \int_0^\infty \rho\, \mathrm{e}^{-\rho^2/2} I_{\{\rho \sin(\varphi) \le |a|,\, \rho \sin(\varphi + \theta) > |a|\}} d\varphi d\rho \\ &= \frac{\theta}{2\pi} \mathrm{e}^{-a^2/2} - \frac{1}{2\pi} \int_{|a|}^{|a|/\sin((\pi-\theta)/2)} \big(2 \arcsin\big(\rho^{-1}|a|\big) + \theta - \pi\big) \rho\, \mathrm{e}^{-\rho^2/2} d\rho \end{aligned}$$

where $\theta = \arccos(\mathrm{e}^{-t})$. The absolute value of the second term of the latter may be bounded by

$$\frac{\theta}{2\pi}\left(\mathrm{e}^{-a^2/2} - \mathrm{e}^{-a^2/2\cos^2(\theta/2)}\right) \le \frac{\theta}{2\pi} \cdot \frac{a^2}{2} \tan^2\left(\frac{\theta}{2}\right) \mathrm{e}^{-a^2/2} \le \frac{\theta^3}{2\pi}\, a^2\, \mathrm{e}^{-a^2/2}$$

at least for all θ small enough. In particular, since $\theta = \arccos(\mathrm{e}^{-t})$ and thus $\theta \sim \sqrt{2t}$ when $t \to 0$, it follows that

$$\lim_{t \to 0} (2\pi)^{1/2} \left[\arccos(\mathrm{e}^{-t})\right]^{-1} K_t(H, H^c) = \mathcal{O}_{n-1}(\partial H). \tag{8.26}$$

On the basis of Proposition 8.5, we now would need lower estimates of the functional $K_t(A, A^c)$ for the small values of t. The typical isoperimetric approach would be to use a symmetrization result of C. Borell [Bo10], analogous to (8.7), asserting that if H is a half-space with the same measure as A, then for every $t \ge 0$,

$$K_t(A, A) \le K_t(H, H). \tag{8.27}$$

Hence $K_t(A, A^c) \ge K_t(H, H^c)$ and we would conclude from Proposition 8.5 and (8.26) that

$$\mathcal{O}_{n-1}(\partial A) \ge \mathcal{O}_{n-1}(\partial H).$$

In particular, and as in the classical case, isoperimetry is therefore equivalent to saying that

$$\|P_t(I_A)\|_2 \le \|P_t(I_H)\|_2, \quad t \ge 0, \tag{8.28}$$

for H a half-space with the same measure as A. This inequality is established in [Bo10], extending ideas of [Eh2] on Gaussian symmetrization and based on Baernstein's transformation [Ba] developed in the classical case. Borell's techniques also apply to some other diffusion processes [Bo11], [Bo12]. Inequality (8.28) may also be seen to follow from rearrangement inequalities on the sphere [B-T] via Poincaré's limit as was shown in [Ca-L] following indications of [Be3].

Our approach to bound $K_t(A,A)$ will be to use hypercontractivity as the corresponding semigroup estimate in this Gaussian setting. Namely, we simply write for A a Borel set in $\mathbb{R}^n$ and $p(t) = 1 + e^{-t}$ that

$$K_t(A,A) = \|P_{t/2}(I_A)\|_2^2 \le \|I_A\|_{p(t)}^2, \quad t \ge 0. \tag{8.29}$$

Hence

$$K_t(A,A^c) \ge \gamma_n(A)\big[1 - \gamma_n(A)^{(2/p(t))-1}\big].$$

Therefore, combined with Proposition 8.5,

$$\mathcal{O}_{n-1}(\partial A) \ge (2\pi)^{1/2}\gamma_n(A)\sup_{t>0}\big[\big(\arccos(e^{-t})\big)^{-1}\big(1 - \gamma_n(A)^{(2/p(t))-1}\big)\big].$$

Setting $\theta = \arccos(e^{-t}) \in (0, \frac{\pi}{2}]$ we need to evaluate

$$\sup_{0<\theta\le\frac{\pi}{2}} \frac{1}{\theta}\left[1 - \exp\left(-\frac{1-\cos\theta}{1+\cos\theta}\log\frac{1}{\gamma_n(A)}\right)\right].$$

To this aim, we can note for example that

$$\frac{1-\cos\theta}{1+\cos\theta} \ge \frac{\theta^2}{2\pi},$$

and choosing thus θ of the form

$$\theta = (2\pi)^{1/2}\left(\log\frac{1}{\gamma_n(A)}\right)^{-1/2},$$

provided that $\gamma_n(A) \le e^{-8/\pi}$, we find that

$$\mathcal{O}_{n-1}(\partial A) \ge \left(1 - \frac{1}{e}\right)\gamma_n(A)\left(\log\frac{1}{\gamma_n(A)}\right)^{1/2}.$$

Due to the equivalence (8.16), there exists $\delta > 0$ such that when $\gamma_n(A) \le \delta$,

$$\mathcal{O}_{n-1}(\partial A) \ge \frac{1}{3}\varphi_1 \circ \Phi^{-1}\big(\gamma_n(A)\big).$$

When $\delta < \gamma_n(A) \le 1/2$, we can always use (8.25) to get

$$\mathcal{O}_{n-1}(\partial A) \ge \left(\frac{\pi}{2}\right)\gamma_n(A) \ge c(\delta)\varphi_1 \circ \Phi^{-1}\big(\gamma_n(A)\big)$$

for some $c(\delta) > 0$. These two inequalities, together with symmetry, yield that, for some numerical constant $0 < c < 1$ and all subsets A in $\mathbb{R}^n$ with smooth boundary,

$$\mathcal{O}_{n-1}(\partial A) \ge c\,\varphi_1 \circ \Phi^{-1}\big(\gamma_n(A)\big). \tag{8.30}$$

One may try to tighten the preceding computations to reach the value $c = 1$ in (8.30). This however does not seem likely and it is certainly in the hypercontractive estimate (8.29) that a good deal of the best constant is lost. One may wonder why this is the case. It seems that hypercontractivity, while an equality on exponential functions, is perhaps not that sharp on indicator functions. This would have to be understood in connection with (8.28). Note finally that one may easily integrate back (8.30) to obtain, with these functional tools, the following analogue of the Gaussian isoperimetric inequality: if $\gamma_n(A) \geq \Phi(a)$, for every $r \geq 0$,

$$\gamma_n(A_r) \geq \Phi(a + cr). \tag{8.31}$$

We briefly sketch one argument taken from [Bob1] (where a related equivalent functional formulation of the Gaussian isoperimetry is studied) that works for arbitrary measurable sets A. First, if $f_r(x) = (1 - d(x, A_r)/2r)^+$, $r > 0$,

$$\lim_{r \to 0} \int |\nabla f_r| d\gamma_n = \mathcal{O}_{n-1}(\partial A) = \liminf_{r \to 0} \frac{1}{r}\left[\gamma_n(A_r) - \gamma_n(A)\right]$$

if A is closed. In general, one may note that $\mathcal{O}_{n-1}(\partial A) = \mathcal{O}_{n-1}(\partial \bar{A})$ if $\gamma_n(A) = \gamma_n(\bar{A})$ and $\mathcal{O}_{n-1}(\partial A) = \infty$ if not. Now, the family of functions

$$R_r(p) = \Phi\big(\Phi^{-1}(p) + r\big), \quad 0 \leq p \leq 1, \quad r \geq 0,$$

satisfy $R_{r_1} \circ R_{r_2} = R_{r_1 + r_2}$, $r_1, r_2 \geq 0$. Similarly, $(A_{r_1})_{r_2} = A_{r_1 + r_2}$. Therefore, if (8.31) holds for r_1 and r_2, then it also holds for $r_1 + r_2$. Hence, (8.31) is satisfied as soon as it is satisfied for all $r > 0$ small enough and this is actually given by (8.30) since the derivative of $\Phi^{-1}(\gamma_n(A_r))$ is

$$\mathcal{O}_{n-1}(\partial A_r) / \varphi_1 \circ \Phi^{-1}\big(\gamma_n(A_r)\big).$$

To be more precise, one should actually work out this argument with the functions $\Phi(x/\sigma)$, $\sigma > 1$, and let then σ tend to one. We refer to [Bob1] for all the details.

It is likely that the preceding approach has some interesting consequences in more abstract settings.

It might be worthwhile noting finally that Ehrhard's tensorization argument together with symmetrization may also be used to establish directly hypercontractivity, a comment we learned from C. Borell. One approach through logarithmic Sobolev inequalities is developed in [Eh4]. Alternatively, by the result of [Bo10],

$$\int g P_t f d\gamma_n \leq \int g^* P_t f^* d\gamma_1$$

for every $t \geq 0$ and every f, g say in $\mathrm{L}^2(\gamma_n)$ where f^* denotes the (one-dimensional) nonincreasing rearrangement of f with respect to the Gaussian measure γ_n (see [Eh3], [Bo10]). If $1 < p < q < \infty$ and $q < 1 + (p-1)e^{2t}$, a trivial application of Hölder's inequality shows that, for every φ in $\mathrm{L}^p(\gamma_1)$,

$$\|P_t \varphi\|_q \leq C \|\varphi\|_p$$

for some numerical $C > 0$. Now, if q' is the conjugate of q,

$$\begin{aligned}\int gP_t f d\gamma_n &\leq \int g^* P_t f^* d\gamma_1 \\ &\leq \|g^*\|_{q'} \|P_t f^*\|_q \\ &\leq C\|g^*\|_{q'} \|f^*\|_p \leq C\|g\|_{q'} \|f\|_p\end{aligned}$$

so that, by duality,

$$\|P_t f\|_q \leq C\|f\|_p.$$

Applying this inequality to $f^{\otimes k}$ on $(\mathbb{R}^n)^k = \mathbb{R}^{nk}$ yields

$$\|P_t f\|_q \leq C^{1/k}\|f\|_p.$$

Letting k tend to infinity, and q to its optimal value $1 + (p-1)e^{2t}$ concludes the proof of the claim.

REFERENCES

[A-C] R. A. Adams, F. H. Clarke. Gross's logarithmic Sobolev inequality: a simple proof. Amer. J. Math. 101, 1265–1269 (1979).

[A-K-S] S. Aida, S. Kusuoka, D. Stroock. On the support of Wiener functionals. Asymptotic problems in probability theory: Wiener functionals and asymptotics. Pitman Research Notes in Math. Series 284, 1–34 (1993). Longman.

[A-M-S] S. Aida, T. Masuda, I. Shigekawa. Logarithmic Sobolev inequalities and exponential integrability. J. Funct. Anal. 126, 83–101 (1994).

[A-L-R] M. Aizenman, J. L. Lebowitz, D. Ruelle. Some rigorous results on the Sherrington-Kirkpatrick spin glass model. Comm. Math. Phys. 112, 3–20 (1987).

[An] T. W. Anderson. The integral of a symmetric unimodal function over a symmetric convex set and some probability inequalities. Proc. Amer. Math. Soc. 6, 170–176 (1955).

[A-G] M. Arcones, E. Giné. On decoupling, series expansions and tail behavior of chaos processes. J. Theoretical Prob. 6, 101–122 (1993).

[Az] R. Azencott. Grandes déviations et applications. École d'Été de Probabilités de St-Flour 1978. Lecture Notes in Math. 774, 1–176 (1978). Springer-Verlag.

[Azu] K. Azuma. Weighted sums of certain dependent random variables. Tohoku Math. J. 19, 357–367 (1967).

[B-C] A. Badrikian, S. Chevet. Mesures cylindriques, espaces de Wiener et fonctions aléatoires gaussiennes. Lecture Notes in Math. 379, (1974). Springer-Verlag.

[Ba] A. Baernstein II. Integral means, univalent functions and circular symmetrization. Acta Math. 133, 139–169 (1974).

[B-T] A. Baernstein II, B. A. Taylor. Spherical rearrangements, subharmonic functions and *-functions in n-space. Duke Math. J. 43, 245–268 (1976).

[Bak] D. Bakry. L'hypercontractivité et son utilisation en théorie des semigroupes. École d'Été de Probabilités de St-Flour. Lecture Notes in Math. 1581, 1–114 (1994). Springer-Verlag.

[B-É] D. Bakry, M. Émery. Diffusions hypercontractives. Séminaire de Probabilités XIX. Lecture Notes in Math. 1123, 175–206 (1985). Springer-Verlag.

[B-R] P. Baldi, B. Roynette. Some exact equivalents for Brownian motion in Hölder norm. Prob. Th. Rel. Fields 93, 457–484 (1992).

[B-BA-K] P. Baldi, G. Ben Arous, G. Kerkyacharian. Large deviations and the Strassen theorem in Hölder norm. Stochastic Processes and Appl. 42, 171–180 (1992).

[Bas] R. Bass. Probability estimates for multiparameter Brownian processes. Ann. Probability 16, 251–264 (1988).

[Be1] W. Beckner. Inequalities in Fourier analysis. Ann. Math. 102, 159–182 (1975).

[Be2] W. Beckner. Unpublished (1982).

[Be3] W. Beckner. Sobolev inequalities, the Poisson semigroup and analysis on the sphere S^n. Proc. Nat. Acad. Sci. 89, 4816–4819 (1992).

[Bel] D. R. Bell. The Malliavin calculus. Pitman Monographs 34. Longman (1987).

[BA-L1] G. Ben Arous, M. Ledoux. Schilder's large deviation principle without topology. Asymptotic problems in probability theory: Wiener functionals and asymptotics. Pitman Research Notes in Math. Series 284, 107–121 (1993). Longman.

[BA-L2] G. Ben Arous, M. Ledoux. Grandes déviations de Freidlin-Wentzell en norme hölderienne. Séminaire de Probabilités XXVIII. Lecture Notes in Math. 1583, 293–299 (194). Springer-Verlag.

[BA-G-L] G. Ben Arous, M. Gradinaru, M. Ledoux. Hölder norms and the support theorem for diffusions. Ann. Inst. H. Poincaré 30, 415–436 (1994).

[Bob1] S. Bobkov. A functional form of the isoperimetric inequality for the Gaussian measure (1993). To appear in J. Funct. Anal..

[Bob2] S. Bobkov. An isoperimetric inequality on the discrete cube and an elementary proof of the isoperimetric inequality in Gauss space. Preprint (1994).

[Bog] V. I. Bogachev. Gaussian measures on linear spaces (1994). To appear.

[Bon] A. Bonami. Etude des coefficients de Fourier des fonctions de $L^p(G)$. Ann. Inst. Fourier 20, 335–402 (1970).

[Bo1] C. Borell. Convex measures on locally convex spaces. Ark. Mat. 12, 239–252 (1974).

[Bo2] C. Borell. The Brunn-Minskowski inequality in Gauss space. Invent. Math. 30, 207–216 (1975).

[Bo3] C. Borell. Gaussian Radon measures on locally convex spaces. Math. Scand. 38, 265–284 (1976).

[Bo4] C. Borell. A note on Gauss measures which agree on small balls. Ann. Inst. H. Poincaré 13, 231–238 (1977).

[Bo5] C. Borell. Tail probabilities in Gauss space. Vector Space Measures and Applications, Dublin 1977. Lecture Notes in Math. 644, 71–82 (1978). Springer-Verlag.

[Bo6] C. Borell. On the integrability of Banach space valued Walsh polynomials. Séminaire de Probabilités XIII. Lecture Notes in Math. 721, 1–3 (1979). Springer-Verlag.

[Bo7] C. Borell. A Gaussian correlation inequality for certain bodies in $\mathbb{R}^n$. Math. Ann. 256, 569–573 (1981).

[Bo8] C. Borell. On polynomials chaos and integrability. Prob. Math. Statist. 3, 191–203 (1984).

[Bo9] C. Borell. On the Taylor series of a Wiener polynomial. Seminar Notes on multiple stochastic integration, polynomial chaos and their integration. Case Western Reserve University, Cleveland (1984).

[Bo10] C. Borell. Geometric bounds on the Ornstein-Uhlenbeck process. Z. Wahrscheinlichkeitstheor. verw. Gebiete 70, 1–13 (1985).

[Bo11] C. Borell. Intrinsic bounds on some real-valued stationary random functions. Probability in Banach spaces V. Lecture Notes in Math. 1153, 72–95 (1985). Springer-Verlag.

[Bo12] C. Borell. Analytic and empirical evidences of isoperimetric processes. Probability in Banach spaces 6. Progress in Probability 20, 13–40 (1990). Birkhäuser.

[B-M] A. Borovkov, A. Mogulskii. On probabilities of small deviations for stochastic processes. Siberian Adv. Math. 1, 39–63 (1991).

[B-L-L] H. Brascamp, E. H. Lieb, J. M. Luttinger. A general rearrangement inequality for multiple integrals. J. Funct. Anal. 17, 227–237 (1974).

[B-Z] Y. D. Burago, V. A. Zalgaller. Geometric inequalities. Springer-Verlag (1988). First Edition (russian): Nauka (1980).

[C-M] R. H. Cameron, W. T. Martin. Transformations of Wiener integrals under translations. Ann. Math. 45, 386–396 (1944).

[Ca] M. Capitaine. Onsager-Machlup functional for some smooth norms on Wiener space (1994). To appear in Prob. Th. Rel. Fields.

[Ca-L] E. Carlen, M. Loss. Extremals of functionals with competing symmetries. J. Funct. Anal. 88, 437–456 (1990).

[C-F] I. Chavel, E. Feldman. Modified isoperimetric constants, and large time heat diffusion in Riemannian manifold. Duke Math. J. 64, 473–499 (1991).

[Ch] J. Cheeger. A lower bound for the smallest eigenvalue of the Laplacian. Problems in Analysis, Symposium in honor of S. Bochner, 195–199, Princeton Univ. Press, Princeton (1970).

[C-L-Y] S. Cheng, P. Li, S.-T. Yau. On the upper estimate of the heat kernel on a complete Riemannian manifold. Amer. J. Math. 156, 153–201 (1986).

[Che] S. Chevet. Gaussian measures and large deviations. Probability in Banach spaces IV. Lecture Notes in Math. 990, 30–46 (1983). Springer-Verlag.

[Ci1] Z. Ciesielski. On the isomorphisms of the spaces H_α and m. Bull. Acad. Pol. Sc. 8, 217–222 (1960).

[Ci2] Z. Ciesielski. Orlicz spaces, spline systems and brownian motion. Constr. Approx. 9, 191–208 (1993).

[Co] F. Comets. A spherical bound for the Sherrington-Kirkpatrick model. Preprint (1994).

[C-N] F. Comets, J. Neveu. The Sherrington-Kirkpatrick model of spin glasses and stochastic calculus: the high temperature case. Preprint (1993).

[C-L] T. Coulhon, M. Ledoux. Isopérimétrie, décroissance du noyau de la chaleur et transformations de Riesz: un contre-exemple. Ark. Mat. 32, 63–77 (1994).

[DG-E-...] S. Das Gupta, M. L. Eaton, I. Olkin, M. Perlman, L. J. Savage, M. Sobel. Inequalities on the probability content of convex regions for elliptically contoured distributions. Proc. Sixth Berkeley Symp. Math. Statist. Prob. 2, 241–264 (1972). Univ. of California Press.

[Da] E. B. Davies. Heat kernels and spectral theory. Cambridge Univ. Press (1989).

[Da-S] E. B. Davies, B. Simon. Ultracontractivity and the heat kernel for Schrödinger operators and Dirichlet Laplacians. J. Funct. Anal. 59, 335–395 (1984).

[D-L] P. Deheuvels, M. A. Lifshits. Strassen-type functional laws for strong topologies. Prob. Th. Rel. Fields 97, 151–167 (1993).

[De] J. Delporte. Fonctions aléatoires presque sûrement continues sur un intervalle fermé. Ann. Inst. H. Poincaré 1, 111–215 (1964).

[D-S] J.-D. Deuschel, D. Stroock. Large deviations. Academic Press (1989).

[D-F] P. Diaconis, D. Freedman. A dozen de Finetti-style results in search of a theory. Ann. Inst. H. Poincaré 23, 397–423 (1987).

[D-V] M. D. Donsker, S. R. S. Varadhan. Asymptotic evaluation of certain Markov process expectations for large time III. Comm. Pure Appl. Math. 29, 389–461 (1976).

[Du1] R. M. Dudley. The sizes of compact subsets of Hilbert space and continuity of Gaussian processes. J. Funct. Anal. 1, 290–330 (1967).

[Du2] R. M. Dudley. Sample functions of the Gaussian process. Ann. Probability 1, 66–103 (1973).

[D-HJ-S] R. M. Dudley, J. Hoffmann-Jorgensen, L. A. Shepp. On the lower tail of Gaussian seminorms. Ann. Probability 7, 319–342 (1979).

[Dv] A. Dvoretzky. Some results on convex bodies and Banach spaces. Proc. Symp. on Linear Spaces, Jerusalem, 123–160 (1961).

[Eh1] A. Ehrhard. Une démonstration de l'inégalité de Borell. Ann. Scientifiques de l'Université de Clermont-Ferrand 69, 165–184 (1981).

[Eh2] A. Ehrhard. Symétrisation dans l'espace de Gauss. Math. Scand. 53, 281–301 (1983).

[Eh3] A. Ehrhard. Inégalités isopérimétriques et intégrales de Dirichlet gaussiennes. Ann. scient. Éc. Norm. Sup. 17, 317–332 (1984).

[Eh4] A. Ehrhard. Sur l'inégalité de Sobolev logarithmique de Gross. Séminaire de Probabilités XVIII. Lecture Notes in Math. 1059, 194–196 (1984). Springer-Verlag.

[Eh5] A. Ehrhard. Eléments extrémaux pour les inégalités de Brunn-Minkowski gaussiennes. Ann. Inst. H. Poincaré 22, 149–168 (1986).

[E-S] O. Enchev, D. Stroock. Rademacher's theorem for Wiener functionals. Ann. Probability 21, 25–33 (1993).

[Fa] S. Fang. On the Ornstein-Uhlenbeck process. Stochastics and Stochastic Reports 46, 141–159 (1994).

[Fed] H. Federer. Geometric measure theory. Springer-Verlag (1969).

[F-F] H. Federer, W. H. Fleming. Normal and integral current. Ann. Math. 72, 458–520 (1960).

[Fe1] X. Fernique. Continuité des processus gaussiens. C. R. Acad. Sci. Paris 258, 6058–6060 (1964).

[Fe2] X. Fernique. Intégrabilité des vecteurs gaussiens. C. R. Acad. Sci. Paris 270, 1698–1699 (1970).

[Fe3] X. Fernique. Régularité des processus gaussiens. Invent. Math. 12, 304–320 (1971).

[Fe4] X. Fernique. Régularité des trajectoires des fonctions aléatoires gaussiennes. École d'Été de Probabilités de St-Flour 1974. Lecture Notes in Math. 480, 1–96 (1975). Springer-Verlag.

[Fe5] X. Fernique. Gaussian random vectors and their reproducing kernel Hilbert spaces. Technical report, University of Ottawa (1985).

[F-L-M] T. Figiel, J. Lindenstrauss, V. D. Milman. The dimensions of almost spherical sections of convex bodies. Acta Math. 139, 52–94 (1977).

[F-W1] M. Freidlin, A. Wentzell. On small random perturbations of dynamical systems. Russian Math. Surveys 25, 1–55 (1970).

[F-W2] M. Freidlin, A. Wentzell. Random perturbations of dynamical systems. Springer-Verlag (1984).

[Ga] E. Gagliardo. Proprieta di alcune classi di funzioni in piu variabili. Ricerche Mat. 7, 102–137 (1958).

[G-R-R] A. M. Garsia, E. Rodemich, H. Rumsey Jr.. A real variable lemma and the continuity of paths of some Gaussian processes. Indiana Math. J. 20, 565–578 (1978).

[Gal] L. Gallardo. Au sujet du contenu probabiliste d'un lemme d'Henri Poincaré. Ann. Scientifiques de l'Université de Clermont-Ferrand 69, 185–190 (1981).

[G-H-L] S. Gallot, D. Hulin, J. Lafontaine. Riemannian Geometry. Second Edition. Springer-Verlag (1990).

[Go1] V. Goodman. Characteristics of normal samples. Ann. Probability 16, 1281–1290 (1988).

[Go2] V. Goodman. Some probability and entropy estimates for Gaussian measures. Probability in Banach spaces 6. Progress in Probability 20, 150–156 (1990). Birkhäuser.

[G-K1] V. Goodman, J. Kuelbs. Cramér functional estimates for Gaussian measures. Diffusion processes and related topics in Analysis. Progress in Probability 22, 473–495 (1990). Birkhäuser.

[G-K2] V. Goodman, J. Kuelbs. Gaussian chaos and functional laws of the iterated logarithm for Ito-Wiener integrals. Ann. Inst. H. Poincaré 29, 485–512 (1993).

[Gri] K. Grill. Exact convergence rate in Strassen's law of the iterated logarithm. J. Theoretical Prob. 5, 197–204 (1991).

[Gro] M. Gromov. Paul Lévy's isoperimetric inequality. Preprint I.H.E.S. (1980).

[G-M] M. Gromov, V. D. Milman. A topological application of the isoperimetric inequality. Amer. J. Math. 105, 843–854 (1983).

[Gr1] L. Gross. Abstract Wiener spaces. Proc. 5th Berkeley Symp. Math. Stat. Prob. 2, 31–42 (1965).

[Gr2] L. Gross. Potential theory on Hilbert space. J. Funct. Anal. 1, 123–181 (1967).

[Gr3] L. Gross. Logarithmic Sobolev inequalities. Amer. J. Math. 97, 1061–1083 (1975).

[Gr4] L. Gross. Logarithmic Sobolev inequalities and contractive properties of semigroups. Dirichlet forms, Varenna (Italy) 1992. Lecture Notes in Math. 1563, 54–88 (1993). Springer-Verlag.

[G-N-SS] I. Gyöngy, D. Nualart, M. Sanz-Solé. Approximation and support theorems in modulus spaces (1994). To appear in Prob. Th. Rel. Fields.

[Ha] H. Hadwiger. Vorlesungen über Inhalt, Oberfläche und Isoperimetrie. Springer-Verlag (1957).

[Har] L. H. Harper. Optimal numbering and isoperimetric problems on graphs. J. Comb. Th. 1, 385–393 (1966).

[He] B. Heinkel. Mesures majorantes et régularité de fonctions aléatoires. Aspects Statistiques et Aspects Physiques des Processus Gaussiens, St-Flour 1980. Colloque C.N.R.S. 307, 407–434 (1980).

[I-S-T] I. A. Ibragimov, V. N. Sudakov, B. S. Tsirel'son. Norms of Gaussian sample functions. Proceedings of the third Japan-USSR Symposium on Probability Theory. Lecture Notes in Math. 550, 20–41 (1976). Springer-Verlag.

[I-W] N. Ikeda, S. Watanabe. Stochastic differential equations and diffusion processes. North-Holland (1989).

[It] K. Itô. Multiple Wiener integrals. J. Math. Soc. Japan 3, 157–164 (1951).

[Ka1] J.-P. Kahane. Sur les sommes vectorielles $\sum \pm u_n$. C. R. Acad. Sci. Paris 259, 2577–2580 (1964).

[Ka2] J.-P. Kahane. Some random series of functions. Heath Math. Monographs (1968). Second Edition: Cambridge Univ. Press (1985).

[Ke] H. Kesten. On the speed of convergence in first-passage percolation. Ann. Appl. Probability 3, 296–338 (1993).

[Kh] C. Khatri. On certain inequalities for normal distributions and their applications to simultaneous confidence bounds. Ann. Math. Statist. 38, 1853–1867 (1967).

[K-L1] J. Kuelbs, W. Li. Small ball probabilities for Brownian motion and the Brownian sheet. J. Theoretical Prob. 6, 547–577 (1993).

[K-L2] J. Kuelbs, W. Li. Metric entropy and the small ball problem for Gaussian measures J. Funct. Anal. 116, 133-157 (1993).

[K-L-L] J. Kuelbs, W. Li, W. Linde. The Gaussian measure of shifted balls. Prob. Th. Rel. Fields 98, 143–162 (1994).

[K-L-S] J. Kuelbs, W. Li, Q.-M. Shao. Small ball probabilities for Gaussian processes with stationary increments under Hölder norms (1993). To appear in J. Theoretical Prob..

[K-L-T] J. Kuelbs, W. Li, M. Talagrand. Liminf results for Gaussian samples and Chung's functional LIL. Ann. Probability 22, 1879–1903 (1994).

[Ku] H.-H. Kuo. Gaussian measures in Banach spaces. Lecture Notes in Math. 436 (1975). Springer-Verlag.

[Kus] S. Kusuoka. A diffusion process on a fractal. Probabilistic methods in mathematical physics. Proc. of Taniguchi International Symp. 1985, 251–274. Kinokuniga, Tokyo (1987).

[Kw] S. Kwapień. A theorem on the Rademacher series with vector valued coefficients. Probability in Banach Spaces, Oberwolfach 1975. Lecture Notes in Math. 526, 157–158 (1976). Springer-Verlag.

[K-S] S. Kwapień, J. Sawa. On some conjecture concerning Gaussian measures of dilatations of convex symmetric sets. Studia Math. 105, 173–187 (1993).

[L-S] H. J. Landau, L. A. Shepp. On the supremum of a Gaussian process. Sankhyà A32, 369–378 (1970).

[La] R. Latała. A note on the Ehrhard inequality. Preprint (1994).

[L-O] R. Latała, K. Oleszkiewicz. On the best constant in the Khintchine-Kahane inequality. Studia Math. 109, 101–104 (1994).

[Led1] M. Ledoux. Isopérimétrie et inégalités de Sobolev logarithmiques gaussiennes. C. R. Acad. Sci. Paris 306, 79–82 (1988).

[Led2] M. Ledoux. A note on large deviations for Wiener chaos. Séminaire de Probabilités XXIV, Lecture Notes in Math. 1426, 1–14 (1990). Springer-Verlag.

[Led3] M. Ledoux. On an integral criterion for hypercontractivity of diffusion semigroups and extremal functions. J. Funct. Anal. 105, 444–465 (1992).

[Led4] M. Ledoux. A heat semigroup approach to concentration on the sphere and on a compact Riemannian manifold. Geom. and Funct. Anal. 2, 221–224 (1992).

[Led5] M. Ledoux. Semigroup proofs of the isoperimetric inequality in Euclidean and Gauss space. Bull. Sci. math. 118, 485–510 (1994).

[L-T1] M. Ledoux, M. Talagrand. Characterization of the law of the iterated logarithm in Banach spaces. Ann. Probability 16, 1242–1264 (1988).

[L-T2] M. Ledoux, M. Talagrand. Probability in Banach spaces (Isoperimetry and processes). Ergebnisse der Mathematik und ihrer Grenzgebiete. Springer-Verlag (1991).

[Lé] P. Lévy. Problèmes concrets d'analyse fonctionnelle. Gauthier-Villars (1951).

[Li] W. Li. Comparison results for the lower tail of Gaussian semi-norms. J. Theoretical Prob. 5, 1–31 (1992).

[Li-S] W. Li, Q.-M. Shao. Small ball estimates for Gaussian processes under Sobolev type norms. Preprint (1994).

[L-Y] P. Li, S.-T. Yau. On the parabolic kernel of the Schrödinger operator. Acta Math. 156, 153–201 (1986).

[Lif1] M. A. Lifshits. On the distribution of the maximum of a Gaussian process. Probability Theory and its Appl. 31, 125-132 (1987).

[Lif2] M. A. Lifshits. Tail probabilities of Gaussian suprema and Laplace transform. Ann. Inst. H. Poincaré 30, 163–180 (1994).

[Lif3] M. A. Lifshits. Gaussian random functions (1994). Kluwer, to appear.

[Lif-T] M. A. Lifshits, B. S. Tsirel'son. Small deviations of Gaussian fields. Probability Theory and its Appl. 31, 557-558 (1987).

[L-Z] T. Lyons, W. Zheng. A crossing estimate for the canonical process on a Dirichlet space and tightness result. Colloque Paul Lévy, Astérisque 157-158, 249–272 (1988).

[MD] C. J. H. McDiarmid. On the method of bounded differences. Twelfth British Combinatorial Conference. Surveys in Combinatorics, 148–188 (1989). Cambrige Univ. Press.

[MK] H. P. McKean. Geometry of differential space. Ann. Probability 1, 197–206 (1973).

[M-P] M. B. Marcus, G. Pisier. Random Fourier series with applications to harmonic analysis. Ann. Math. Studies, vol. 101 (1981). Princeton Univ. Press.

[M-S] M. B. Marcus, L. A. Shepp. Sample behavior of Gaussian processes. Proc. of the Sixth Berkeley Symposium on Math. Statist. and Prob. 2, 423–441 (1972).

[Ma1] B. Maurey. Constructions de suites symétriques. C. R. Acad. Sci. Paris 288, 679–681 (1979).

[Ma2] B. Maurey. Sous-espaces ℓ^p des espaces de Banach. Séminaire Bourbaki, exp. 608. Astérisque 105-106, 199–216 (1983).

[Ma3] B. Maurey. Some deviations inequalities. Geometric and Funct. Anal. 1, 188–197 (1991).

[MW-N-PA] E. Mayer-Wolf, D. Nualart, V. Perez-Abreu. Large deviations for multiple Wiener-Itô integrals. Séminaire de Probabilités XXVI. Lecture Notes in Math. 1526, 11–31 (1992). Springer-Verlag.

[Maz1] V. G. Maz'ja. Classes of domains and imbedding theorems for function spaces. Soviet Math. Dokl. 1, 882–885 (1960).

[Maz2] V. G. Maz'ja. Sobolev spaces. Springer-Verlag (1985).

[Me] M. Mellouk. Support des diffusions dans les espaces de Besov-Orlicz. C. R. Acad. Sci. Paris 319, 261–266 (1994).

[M-SS] A. Millet, M. Sanz-Solé. A simple proof of the support theorem for diffusion processes. Séminaire de Probabilités XXVIII, Lecture Notes in Math. 1583, 36–48 (1994). Springer-Verlag.

[Mi1] V. D. Milman. New proof of the theorem of Dvoretzky on sections of convex bodies. Funct. Anal. Appl. 5, 28–37 (1971).

[Mi2] V. D. Milman. The heritage of P. Lévy in geometrical functional analysis. Colloque Paul Lévy sur les processus stochastiques. Astérisque 157-158, 273–302 (1988).

[Mi3] V. D. Milman. Dvoretzky's theorem - Thirty years later (Survey). Geometric and Funct. Anal. 2, 455–479 (1992).

[Mi-S] V. D. Milman, G. Schechtman. Asymptotic theory of finite dimensional normed spaces. Lecture Notes in Math. 1200 (1986). Springer-Verlag.

[M-R] D. Monrad, H. Rootzén. Small values of Gaussian processes and functional laws of the iterated logarithm (1993). To appear in Prob. Th. Rel. Fields.

[Mo] J. Moser. On Harnack's theorem for elliptic differential equations. Comm. Pure Appl. Math. 14, 557–591 (1961).

[Na] J. Nash. Continuity of solutions of parabolic and elliptic equations. Amer. J. Math. 80, 931–954 (1958).

[Nel] E. Nelson. The free Markov field. J. Funct. Anal. 12, 211–227 (1973).

[Ne1] J. Neveu. Processus aléatoires gaussiens. Presses de l'Université de Montréal (1968).

[Ne2] J. Neveu. Martingales à temps discret. Masson (1972).

[Ne3] J. Neveu. Sur l'espérance conditionnelle par rapport à un mouvement brownien. Ann. Inst. H. Poincaré 2, 105–109 (1976).

[Ni] L. Nirenberg. On elliptic partial differential equations. Ann. Sc. Norm. Sup. Pisa 13, 116–162 (1959).

[Nu] D. Nualart. The Malliavin calculs and related topics (1994). To appear.

[Os] R. Osserman. The isoperimetric inequality. Bull. Amer. Math. Soc. 84, 1182–1238 (1978).

[Pi1] G. Pisier. Probabilistic methods in the geometry of Banach spaces. Probability and Analysis, Varenna (Italy) 1985. Lecture Notes in Math. 1206, 167–241 (1986). Springer-Verlag.

[Pi2] G. Pisier. Riesz transforms : a simpler analytic proof of P. A. Meyer inequality. Séminaire de Probabilités XXII. Lecture Notes in Math. 1321, 485–501, Springer-Verlag (1988).

[Pi3] G. Pisier. The volume of convex bodies and Banach space geometry. Cambridge Univ. Press (1989).

[Pit] L. Pitt. A Gaussian correlation inequality for symmetric convex sets. Ann. Probability 5, 470–474 (1977).

[Pr1] C. Preston. Banach spaces arising from some integral inequalities. Indiana Math. J. 20, 997–1015 (1971).

[Pr2] C. Preston. Continuity properties of some Gaussian processes. Ann. Math. Statist. 43, 285–292 (1972).

[Sc] M. Schilder. Asymptotic formulas for Wiener integrals. Trans. Amer. Math. Soc. 125, 63–85 (1966).

[Sch] E. Schmidt. Die Brunn-Minkowskische Ungleichung und ihr Spiegelbild sowie die isoperime- trische Eigenschaft der Kugel in der euklidischen und nichteuklidischen Geometrie. Math. Nach. 1, 81–157 (1948).

[Sco] A. Scott. A note on conservative confidence regions for the mean value of multivariate normal. Ann. Math. Statist. 38, 278–280 (1967).

[S-Z1] L. A. Shepp, O. Zeitouni. A note on conditional exponential moments and Onsager-Machlup functionals. Ann. Probability 20, 652–654 (1992).

[S-Z2] L. A. Shepp, O. Zeitouni. Exponential estimates for convex norms and some applications. Barcelona seminar on Stochastic Analysis, St Feliu de Guixols 1991. Progress in Probability 32, 203–215 (1993). Birkhäuser.

[Sh] Q.-M. Shao. A note on small ball probability of a Gaussian process with stationary increments. J. Theoretical Prob. 6, 595–602 (1993).

[S-W] Q.-M. Shao, D. Wang. Small ball probabilities of Gaussian fields. Preprint (1994).

[Si] Z. Sidak. Rectangular confidence regions for the means of multivariate normal distributions. J. Amer. Statist. Assoc. 62, 626–633 (1967).

[Sk] A. V. Skorohod. A note on Gaussian measures in a Banach space. Theor. Probability Appl. 15, 519–520 (1970).

[Sl] D. Slepian. The one-sided barrier problem for Gaussian noise. Bell. System Tech. J. 41, 463–501 (1962).

[So] S. L. Sobolev. On a theorem in functional analysis. Amer. Math. Soc. Translations (2) 34, 39–68 (1963); translated from Mat. Sb. (N.S.) 4 (46), 471–497 (1938).

[St1] W. Stolz. Une méthode élémentaire pour l'évaluation de petites boules browniennes. C. R. Acad. Sci. Paris, 316, 1217–1220 (1993).

[St2] W. Stolz. Some small ball probabilities for Gaussian processes under non-uniform norms (1994). To appear in J. Theoretical Prob..

[Str] D. Stroock. Homogeneous chaos revisited. Séminaire de Probabilités XXI. Lecture Notes in Math. 1247, 1–7 (1987). Springer-Verlag.

[Su1] V. N. Sudakov. Gaussian measures, Cauchy measures and ε-entropy. Soviet Math. Dokl. 10, 310–313 (1969).

[Su2] V. N. Sudakov. Gaussian random processes and measures of solid angles in Hilbert spaces. Soviet Math. Dokl. 12, 412–415 (1971).

[Su3] V. N. Sudakov. A remark on the criterion of continuity of Gaussian sample functions. Proceedings of the Second Japan-USSR Symposium on Probability Theory . Lecture Notes in Math. 330, 444–454 (1973). Springer-Verlag.

[Su4] V. N. Sudakov. Geometric problems of the theory of infinite-dimensional probability distributions. Trudy Mat. Inst. Steklov 141 (1976).

[S-T] V. N. Sudakov, B. S. Tsirel'son. Extremal properties of half-spaces for spherically invariant measures. J. Soviet. Math. 9, 9–18 (1978); translated from Zap. Nauch. Sem. L.O.M.I. 41, 14–24 (1974).

[Sy] G. N. Sytaya. On some asymptotic representation of the Gaussian measure in a Hilbert space. Theory of Stochastic Processes (Kiev) 2, 94-104 (1974).

[Sz] S. Szarek. On the best constant in the Khintchine inequality. Studia Math. 58, 197–208 (1976).

[Tak] M. Takeda. On a martingale method for symmetric diffusion processes and its applications. Osaka J. Math. 26, 605–623 (1989).

[Ta1] M. Talagrand. Sur l'intégrabilité des vecteurs gaussiens. Z. Wahrscheinlichkeitstheor. verw. Gebiete 68, 1–8 (1984).

[Ta2] M. Talagrand. Regularity of Gaussian processes. Acta Math. 159, 99–149 (1987).

[Ta3] M. Talagrand. An isoperimetric theorem on the cube and the Khintchin-Kahane inequalities. Proc. Amer. Math. Soc. 104, 905–909 (1988).

[Ta4] M. Talagrand. Small tails for the supremum of a Gaussian process. Ann. Inst. H. Poincaré 24, 307–315 (1988).

[Ta5] M. Talagrand. Isoperimetry and integrability of the sum of independent Banach space valued random variables. Ann. Probability 17, 1546–1570 (1989).

[Ta6] M. Talagrand. A new isoperimetric inequality for product measure and the tails of sums of independent random variables. Geometric and Funct. Anal. 1, 211-223 (1991).

[Ta7] M. Talagrand. Simple proof of the majorizing measure theorem. Geometric and Funct. Anal. 2, 118–125 (1992).

[Ta8] M. Talagrand. On the rate of clustering in Strassen's law of the iterated logarithm. Probability in Banach spaces 8. Progress in Probability 30, 339–351 (1992). Birkhäuser.

[Ta9] M. Talagrand. New Gaussian estimates for enlarged balls. Geometric and Funct. Anal. 3, 502–526 (1993).

[Ta10] M. Talagrand. Regularity of infinitely divisible processes. Ann. Probability 21, 362–432 (1993).

[Ta11] M. Talagrand. Isoperimetry, logarithmic Sobolev inequalities on the discrete cube, and Margulis' graph connectivity theorem. Geometric and Funct. Anal. 3, 295-314 (1993).

[Ta12] M. Talagrand. The supremum of some canonical processes. Amer. Math. J. 116, 283–325 (1994).

[Ta13] M. Talagrand. Sharper bounds for Gaussian and empirical processes. Ann. Probability 22, 28–76 (1994).

[Ta14] M. Talagrand. Constructions of majorizing measures. Bernoulli processes and cotype. Geometric and Funct. Anal. 4, 660–717 (1994).

[Ta15] M. Talagrand. The small ball problem for the Brownian sheet. Ann. Probability 22, 1331–1354 (1994).

[Ta16] M. Talagrand. Concentration of measure and isoperimetric inequalities in product spaces (1994). To appear in Publ. de l'IHES.

[Ta17] M. Talagrand. Isoperimetry in product spaces: higher level, large sets. Preprint (1994).

[Ta18] M. Talagrand. Majorizing measures: the generic chaining. Preprint (1994).

[TJ] N. Tomczak-Jaegermann. Dualité des nombres d'entropie pour des opérateurs à valeurs dans un espace de Hilbert. C. R. Acad. Sci. Paris 305, 299–301 (1987).

[Var] S. R. S. Varadhan. Large deviations and applications. S. I. A. M. Philadelphia (1984).

[Va1] N. Varopoulos. Une généralisation du théorème de Hardy-Littlewood-Sobolev pour les espaces de Dirichlet. C. R. Acad. Sci. Paris 299, 651–654 (1984).

[Va2] N. Varopoulos. Hardy-Littlewood theory for semigroups. J. Funct. Anal. 63, 240–260 (1985).

[Va3] N. Varopoulos. Isoperimetric inequalities and Markov chains. J. Funct. Anal. 63, 215–239 (1985).

[Va4] N. Varopoulos. Small time Gaussian estimates of heat diffusion kernels. Part I: The semigroup technique. Bull. Sc. math. 113, 253–277 (1989).

[Va5] N. Varopoulos. Analysis and geometry on groups. Proceedings of the International Congress of Mathematicians, Kyoto (1990), vol. II, 951–957 (1991). Springer-Verlag.

[V-SC-C] N. Varopoulos, L. Saloff-Coste, T. Coulhon. Analysis and geometry on groups. Cambridge Univ. Press (1992).

[Wa] S. Watanabe. Lectures on stochastic differential equations and Malliavin calculus. Tata Institute of Fundamental Research Lecture Notes. Springer-Verlag (1984).

[W-W] D. L. Wang, P. Wang. Extremal configurations on a discrete torus and a generalization of the generalized Macaulay theorem. Siam J. Appl. Math. 33, 55–59 (1977).

[Wi] N. Wiener. The homogeneous chaos. Amer. Math. J. 60, 897–936 (1930).

[Ya] S.-T. Yau. Isoperimetric constants and the first eigenvalue of a compact Riemannian manifold. Ann. scient. Éc. Norm. Sup. 8, 487–507 (1975).

[Zo] V. M. Zolotarev. Asymptotic behavior of the Gaussian measure in ℓ^2. J. Sov. Math. 24, 2330-2334 (1986).

LISTE DES AUDITEURS

Mr	AMGHIBECH Said	Analyse, Modèles Stochastiques, Université de Rouen
Mr	AZEMA Jacques	Laboratoire de Probabilités, Université PARIS VI
Mr	BADRIKIAN Albert	Université Blaise Pascal, CLERMONT-FD
Mr	BALDI Paolo	Mathématiques, Université de ROME (Italie)
Mr	BARAUD Yannick	Probabilités-Statistiques, Université Paris-Sud (ORSAY)
Mr	BENAROUS Gérard	Mathématiques, Université Paris-Sud, ORSAY
Mr	BENASSI Albert	Université Blaise Pascal, CLERMONT-FD
Mr	BERNARD Pierre	Université Blaise Pascal, CLERMONT-FD
Mr	BIRGE Lucien	L.S.T.A., Université PARIS VI
Mr	BITOUZE Denis	Mathématiques, Université Paris-Sud (ORSAY)
Mr	BODINEAU Thierry	Mathématiques, Université PARIS VII
Mr	BRIAND Philippe	Université Blaise Pascal, CLERMONT-FD
Mr	BRUNAUD Marc	UFR Mathématiques, Université PARIS VII
Mme	CAPITAINE Mireille	Stat. et Proba, Université Paul Sabatier, TOULOUSE
Mr	CARAMELLINO Lucia	Università degli Studi di Roma, ROME (Italie)
Melle	CASTELL Fabienne	Mathématiques, Université Paris-Sud (ORSAY)
Mr	CATONI Olivier	Ecole Normale Supérieure à PARIS
Mr	CATTIAUX Patrick	Mathématiques, Université Paris-Sud (ORSAY)
Mr	CERF Raphaël	Mathématiques, Université MONTPELLIER II
Mme	CHALEYAT-MAUREL Mireille	Laboratoire de Probabilités, Université PARIS VI
Melle	CHEVET Simone	Université Blaise Pascal, CLERMONT-FD
Mr	DELLACHERIE Claude	Analyse,Modèles Stochastiques,Université de ROUEN
Mr	DONOHO David	Department of Statistics, Stanford University (USA)
Mme	ECHERBAULT Mireille	Stat. Université Paul Sabatier, TOULOUSE
Mr	EDDAHBI M'hamed	Faculté des Sciences de MARRAKECH
Mr	EL-NOUTY Charles	L.S.T.A., Université PARIS VI
Mr	EMERY Michel	IRMA, Université Louis Pasteur, STRASBOURG I
Melle	ESTRADE Anne	Mathématiques, Université d'Orléans
Melle	FORBES Florence	Statistique, Université de GRENOBLE
Melle	FRADON Myriam	Mathématiques, Université Paris-Sud (ORSAY)
Mr	GOBRON Thierry	Ecole Polytechnique, PALAISEAU
Mr	GRADINARU Mihai	Mathématiques, Université Paris-Sud (ORSAY)
Mr	GRISHIN Stanislav	Mathematics, University of California, IRVINE (USA)
Mme	GRUET Marie-Anne	Labo de Biométrie, I.N.R.A., JOUY-EN-JOSAS
Mr	GRUNBAUM Francisco	Department of Mathematics, University of California BERKELEY (USA)
Mr	GUERME Emmanuel	L.S.T.A., Université PARIS VI
Mr	GUIMIER Alain	Université de NOUAKCHOTT, Mauritanie
Melle	GUIONNET Alice	Mathématiques, Université Paris-Sud (ORSAY)
Mr	HENNEQUIN Paul-Louis	Université Blaise Pascal, CLERMONT-FD
Mr	HOFFMANN Marc	Mathématiques, Université PARIS VII
Mr	HU Ying	Université Blaise Pascal, CLERMONT-FD
Mr	IOFFE Dima	Northwestern University, EVANSTON (USA)
Mr	IOUDITSKI Anatoli	IRISA, Université de RENNES
Mr	JOLIVET Emmanuel	Labo de Biométrie, I.N.R.A., JOUY-EN-JOSAS
Mr	KERKYACHARIAN Gérard	Faculté de Mathématiques, Université d'AMIENS
Mr	KONGTCHEU Philibert	Mathématiques, Université de Provence, MARSEILLE
Mme	LAREDO Catherine	Labo de Biométrie, I.N.R.A., JOUY-EN-JOSAS
Mr	LAROCHE Etienne	Stat. et Proba, Université Paul Sabatier, TOULOUSE
Mr	LATALA Rafal	Department of Mathematics, Warsaw University, WARSZAWA (Pologne)
Mme	LAURENT Béatrice	Mathématiques, Université Paris-Sud (ORSAY)

Mr	LE BORGNE Stéphane	IRMAR, Université de RENNES I
Mr	LE GALL Jean-François	Laboratoire de Probabilités, Université PARIS VI
Mr	LE JAN Yves	Mathématiques, Université Paris-Sud (ORSAY)
Mr	LEMDANI Mohamed	Labo de Biométrie, I.N.R.A., JOUY-EN-JOSAS
Mr	LEONARD Christian	Mathématiques, Université Paris-Sud (ORSAY)
Mr	LEPINGLE Dominique	Mathématiques, Université d'ORLEANS
Melle	LI Xue-Mei	Mathematics Institute, University of Warwick, COVENTRY (USA)
Mr	LLOPIS Jaume	Statistique, Faculté de Biologie, BARCELONE
Mme	MAILLE Sophie	Stat. et Proba, Université Paul Sabatier, TOULOUSE
Mr	MATHIEU Pierre	Mathématiques, Université de Provence, MARSEILLE
Mr	MICLO Laurent	IRMA, Université Louis Pasteur, STRASBOURG I
Mme	MILLET Annie	Laboratoire de Probabilités, Université PARIS VI
Mme	MONAT Pascale	Mathématiques, Université La Bouloie, BESANCON
Mr	NOBLE John	Department of Statistics, University College, CORK (Irlande)
Mr	NUALART David	Université de BARCELONE (Espagne)
Mr.	OLESZKIEWICZ Krzysztof	Department of Mathematics, Warsaw University, WARSZAWA (Pologne)
Mr	PARDOUX Etienne	Mathématiques, Université de Provence, MARSEILLE
Mme	PELIGRAD Magda	Mathematical Sciences, University of Cincinnati OHIO (USA)
Mr	PERMAN Mihael	Department of Mathematics, University of Southern California, LOS ANGELES (USA)
Mme	PICARD Dominique	U.F.R. de Mathématiques, Université PARIS VII
Mr	PICARD Jean	Université Blaise Pascal, CLERMONT-FD
Mr	RAMPONI Alessandro	Dipartimento di Matematica Università degli Studi di Roma, ROME (Italie)
Mr	RIO Emmanuel	Mathématiques, Université Paris-Sud (ORSAY)
Mr	RIOS Ricardo	Mathématiques, Université Paris-Sud (ORSAY)
Mr	ROLSKI Tomasz	Department of Mathematics University of WROCLAW (Pologne)
Mr	ROUAULT Alain	Mathématiques, Université de VERSAILLES
Mr	ROVIRA ESCOFET Carles	Facultat de Matematiques Université de BARCELONE (Espagne)
Mr	ROZENHOLC Yves	Mathématiques, Université PARIS VII
Melle	SAVONA Catherine	Université Blaise Pascal, CLERMONT-FD
Mme	SIRI Paola	Dipartimento di Matematica Università di Genova (ITALIE)
Mr	SOLE I CLIVILLES Josep	ETSECCP, Université de BARCELONE (Espagne)
Mme	TRIBOULEY Karine	Mathématiques, Université PARIS VII
Mr	TROUVE Alain	Ecole Normale Supérieure, Rue d'Ulm, PARIS
Mme	TROUVE Isabelle	Ecole Normale Supérieure, CACHAN
Mr	UTZET Frédéric	Departement de Mathématiques Université de BARCELONE (Espagne)
Mr	VAN CASTEREN Jan	Mathématiques et Informatique Université de WILRIJK (Belgique)
Melle	VIENNET Gabrielle	Mathématiques, Université PARIS VII
Mr	VIENS Frederi	Department of Mathematics University of California, IRVINE (USA)
Mr	WANG Mei	Department of Mathematics Illinois Institute, CHICAGO (USA)
Mr	WELLNER Jon	Department of Mathematics University of Washington (USA)
Mr	ZAMBRINI Jean-Claude	CFMC, Université de LISBONNE (Portugal)

LIST OF PREVIOUS VOLUMES OF THE "Ecole d'Eté de Probabilités"

1971 - J.L. Bretagnolle (LNM 307)
"Processus à accroissements indépendants"
S.D. Chatterji
"Les martingales et leurs applications analytiques"
P.A. MEYER
"Présentation des processus de Markov"

1973 - P.A. MEYER (LNM 390)
"Transformation des processus de Markov"
P. PRIOURET
"Processus de diffusion et équations différentielles stochastiques"
F. SPITZER
"Introduction aux processus de Markov à paramètres dans Z_V"

1974 - X. FERNIQUE (LNM 480)
"Régularité des trajectoires des fonctions aléatoires gaussiennes"
J.P. CONZE
"Systèmes topologiques et métriques en théorie ergodique"
J. GANI
"Processus stochastiques de population"

1975 A. BADRIKIAN (LNM 539)
"Prolégomènes au calcul des probabilités dans les Banach"
J.F.C. KINGMAN
"Subadditive processes"
J. KUELBS
"The law of the iterated logarithm and related strong convergence theorems for Banach space valued random variables"

1976 J. HOFFMANN-JORGENSEN (LNM 598)
"Probability in Banach space"
T.M. LIGGETT
"The stochastic evolution of infinite systems of interacting particles"
J. NEVEU
"Processus ponctuels"

1977 D. DACUNHA-CASTELLE (LNM 678)
"Vitesse de convergence pour certains problèmes statistiques"
H. HEYER
"Semi-groupes de convolution sur un groupe localement compact et applications à la théorie des probabilités"
B. ROYNETTE
"Marches aléatoires sur les groupes de Lie"

1978	R. AZENCOTT "Grandes déviations et applications" Y. GUIVARC'H "Quelques propriétés asymptotiques des produits de matrices aléatoires" R.F. GUNDY "Inégalités pour martingales à un et deux indices : l'espace H^p"	(LNM 774)
1979	J.P. BICKEL "Quelques aspects de la statistique robuste" N. EL KAROUI "Les aspects probabilistes du contrôle stochastique" M. YOR "Sur la théorie du filtrage"	(LNM 876)
1980	J.M. BISMUT "Mécanique aléatoire" L. GROSS "Thermodynamics, statistical mechanics and random fields" K. KRICKEBERG "Processus ponctuels en statistique"	(LNM 929)
1981	X. FERNIQUE "Régularité de fonctions aléatoires non gaussiennes" P.W. MILLAR "The minimax principle in asymptotic statistical theory" D.W. STROOCK "Some application of stochastic calculus to partial differential equations" M. WEBER "Analyse infinitésimale de fonctions aléatoires"	(LNM 976)
1982	R.M. DUDLEY "A course on empirical processes" H. KUNITA "Stochastic differential equations and stochastic flow of diffeomorphisms" F. LEDRAPPIER "Quelques propriétés des exposants caractéristiques"	(LNM 1097)
1983	D.J. ALDOUS "Exchangeability and related topics" I.A. IBRAGIMOV "Théorèmes limites pour les marches aléatoires" J. JACOD "Théorèmes limite pour les processus"	(LNM 1117)
1984	R. CARMONA "Random Schrödinger operators" H. KESTEN "Aspects of first passage percolation" J.B. WALSH "An introduction to stochastic partial differential equations"	(LNM 1180)

1985-87	S.R.S. VARADHAN "Large deviations" P. DIACONIS "Applications of non-commutative Fourier analysis to probability theorems H. FÖLLMER "Random fields and diffusion processes" G.C. PAPANICOLAOU "Waves in one-dimensional random media" D. ELWORTHY Geometric aspects of diffusions on manifolds" E. NELSON "Stochastic mechanics and random fields"	(LNM 1362)
1986	O.E. BARNDORFF-NIELSEN "Parametric statistical models and likelihood"	(LNS M50)
1988	A. ANCONA "Théorie du potentiel sur les graphes et les variétés" D. GEMAN "Random fields and inverse problems in imaging" N. IKEDA "Probabilistic methods in the study of asymptotics"	(LNM 1427)
1989	D.L. BURKHOLDER "Explorations in martingale theory and its applications" E. PARDOUX "Filtrage non linéaire et équations aux dérivées partielles stochastiques associées" A.S. SZNITMAN "Topics in propagation of chaos"	(LNM 1464)
1990	M.I. FREIDLIN "Semi-linear PDE's and limit theorems for large deviations" J.F. LE GALL "Some properties of planar Brownian motion"	(LNM 1527)
1991	D.A. DAWSON "Measure-valued Markov processes" B. MAISONNEUVE "Processus de Markov : Naissance, Retournement, Régénération" J. SPENCER "Nine Lectures on Random Graphs"	(LNM 1541)
1992	D. BAKRY "L'hypercontractivité et son utilisation en théorie des semigroupes" R.D. GILL "Lectures on Survival Analysis" S.A. MOLCHANOV "Lectures on the Random Media"	(LNM 1581)

1993	P. BIANE "Calcul stochastique non-commutatif" R. DURRETT "Ten Lectures on Particle Systems"	(LNM 1608)
1994	R. DOBRUSHIN "Perturbation methods of the theory of Gibbsian fields" P. GROENEBOOM "Lectures on inverse problems" M. LEDOUX "Isoperimetry and gaussian analysis"	(LNM 1648)

Lecture Notes in Mathematics

For information about Vols. 1–1459
please contact your bookseller or Springer-Verlag

Vol. 1460: G. Toscani, V. Boffi, S. Rionero (Eds.), Mathematical Aspects of Fluid Plasma Dynamics. Proceedings, 1988. V, 221 pages. 1991.

Vol. 1461: R. Gorenflo, S. Vessella, Abel Integral Equations. VII, 215 pages. 1991.

Vol. 1462: D. Mond, J. Montaldi (Eds.), Singularity Theory and its Applications. Warwick 1989, Part I. VIII, 405 pages. 1991.

Vol. 1463: R. Roberts, I. Stewart (Eds.), Singularity Theory and its Applications. Warwick 1989, Part II. VIII, 322 pages. 1991.

Vol. 1464: D. L. Burkholder, E. Pardoux, A. Sznitman, Ecole d'Eté de Probabilités de Saint- Flour XIX-1989. Editor: P. L. Hennequin. VI, 256 pages. 1991.

Vol. 1465: G. David, Wavelets and Singular Integrals on Curves and Surfaces. X, 107 pages. 1991.

Vol. 1466: W. Banaszczyk, Additive Subgroups of Topological Vector Spaces. VII, 178 pages. 1991.

Vol. 1467: W. M. Schmidt, Diophantine Approximations and Diophantine Equations. VIII, 217 pages. 1991.

Vol. 1468: J. Noguchi, T. Ohsawa (Eds.), Prospects in Complex Geometry. Proceedings, 1989. VII, 421 pages. 1991.

Vol. 1469: J. Lindenstrauss, V. D. Milman (Eds.), Geometric Aspects of Functional Analysis. Seminar 1989-90. XI, 191 pages. 1991.

Vol. 1470: E. Odell, H. Rosenthal (Eds.), Functional Analysis. Proceedings, 1987-89. VII, 199 pages. 1991.

Vol. 1471: A. A. Panchishkin, Non-Archimedean L-Functions of Siegel and Hilbert Modular Forms. VII, 157 pages. 1991.

Vol. 1472: T. T. Nielsen, Bose Algebras: The Complex and Real Wave Representations. V, 132 pages. 1991.

Vol. 1473: Y. Hino, S. Murakami, T. Naito, Functional Differential Equations with Infinite Delay. X, 317 pages. 1991.

Vol. 1474: S. Jackowski, B. Oliver, K. Pawałowski (Eds.), Algebraic Topology, Poznań 1989. Proceedings. VIII, 397 pages. 1991.

Vol. 1475: S. Busenberg, M. Martelli (Eds.), Delay Differential Equations and Dynamical Systems. Proceedings, 1990. VIII, 249 pages. 1991.

Vol. 1476: M. Bekkali, Topics in Set Theory. VII, 120 pages. 1991.

Vol. 1477: R. Jajte, Strong Limit Theorems in Noncommutative L_2-Spaces. X, 113 pages. 1991.

Vol. 1478: M.-P. Malliavin (Ed.), Topics in Invariant Theory. Seminar 1989-1990. VI, 272 pages. 1991.

Vol. 1479: S. Bloch, I. Dolgachev, W. Fulton (Eds.), Algebraic Geometry. Proceedings, 1989. VII, 300 pages. 1991.

Vol. 1480: F. Dumortier, R. Roussarie, J. Sotomayor, H. Żołądek, Bifurcations of Planar Vector Fields: Nilpotent Singularities and Abelian Integrals. VIII, 226 pages. 1991.

Vol. 1481: D. Ferus, U. Pinkall, U. Simon, B. Wegner (Eds.), Global Differential Geometry and Global Analysis. Proceedings, 1991. VIII, 283 pages. 1991.

Vol. 1482: J. Chabrowski, The Dirichlet Problem with L^2-Boundary Data for Elliptic Linear Equations. VI, 173 pages. 1991.

Vol. 1483: E. Reithmeier, Periodic Solutions of Nonlinear Dynamical Systems. VI, 171 pages. 1991.

Vol. 1484: H. Delfs, Homology of Locally Semialgebraic Spaces. IX, 136 pages. 1991.

Vol. 1485: J. Azéma, P. A. Meyer, M. Yor (Eds.), Séminaire de Probabilités XXV. VIII, 440 pages. 1991.

Vol. 1486: L. Arnold, H. Crauel, J.-P. Eckmann (Eds.), Lyapunov Exponents. Proceedings, 1990. VIII, 365 pages. 1991.

Vol. 1487: E. Freitag, Singular Modular Forms and Theta Relations. VI, 172 pages. 1991.

Vol. 1488: A. Carboni, M. C. Pedicchio, G. Rosolini (Eds.), Category Theory. Proceedings, 1990. VII, 494 pages. 1991.

Vol. 1489: A. Mielke, Hamiltonian and Lagrangian Flows on Center Manifolds. X, 140 pages. 1991.

Vol. 1490: K. Metsch, Linear Spaces with Few Lines. XIII, 196 pages. 1991.

Vol. 1491: E. Lluis-Puebla, J.-L. Loday, H. Gillet, C. Soulé, V. Snaith, Higher Algebraic K-Theory: an overview. IX, 164 pages. 1992.

Vol. 1492: K. R. Wicks, Fractals and Hyperspaces. VIII, 168 pages. 1991.

Vol. 1493: E. Benoît (Ed.), Dynamic Bifurcations. Proceedings, Luminy 1990. VII, 219 pages. 1991.

Vol. 1494: M.-T. Cheng, X.-W. Zhou, D.-G. Deng (Eds.), Harmonic Analysis. Proceedings, 1988. IX, 226 pages. 1991.

Vol. 1495: J. M. Bony, G. Grubb, L. Hörmander, H. Komatsu, J. Sjöstrand, Microlocal Analysis and Applications Montecatini Terme, 1989. Editors: L. Cattabriga, L. Rodino. VII, 349 pages. 1991.

Vol. 1496: C. Foias, B. Francis, J. W. Helton, H. Kwakernaak, J. B. Pearson, H_∞-Control Theory. Como, 1990 Editors: E. Mosca, L. Pandolfi. VII, 336 pages. 1991.

Vol. 1497: G. T. Herman, A. K. Louis, F. Natterer (Eds.), Mathematical Methods in Tomography. Proceedings 1990. X, 268 pages. 1991.

Vol. 1498: R. Lang, Spectral Theory of Random Schrödinger Operators. X, 125 pages. 1991.

Vol. 1499: K. Taira, Boundary Value Problems and Markov Processes. IX, 132 pages. 1991.

Vol. 1500: J.-P. Serre, Lie Algebras and Lie Groups. VII, 168 pages. 1992.

Vol. 1501: A. De Masi, E. Presutti, Mathematical Methods for Hydrodynamic Limits. IX, 196 pages. 1991.

Vol. 1502: C. Simpson, Asymptotic Behavior of Monodromy. V, 139 pages. 1991.

Vol. 1503: S. Shokranian, The Selberg-Arthur Trace Formula (Lectures by J. Arthur). VII, 97 pages. 1991.

Vol. 1504: J. Cheeger, M. Gromov, C. Okonek, P. Pansu, Geometric Topology: Recent Developments. Editors: P. de Bartolomeis, F. Tricerri. VII, 197 pages. 1991.

Vol. 1505: K. Kajitani, T. Nishitani, The Hyperbolic Cauchy Problem. VII, 168 pages. 1991.

Vol. 1506: A. Buium, Differential Algebraic Groups of Finite Dimension. XV, 145 pages. 1992.

Vol. 1507: K. Hulek, T. Peternell, M. Schneider, F.-O. Schreyer (Eds.), Complex Algebraic Varieties. Proceedings, 1990. VII, 179 pages. 1992.

Vol. 1508: M. Vuorinen (Ed.), Quasiconformal Space Mappings. A Collection of Surveys 1960-1990. IX, 148 pages. 1992.

Vol. 1509: J. Aguadé, M. Castellet, F. R. Cohen (Eds.), Algebraic Topology - Homotopy and Group Cohomology. Proceedings, 1990. X, 330 pages. 1992.

Vol. 1510: P. P. Kulish (Ed.), Quantum Groups. Proceedings, 1990. XII, 398 pages. 1992.

Vol. 1511: B. S. Yadav, D. Singh (Eds.), Functional Analysis and Operator Theory. Proceedings, 1990. VIII, 223 pages. 1992.

Vol. 1512: L. M. Adleman, M.-D. A. Huang, Primality Testing and Abelian Varieties Over Finite Fields. VII, 142 pages. 1992.

Vol. 1513: L. S. Block, W. A. Coppel, Dynamics in One Dimension. VIII, 249 pages. 1992.

Vol. 1514: U. Krengel, K. Richter, V. Warstat (Eds.), Ergodic Theory and Related Topics III, Proceedings, 1990. VIII, 236 pages. 1992.

Vol. 1515: E. Ballico, F. Catanese, C. Ciliberto (Eds.), Classification of Irregular Varieties. Proceedings, 1990. VII, 149 pages. 1992.

Vol. 1516: R. A. Lorentz, Multivariate Birkhoff Interpolation. IX, 192 pages. 1992.

Vol. 1517: K. Keimel, W. Roth, Ordered Cones and Approximation. VI, 134 pages. 1992.

Vol. 1518: H. Stichtenoth, M. A. Tsfasman (Eds.), Coding Theory and Algebraic Geometry. Proceedings, 1991. VIII, 223 pages. 1992.

Vol. 1519: M. W. Short, The Primitive Soluble Permutation Groups of Degree less than 256. IX, 145 pages. 1992.

Vol. 1520: Yu. G. Borisovich, Yu. E. Gliklikh (Eds.), Global Analysis – Studies and Applications V. VII, 284 pages. 1992.

Vol. 1521: S. Busenberg, B. Forte, H. K. Kuiken, Mathematical Modelling of Industrial Process. Bari, 1990. Editors: V. Capasso, A. Fasano. VII, 162 pages. 1992.

Vol. 1522: J.-M. Delort, F. B. I. Transformation. VII, 101 pages. 1992.

Vol. 1523: W. Xue, Rings with Morita Duality. X, 168 pages. 1992.

Vol. 1524: M. Coste, L. Mahé, M.-F. Roy (Eds.), Real Algebraic Geometry. Proceedings, 1991. VIII, 418 pages. 1992.

Vol. 1525: C. Casacuberta, M. Castellet (Eds.), Mathematical Research Today and Tomorrow. VII, 112 pages. 1992.

Vol. 1526: J. Azéma, P. A. Meyer, M. Yor (Eds.), Séminaire de Probabilités XXVI. X, 633 pages. 1992.

Vol. 1527: M. I. Freidlin, J.-F. Le Gall, Ecole d'Eté de Probabilités de Saint-Flour XX – 1990. Editor: P. L. Hennequin. VIII, 244 pages. 1992.

Vol. 1528: G. Isac, Complementarity Problems. VI, 297 pages. 1992.

Vol. 1529: J. van Neerven, The Adjoint of a Semigroup of Linear Operators. X, 195 pages. 1992.

Vol. 1530: J. G. Heywood, K. Masuda, R. Rautmann, S. A. Solonnikov (Eds.), The Navier-Stokes Equations II – Theory and Numerical Methods. IX, 322 pages. 1992.

Vol. 1531: M. Stoer, Design of Survivable Networks. IV, 206 pages. 1992.

Vol. 1532: J. F. Colombeau, Multiplication of Distributions. X, 184 pages. 1992.

Vol. 1533: P. Jipsen, H. Rose, Varieties of Lattices. X, 162 pages. 1992.

Vol. 1534: C. Greither, Cyclic Galois Extensions of Commutative Rings. X, 145 pages. 1992.

Vol. 1535: A. B. Evans, Orthomorphism Graphs of Groups. VIII, 114 pages. 1992.

Vol. 1536: M. K. Kwong, A. Zettl, Norm Inequalities for Derivatives and Differences. VII, 150 pages. 1992.

Vol. 1537: P. Fitzpatrick, M. Martelli, J. Mawhin, R. Nussbaum, Topological Methods for Ordinary Differential Equations. Montecatini Terme, 1991. Editors: M. Furi, P. Zecca. VII, 218 pages. 1993.

Vol. 1538: P.-A. Meyer, Quantum Probability for Probabilists. X, 287 pages. 1993.

Vol. 1539: M. Coornaert, A. Papadopoulos, Symbolic Dynamics and Hyperbolic Groups. VIII, 138 pages. 1993.

Vol. 1540: H. Komatsu (Ed.), Functional Analysis and Related Topics, 1991. Proceedings. XXI, 413 pages. 1993.

Vol. 1541: D. A. Dawson, B. Maisonneuve, J. Spencer, Ecole d' Eté de Probabilités de Saint-Flour XXI - 1991. Editor: P. L. Hennequin. VIII, 356 pages. 1993.

Vol. 1542: J.Fröhlich, Th.Kerler, Quantum Groups, Quantum Categories and Quantum Field Theory. VII, 431 pages. 1993.

Vol. 1543: A. L. Dontchev, T. Zolezzi, Well-Posed Optimization Problems. XII, 421 pages. 1993.

Vol. 1544: M.Schürmann, White Noise on Bialgebras. VII, 146 pages. 1993.

Vol. 1545: J. Morgan, K. O'Grady, Differential Topology of Complex Surfaces. VIII, 224 pages. 1993.

Vol. 1546: V. V. Kalashnikov, V. M. Zolotarev (Eds.), Stability Problems for Stochastic Models. Proceedings, 1991. VIII, 229 pages. 1993.

Vol. 1547: P. Harmand, D. Werner, W. Werner, M-ideals in Banach Spaces and Banach Algebras. VIII, 387 pages. 1993.

Vol. 1548: T. Urabe, Dynkin Graphs and Quadrilateral Singularities. VI, 233 pages. 1993.

Vol. 1549: G. Vainikko, Multidimensional Weakly Singular Integral Equations. XI, 159 pages. 1993.

Vol. 1550: A. A. Gonchar, E. B. Saff (Eds.), Methods of Approximation Theory in Complex Analysis and Mathematical Physics IV, 222 pages, 1993.

Vol. 1551: L. Arkeryd, P. L. Lions, P.A. Markowich, S.R. S. Varadhan. Nonequilibrium Problems in Many-Particle Systems. Montecatini, 1992. Editors: C. Cercignani, M. Pulvirenti. VII, 158 pages 1993.

Vol. 1552: J. Hilgert, K.-H. Neeb, Lie Semigroups and their Applications. XII, 315 pages. 1993.

Vol. 1553: J.-L- Colliot-Thélène, J. Kato, P. Vojta. Arithmetic Algebraic Geometry. Trento, 1991. Editor: E. Ballico. VII, 223 pages. 1993.

Vol. 1554: A. K. Lenstra, H. W. Lenstra, Jr. (Eds.), The Development of the Number Field Sieve. VIII, 131 pages. 1993.

Vol. 1555: O. Liess, Conical Refraction and Higher Microlocalization. X, 389 pages. 1993.

Vol. 1556: S. B. Kuksin, Nearly Integrable Infinite-Dimensional Hamiltonian Systems. XXVII, 101 pages. 1993.

Vol. 1557: J. Azéma, P. A. Meyer, M. Yor (Eds.), Séminaire de Probabilités XXVII. VI, 327 pages. 1993.

Vol. 1558: T. J. Bridges, J. E. Furter, Singularity Theory and Equivariant Symplectic Maps. VI, 226 pages. 1993.

Vol. 1559: V. G. Sprindžuk, Classical Diophantine Equations. XII, 228 pages. 1993.

Vol. 1560: T. Bartsch, Topological Methods for Variational Problems with Symmetries. X, 152 pages. 1993.

Vol. 1561: I. S. Molchanov, Limit Theorems for Unions of Random Closed Sets. X, 157 pages. 1993.

Vol. 1562: G. Harder, Eisensteinkohomologie und die Konstruktion gemischter Motive. XX, 184 pages. 1993.

Vol. 1563: E. Fabes, M. Fukushima, L. Gross, C. Kenig, M. Röckner, D. W. Stroock, Dirichlet Forms. Varenna, 1992. Editors: G. Dell'Antonio, U. Mosco. VII, 245 pages. 1993.

Vol. 1564: J. Jorgenson, S. Lang, Basic Analysis of Regularized Series and Products. IX, 122 pages. 1993.

Vol. 1565: L. Boutet de Monvel, C. De Concini, C. Procesi, P. Schapira, M. Vergne. D-modules, Representation Theory, and Quantum Groups. Venezia, 1992. Editors: G. Zampieri, A. D'Agnolo. VII, 217 pages. 1993.

Vol. 1566: B. Edixhoven, J.-H. Evertse (Eds.), Diophantine Approximation and Abelian Varieties. XIII, 127 pages. 1993.

Vol. 1567: R. L. Dobrushin, S. Kusuoka, Statistical Mechanics and Fractals. VII, 98 pages. 1993.

Vol. 1568: F. Weisz, Martingale Hardy Spaces and their Application in Fourier Analysis. VIII, 217 pages. 1994.

Vol. 1569: V. Totik, Weighted Approximation with Varying Weight. VI, 117 pages. 1994.

Vol. 1570: R. deLaubenfels, Existence Families, Functional Calculi and Evolution Equations. XV, 234 pages. 1994.

Vol. 1571: S. Yu. Pilyugin, The Space of Dynamical Systems with the C^0-Topology. X, 188 pages. 1994.

Vol. 1572: L. Göttsche, Hilbert Schemes of Zero-Dimensional Subschemes of Smooth Varieties. IX, 196 pages. 1994.

Vol. 1573: V. P. Havin, N. K. Nikolski (Eds.), Linear and Complex Analysis – Problem Book 3 – Part I. XXII, 489 pages. 1994.

Vol. 1574: V. P. Havin, N. K. Nikolski (Eds.), Linear and Complex Analysis – Problem Book 3 – Part II. XXII, 507 pages. 1994.

Vol. 1575: M. Mitrea, Clifford Wavelets, Singular Integrals, and Hardy Spaces. XI, 116 pages. 1994.

Vol. 1576: K. Kitahara, Spaces of Approximating Functions with Haar-Like Conditions. X, 110 pages. 1994.

Vol. 1577: N. Obata, White Noise Calculus and Fock Space. X, 183 pages. 1994.

Vol. 1578: J. Bernstein, V. Lunts, Equivariant Sheaves and Functors. V, 139 pages. 1994.

Vol. 1579: N. Kazamaki, Continuous Exponential Martingales and *BMO*. VII, 91 pages. 1994.

Vol. 1580: M. Milman, Extrapolation and Optimal Decompositions with Applications to Analysis. XI, 161 pages. 1994.

Vol. 1581: D. Bakry, R. D. Gill, S. A. Molchanov, Lectures on Probability Theory. Editor: P. Bernard. VIII, 420 pages. 1994.

Vol. 1582: W. Balser, From Divergent Power Series to Analytic Functions. X, 108 pages. 1994.

Vol. 1583: J. Azéma, P. A. Meyer, M. Yor (Eds.), Séminaire de Probabilités XXVIII. VI, 334 pages. 1994.

Vol. 1584: M. Brokate, N. Kenmochi, I. Müller, J. F. Rodriguez, C. Verdi, Phase Transitions and Hysteresis. Montecatini Terme, 1993. Editor: A. Visintin. VII. 291 pages. 1994.

Vol. 1585: G. Frey (Ed.), On Artin's Conjecture for Odd 2-dimensional Representations. VIII, 148 pages. 1994.

Vol. 1586: R. Nillsen, Difference Spaces and Invariant Linear Forms. XII, 186 pages. 1994.

Vol. 1587: N. Xi, Representations of Affine Hecke Algebras. VIII, 137 pages. 1994.

Vol. 1588: C. Scheiderer, Real and Étale Cohomology. XXIV, 273 pages. 1994.

Vol. 1589: J. Bellissard, M. Degli Esposti, G. Forni, S. Graffi, S. Isola, J. N. Mather, Transition to Chaos in Classical and Quantum Mechanics. Montecatini Terme, 1991. Editor: S. Graffi. VII, 192 pages. 1994.

Vol. 1590: P. M. Soardi, Potential Theory on Infinite Networks. VIII, 187 pages. 1994.

Vol. 1591: M. Abate, G. Patrizio, Finsler Metrics – A Global Approach. IX, 180 pages. 1994.

Vol. 1592: K. W. Breitung, Asymptotic Approximations for Probability Integrals. IX, 146 pages. 1994.

Vol. 1593: J. Jorgenson & S. Lang, D. Goldfeld, Explicit Formulas for Regularized Products and Series. VIII, 154 pages. 1994.

Vol. 1594: M. Green, J. Murre, C. Voisin, Algebraic Cycles and Hodge Theory. Torino, 1993. Editors: A. Albano, F. Bardelli. VII, 275 pages. 1994.

Vol. 1595: R.D.M. Accola, Topics in the Theory of Riemann Surfaces. IX, 105 pages. 1994.

Vol. 1596: L. Heindorf, L. B. Shapiro, Nearly Projective Boolean Algebras. X, 202 pages. 1994.

Vol. 1597: B. Herzog, Kodaira-Spencer Maps in Local Algebra. XVII, 176 pages. 1994.

Vol. 1598: J. Berndt, F. Tricerri, L. Vanhecke, Generalized

Heisenberg Groups and Damek-Ricci Harmonic Spaces. VIII, 125 pages. 1995.

Vol. 1599: K. Johannson, Topology and Combinatorics of 3-Manifolds. XVIII, 446 pages. 1995.

Vol. 1600: W. Narkiewicz, Polynomial Mappings. VII, 130 pages. 1995.

Vol. 1601: A. Pott, Finite Geometry and Character Theory. VII, 181 pages. 1995.

Vol. 1602: J. Winkelmann, The Classification of Three-dimensional Homogeneous Complex Manifolds. XI, 230 pages. 1995.

Vol. 1603: V. Ene, Real Functions – Current Topics. XIII, 310 pages. 1995.

Vol. 1604: A. Huber, Mixed Motives and their Realization in Derived Categories. XV, 207 pages. 1995.

Vol. 1605: L. B. Wahlbin, Superconvergence in Galerkin Finite Element Methods. XI, 166 pages. 1995.

Vol. 1606: P.-D. Liu, M. Qian, Smooth Ergodic Theory of Random Dynamical Systems. XI, 221 pages. 1995.

Vol. 1607: G. Schwarz, Hodge Decomposition – A Method for Solving Boundary Value Problems. VII, 155 pages. 1995.

Vol. 1608: P. Biane, R. Durrett, Lectures on Probability Theory. Editor: P. Bernard. VII, 210 pages. 1995.

Vol. 1609: L. Arnold, C. Jones, K. Mischaikow, G. Raugel, Dynamical Systems. Montecatini Terme, 1994. Editor: R. Johnson. VIII, 329 pages. 1995.

Vol. 1610: A. S. Üstünel, An Introduction to Analysis on Wiener Space. X, 95 pages. 1995.

Vol. 1611: N. Knarr, Translation Planes. VI, 112 pages. 1995.

Vol. 1612: W. Kühnel, Tight Polyhedral Submanifolds and Tight Triangulations. VII, 122 pages. 1995.

Vol. 1613: J. Azéma, M. Emery, P. A. Meyer, M. Yor (Eds.), Séminaire de Probabilités XXIX. VI, 326 pages. 1995.

Vol. 1614: A. Koshelev, Regularity Problem for Quasilinear Elliptic and Parabolic Systems. XXI, 255 pages. 1995.

Vol. 1615: D. B. Massey, Lê Cycles and Hypersurface Singularities. XI, 131 pages. 1995.

Vol. 1616: I. Moerdijk, Classifying Spaces and Classifying Topoi. VII, 94 pages. 1995.

Vol. 1617: V. Yurinsky, Sums and Gaussian Vectors. XI, 305 pages. 1995.

Vol. 1618: G. Pisier, Similarity Problems and Completely Bounded Maps. VII, 156 pages. 1996.

Vol. 1619: E. Landvogt, A Compactification of the Bruhat-Tits Building. VII, 152 pages. 1996.

Vol. 1620: R. Donagi, B. Dubrovin, E. Frenkel, E. Previato, Integrable Systems and Quantum Groups. Montecatini Terme, 1993. Editors:M. Francaviglia, S. Greco. VIII, 488 pages. 1996.

Vol. 1621: H. Bass, M. V. Otero-Espinar, D. N. Rockmore, C. P. L. Tresser, Cyclic Renormalization and Auto-morphism Groups of Rooted Trees. XXI, 136 pages. 1996.

Vol. 1622: E. D. Farjoun, Cellular Spaces, Null Spaces and Homotopy Localization. XIV, 199 pages. 1996.

Vol. 1623: H.P. Yap, Total Colourings of Graphs. VIII, 131 pages. 1996.

Vol. 1624: V. Brînzănescu, Holomorphic Vector Bundles over Compact Complex Surfaces. X, 170 pages. 1996.

Vol.1625: S. Lang, Topics in Cohomology of Groups. VII, 226 pages. 1996.

Vol. 1626: J. Azéma, M. Emery, M. Yor (Eds.), Séminaire de Probabilités XXX. VIII, 382 pages. 1996.

Vol. 1627: C. Graham, Th. G. Kurtz, S. Méléard, Ph. E. Protter, M. Pulvirenti, D. Talay, Probabilistic Models for Nonlinear Partial Differential Equations. Montecatini Terme, 1995. Editors: D. Talay, L. Tubaro. X, 301 pages. 1996.

Vol. 1628: P.-H. Zieschang, An Algebraic Approach to Association Schemes. XII, 189 pages. 1996.

Vol. 1629: J. D. Moore, Lectures on Seiberg-Witten Invariants. VII, 105 pages. 1996.

Vol. 1630: D. Neuenschwander, Probabilities on the Heisenberg Group: Limit Theorems and Brownian Motion. VIII, 139 pages. 1996.

Vol. 1631: K. Nishioka, Mahler Functions and Transcendence.VIII, 185 pages.1996.

Vol. 1632: A. Kushkuley, Z. Balanov, Geometric Methods in Degree Theory for Equivariant Maps. VII, 136 pages. 1996.

Vol.1633: H. Aikawa, M. Essén, Potential Theory – Selected Topics. IX, 200 pages.1996.

Vol. 1634: J. Xu, Flat Covers of Modules. IX, 161 pages. 1996.

Vol. 1635: E. Hebey, Sobolev Spaces on Riemannian Manifolds. X, 116 pages. 1996.

Vol. 1636: M. A. Marshall, Spaces of Orderings and Abstract Real Spectra. VI, 190 pages. 1996.

Vol. 1637: B. Hunt, The Geometry of some special Arithmetic Quotients. XIII, 332 pages. 1996.

Vol. 1638: P. Vanhaecke, Integrable Systems in the realm of Algebraic Geometry. VIII, 218 pages. 1996.

Vol. 1639: K. Dekimpe, Almost-Bieberbach Groups: Affine and Polynomial Structures. X, 259 pages. 1996.

Vol. 1640: G. Boillat, C. M. Dafermos, P. D. Lax, T. P. Liu, Recent Mathematical Methods in Nonlinear Wave Propagation. Montecatini Terme, 1994. Editor: T. Ruggeri. VII, 142 pages. 1996.

Vol. 1641: P. Abramenko, Twin Buildings and Applications to S-Arithmetic Groups. IX, 123 pages. 1996.

Vol. 1642: M. Puschnigg, Asymptotic Cyclic Cohomology. XXII, 138 pages. 1996.

Vol. 1643: J. Richter-Gebert, Realization Spaces of Polytopes. XI, 187 pages. 1996.

Vol. 1644: A. Adler, S. Ramanan, Moduli of Abelian Varieties. VI, 196 pages. 1996.

Vol. 1645: H. W. Broer, G. B. Huitema, M. B. Sevryuk, Quasi-Periodic Motions in Families of Dynamical Systems. XI, 195 pages. 1996.

Vol. 1646: J.-P. Demailly, T. Peternell, G. Tian, A. N. Tyurin, Transcendental Methods in Algebraic Geometry. Cetraro, 1994. Editors: F. Catanese, C. Ciliberto. VII, 257 pages. 1996.

Vol. 1647: D. Dias, P. Le Barz, Configuration Spaces over Hilbert Schemes and Applications. VII. 143 pages. 1996.

Vol. 1648: R. Dobrushin, P. Groeneboom, M. Ledoux, Lectures on Probability Theory and Statistics. Editor: P. Bernard. VIII, 300 pages. 1996.

GPSR Compliance

The European Union's (EU) General Product Safety Regulation (GPSR) is a set of rules that requires consumer products to be safe and our obligations to ensure this.

If you have any concerns about our products, you can contact us on ProductSafety@springernature.com

In case Publisher is established outside the EU, the EU authorized representative is:

Springer Nature Customer Service Center GmbH
Europaplatz 3
69115 Heidelberg, Germany

Batch number: 09624486

Printed by Printforce, the Netherlands